NON-OXIDE TECHNICAL AND ENGINEERING CERAMICS

Proceedings of the International Conference held at the National Institute for Higher Education, Limerick, Ireland, 10–12 July 1985, co-sponsored by the Commission of the European Communities, European Research Office of the United States Army, Industrial Development Authority of Ireland, Howmedica International Inc., Limerick, National Board for Science and Technology, Dublin, Shannon Free Airport Development Co., and supported by the Irish Durability and Fracture Committee and the Materials Research Centre, NIHE, Limerick.

Conference Chairman
  Dr S. HAMPSHIRE, Materials Research Centre, NIHE, Limerick, Ireland

Irish Organisation and Advisory Group
  M. FARMER, University College, Dublin
  Dr K. HALPIN, Institute for Industrial Research and Standards, Dublin
  Professor E. R. PETTY, NIHE, Limerick
  Dr M. J. POMEROY, NIHE, Limerick
  Professor D. M. R. TAPLIN, Trinity College, Dublin
  Professor S. TIMONEY, University College, Dublin

International Advisory Committee
  Professor M. BILLY, UER des Sciences, Limoges, France
  Professor P. BOCH, Ecole Nationale Supérieure de Céramique Industrielle, Limoges, France
  Professor R. J. BROOK, University of Leeds, UK. (President, Institute of Ceramics)
  Dr D. BROUSSAUD, Ecole Nationale Supérieure des Mines de Paris, Evry, France. (President, Groupe Français de la Céramique)
  Dr R. W. DAVIDGE, AERE, Harwell, UK
  Professor H. HAUSNER, Technische Universität, Berlin, FGR
  Professor K. H. JACK, FRS, Emeritus Professor, University of Newcastle upon Tyne, UK
  Dr A. W. J. M. RAE, Anzon Ltd, Wallsend, UK
  Dr D. P. THOMPSON, University of Newcastle upon Tyne, UK
  Dr G. ZIEGLER, DFVLR, Köln, FGR

NON-OXIDE TECHNICAL AND ENGINEERING CERAMICS

Edited by

STUART HAMPSHIRE

*Materials Research Centre,
National Institute for Higher Education, Limerick, Ireland*

ELSEVIER APPLIED SCIENCE
LONDON and NEW YORK

ELSEVIER APPLIED SCIENCE PUBLISHERS LTD
Crown House, Linton Road, Barking, Essex IG11 8JU, England

Sole Distributor in the USA and Canada
ELSEVIER SCIENCE PUBLISHING CO., INC.
52 Vanderbilt Avenue, New York, NY 10017, USA

WITH 74 TABLES AND 249 ILLUSTRATIONS

© ELSEVIER APPLIED SCIENCE PUBLISHERS LTD 1986

© UNITED KINGDOM ATOMIC ENERGY AUTHORITY 1986—Chapters 22 and 24

British Library Cataloguing in Publication Data

Non-oxide technical and engineering ceramics.
1. Ceramic materials
I. Hampshire, Stuart
620.1'4   TA430

Library of Congress Cataloging in Publication Data

Non-oxide technical and engineering ceramics.

Papers presented at a conference held at the National Institute for Higher Education, Limerick, Ireland, July 10–12, 1985; co-sponsored by the Commission of the European Communities, and others.
Bibliography: p.
Includes index.
1. Ceramics—Congresses. 2. Ceramic materials—Congresses. I. Hampshire, Stuart. II. Commission of the European Communities.
TA430.N66   1986   666   86-16570
ISBN 1-85166-042-9

The selection and presentation of material and the opinions expressed are the sole responsibility of the author(s) concerned.

Special regulations for readers in the USA

This publication has been registered with the Copyright Clearance Center Inc. (CCC), Salem, Massachusetts. Information can be obtained from the CCC about conditions under which photocopies of parts of this publication may be made in the USA. All other copyright questions, including photocopying outside of the USA, should be referred to the publisher.

All rights reserved. No part of this publication may be reproduced, stored in a retrieval system, or transmitted in any form or by any means, electronic, mechanical, photocopying, recording, or otherwise, without the prior written permission of the publisher.

Typeset and printed in Northern Ireland by The Universities Press (Belfast) Ltd

Foreword

Conferences on technical and engineering ceramics are held with increasing frequency, having become fashionable because the potential of ceramics in profitable growth industries is an urgent matter of considerable debate and discussion.

Japanese predictions are that the market value of ceramics will grow at about 10% per annum to reach at least $\$10^{10}$ by the end of the century. Seventy per cent of this market will be in electroceramics, applications for which include insulating substrates in integrated circuits, ferroelectric capacitors, piezoelectric oscillators and transducers, ferrite magnets, and ion-conducting solid electrolytes and sensors. All these are oxides, and so are excluded by the title of the Limerick Conference.

Why 'Non-oxide'? The other major ceramics potential is in structural engineering components and engine applications. Here, the greatest impetus to research and development has been the attempt to produce a ceramic gas turbine. Heat engines become more efficient as their working temperature increases, but nickel-base superalloy engines have about reached their limit. Compared with metals, ceramics have higher strengths at high temperatures, better oxidation and corrosion resistance, and are also less dense. In general, ceramics have better properties above about 1000°C except in one respect—their inherent brittleness. The work of fracture is therefore much smaller than for metals and so the permitted flaw size is also smaller. The fabrication of reliable components thus requires changes and improvements in processing, shaping and densification which have so far been achieved to only a limited extent. The economic costs of more sophisticated and controlled processing must be set against the potential advantages of the products, and in the absence of definitive answers European and particularly UK industry is hesitant.

Of the oxides, only alumina and zirconia are strong enough to be considered for engine applications. But alumina has a high coefficient of thermal expansion and so poor thermal shock resistance. Zirconia also has a high thermal expansion and, additionally, a destructive tetragonal to monoclinic phase change on cooling from above 1200°C. Partially stabilised zirconia, with additions of magnesia or yttria, has remarkably high toughness and strength but the transformation toughening becomes inoperative above the transformation temperature and so useful applications of PSZ are below 1000°C.

This leaves the non-oxides, silicon carbide and silicon nitride, each with high strength and good oxidation resistance. Silicon carbide has a higher thermal expansion than silicon nitride but compensates with a higher thermal conductivity; its intrinsic strength is lower than that of silicon nitride but densification additives do not impair its creep resistance to the same extent as with the nitride. Both the carbide and nitride have been main contenders for high-temperature engine applications and each exists in a variety of types depending on the method of processing, e.g. hot-pressing, reaction bonding, pressureless sintering and so on. Silicon nitride was favoured when the American engine programme started in 1971. Since then, more effort has been directed to the nitride and the concurrent discovery in Japan and at Newcastle of ceramic alloys based upon silicon nitride has widened the field to include the sialons.

The Limerick Conference reflected this background. Out of some thirty papers, two-thirds were concerned with nitrogen ceramics and more than one-half of these included work on sialons. This, of course, results from the particular interests of the participants but it also points to some of the directions for successful commercial development. The Japanese expect that their production of ceramic engine components, wear parts and bearings will reach $2 billion per year in fifteen years' time. Their optimism and willingness to take risks is seldom shared by European industrialists yet the Japanese future is still dependent, in part, on European achievement; more than half-a-dozen Japanese companies are licensed to produce 'Syalon'!

The Conference reported progress in many directions and in doing so I hope it encouraged the faint-hearted.

Due to the efforts of Dr Stuart Hampshire and his colleagues, Limerick has become a new centre for research on nitrogen ceramics and the relationships between structure, properties and processing that are being studied there are complementary to the more direct application of engineering ceramics being made by Professor Timoney's group at Dublin. In the interests of industrial development in Ireland, the UK and the rest of Europe, these and similar efforts must be sustained and expanded.

The Limerick Conference was highly successful. The number of participants, about one hundred, was just right for lively and uninhibited discussion. Facilities provided by the Institute were excellent and the social exchanges, including the memorable mediaeval banquet at Bunratty Castle, were enjoyed by all.

K. H. JACK, FRS
Emeritus Professor
University of Newcastle upon Tyne
UK

Acknowledgements

I wish to express my thanks to all the participants at the International Conference on 'Non-oxide technical and engineering ceramics' for contributing to this important event. In particular, I would like to thank the authors and presenters of the papers for the privilege of working with them to produce what will be an important volume in the area of high performance ceramics.

This conference could not have taken place without financial and moral support from different sources and I would like to express my thanks to the following:

Commission of the European Communities, Brussels;
European Research Office, US Army;
Mr Eamonn Kinsella, National Board for Science and Technology, Dublin;
Ms Kathleen Fitzgerald, Industrial Development Authority of Ireland;
Mr Dermot Whelan, Howmedica International Inc., Limerick;
Mr Bill Moloney, Shannon Free Airport Development Co.;
Dr Aidan Spooner, De Beers Industrial Diamond Division, Shannon;
Irish Durability and Fracture Committee.

I would like to thank the members of the International Advisory Board for their support of this venture and also the session chairmen, Professor Richard Brook, Dr Daniel Broussaud, Dr Roger Davidge, Mr Malcolm Farmer, Professor Ernst Gugel, Professor Ken Jack, Mr Bill Long, Professor Evan Petty, Dr Mike Pomeroy and Dr Derek Thompson for their efficient management of the technical sessions. Thanks are due to Professor Noel Mulcahy who opened the proceedings and welcomed the delegates.

I am especially appreciative of the participation of colleagues at NIHE, Limerick and their dedicated work in making the event such a success. Special mention must be made of Dr Mike Pomeroy, Mr

Killian O'Reilly, Ms Bilge Saruhan, Mr Christian Weber and Mr Martin Redington. Thanks are also due to Ms Helen Sheehan and Mr Michael McInerney for technical assistance during the conference.

Special acknowledgement is due to Ms Charlotte Tuohy, the Conference Coordinator, and to Ms Peggy Hurley, the Conference Registrar, for their dedicated assistance during the stages of planning of the conference and publication of this volume. Thanks are also due to Ms Margaret Bolton and Mr Killian O'Reilly for their help in transcription and compilation of the discussions.

Finally, I would like to express my gratitude to Dr Edward Walsh, President of NIHE, Limerick for use of the facilities.

STUART HAMPSHIRE

Contents

Foreword, by Professor K. H. Jack, FRS v

Acknowledgements ix

1. Sialons: A Study in Materials Development 1
   K. H. JACK

2. Discussion Session: Aspects of Commercialisation of Advanced Ceramics and Future Opportunities 31

3. Analysis of Coarsening and Densification Kinetics During the Heat Treatment of Nitrogen Ceramics 41
   H. PICKUP, U. EISELE, E. GILBART and R. J. BROOK

4. Sinterability of Silicon Nitride Powders and Characterisation of Sintered Materials 53
   V. VANDENEEDE, A. LERICHE, F. CAMBIER, H. PICKUP and R. J. BROOK

5. Microstructural Development and Secondary Phases in Silicon Nitride Sintered with Mixed Neodymia/Magnesia Additions 69
   B. SARUHAN, M. J. POMEROY and S. HAMPSHIRE

6. Microstructural Development, Microstructural Characterization and Relation to Mechanical Properties of Dense Silicon Nitride 83
   G. WÖTTING, B. KANKA and G. ZIEGLER

7. Sialon and Syalon Powders: Production, Properties and Applications 97
   P. FERGUSON and A. W. J. M. RAE

8. Reaction Sequences in the Preparation of Sialon Ceramics . 105
   W. Y. SUN, P. A. WALLS and D. P. THOMPSON

9. Preparation and Densification of Nitrogen Ceramics from Oxides 119
   S. A. SIDDIQI, I. HIGGINS and A. HENDRY

10. Preparation of Sialon Ceramics from Low Cost Raw Materials 133
    C. J. SPACIE

11. Densification Behaviour of Sialon Powders Derived from Aluminosilicate Minerals 149
    M. MOSTAGHACI, Q. FAN, F. L. RILEY, Y. BIGAY and J. P. TORRE

12. Sintering and some Properties of Si_3N_4 Based Ceramics . . 165
    S. BOSKOVIĆ and E. KOSTIĆ

13. Syalon Ceramic for Application at High Temperature and Stress 175
    M. H. LEWIS, S. MASON and A. SZWEDA

14. Structural Evolution Under High Pressure and at High Temperature of a β'-Sialon Phase 191
    G. ROULT, M. BROSSARD, J. C. LABBE and P. GOURSAT

15. Multianion Glasses 203
    W. K. TREDWAY and S. H. RISBUD

16. Non-destructive Evaluation of Ceramic Surfaces and Sub-surfaces 213
    L. MCDONNELL and E. M. CASHELL

17. Preparation and Characterization of Yttrium α'-Sialons . . 223
    S. SLASOR and D. P. THOMPSON

18. Characterisation and Properties of Sialon Materials 231
    T. EKSTRÖM and N. INGELSTRÖM

19. Comparison of CBN, Sialons and Tungsten Carbides in the Machining of Surgical Implant Alloys and Cast Irons. . . . 255
   M. AUSTIN, J. TOOHER, J. MONAGHAN and M. EL-BARADIE

20. The Wear Behaviour of Sialon and Silicon Carbide Ceramics in Sliding Contact 281
   S. A. HORTON, J. DENAPE, D. BROUSSAUD, D. DOWSON, F. L. RILEY and N. WALLBRIDGE

21. Studies on Properties of Low Atomic Number Ceramics as Limiter Materials for Fusion Applications 299
   B. A. THIELE, H. HOVEN, K. KOIZLIK, J. LINKE and E. WALLURE

22. Effect of Microstructural Features on the Mechanical Properties of REFEL Self-bonded Silicon Carbide 301
   P. KENNEDY

23. Failure Probability of Shouldered and Notched Ceramic Components Using Neuber Notch Theory 319
   H. FESSLER and D. C. FRICKER

24. Characterisation and Mechanical Properties of Hot-pressed and Reaction-bonded Silicon Nitride 341
   R. C. PILLER, K. P. BALKWILL, A. BRIGGS and R. W. DAVIDGE

25. Tensile Strength of a Sintered Silicon Nitride 361
   T. SOMA, M. MATSUI and I. ODA

26. Selecting Ceramics for High Temperature Components in IC Engines 375
   M. H. FARMER, M. S. LACEY and J. N. MULCAHY

27. Oxidative Removal of Organic Binders from Injection-molded Ceramics 397
   B. C. MUTSUDDY

28. Silicon Nitride Ceramics and Composites: a View of Reliability Enhancement 409
S. T. BULJAN, J. T. NEIL, A. E. PASTO, J. T. SMITH and G. ZILBERSTEIN

29. Panel Discussion: Aspects of Reliability 433

List of Participants. 439

Index of Contributors 443

Subject Index 445

1

Sialons: A Study in Materials Development

K. H. Jack
Wolfson Laboratory, Department of Metallurgy and Engineering Materials, University of Newcastle upon Tyne, NE1 7RU, UK

ABSTRACT

Although the combination of properties shown by silicon nitride has made it a leading ceramic contender for engineering applications, the technological problem is in fabricating dense components cheaply and to precise dimensional tolerances. One solution is to use the principles of ceramic alloying inherent in the production of 'sialons'—the acronym given to phases in the Si–Al–O–N and related systems that are built up of $(Si, Al)(O, N)_4$ tetrahedra in the same way that the structural units of the mineral silicates are SiO_4 tetrahedra.

The mutual replacement of silicon by aluminium and of nitrogen by oxygen in β-Si_3N_4 gives β'-sialons and allows variations in the covalent and ionic contributions to the interatomic bonding that, in turn, allow variations in properties. The products promise to be outstanding for engine components and are already commercially successful as cutting tools for machining metals. Other applications are in welding and extrusion, in molten metal handling, and for seals, bearings and wear parts.

α'-Sialons are derived from α-Si_3N_4 by replacing Si with Al and effecting valency compensation by accommodating additional metal cations in interstitial sites. Prospects for their technological application are at least as good as for β'-sialons. The replacement of Si and N by Al and O in silicon oxynitride, Si_2N_2O, is more limited but again produces useful and fully dense O'-sialon ceramics by pressureless sintering.

Dual-phase materials in which pairs of the completely compatible phases α', β' and O' are combined have advantages over single-phase sialons.

The sialons include vitreous as well as crystalline materials and, in the glass systems so far studied, up to about one in four oxygen atoms can be replaced by nitrogen. The viscosity, hardness and density all increase with increasing nitrogen content, providing stronger and more refractory glasses.

1. INTRODUCTION

In each of its structural modifications, α and β, silicon nitride (Si_3N_4) has a unique combination of properties. It is strong, hard, wear-resistant, stable to higher than 1800°C, oxidation-resistant and, because of its low coefficient of thermal expansion, has excellent resistance to thermal shock. It is also less than half as dense as steel. These features were pointed out[1] more than 25 years ago, but the greatest impetus to its development was its selection by the Advanced Research Projects Agency of the US Department of Defense in 1971 as the material for the ceramic gas turbine. The latter was specified to run at 1370°C—more than 300°C higher than existing nickel-based super-alloy engines—so giving higher efficiency, the capability of using poorer fuels, and less pollution of the environment.

After 14 years, and government funding in the USA, Germany and Japan of at least $400 million—with perhaps an equal expenditure by private companies in those countries—there is still no commercial ceramic engine. One reason for this is that the desirable properties are achieved only in fully dense silicon nitride, and the main technological difficulty is in fabricating dense, precisely shaped components. The material cannot be densified like an ordinary oxide or silicate ceramic merely by firing it. The strong interatomic covalent bonding between silicon and nitrogen means that self-diffusivity is small and, by the time the high temperature is reached at which the atoms move, the silicon nitride begins to decompose by volatilisation of nitrogen. It can be densified by hot-pressing at about 2 tsi (~30 MPa) in graphite dies heated by induction to 1800°C, but the process is limited to simple shapes and the hot-pressed product (HPSN) is so hard that the final shape must be obtained by expensive diamond grinding. Moreover, so-called 'fluxing agents'—usually oxides, for example of magnesium (MgO) or yttrium (Y_2O_3)—must be added to the silicon nitride powder to achieve densification.

It had not been realised, even when the American engine programme started, that each powder particle of silicon nitride has around it a surface layer of silica (SiO_2) (see Fig. 1). The function of the added metal oxide is to react with this silica and a little silicon nitride to give, at the high hot-pressing temperature, an oxynitride liquid. The latter allows mass transport and densification by 'liquid-phase sintering'.[2,3]

The liquid that is necessary for densification cools and, with the magnesia additive, gives a grain-boundary Mg–Si–O–N glass that softens not much above 1000°C. Although the properties of the product are good at low temperatures, the high-temperature creep resistance is very poor.

With yttrium oxide, the Y–Si–O–N liquid cools to give one or more of the four crystalline quaternary oxynitrides shown on the phase diagram of Fig. 2. The intergranular phases that are formed depend upon the amount of added yttria and the amount of surface silica on the silicon nitride, but all of them oxidise with an increase in volume. In particular, the oxynitride $Y_2Si_3O_3N_4$ is isostructural with the melilite mineral silicate Åkermanite, $Ca_2MgSi_2O_7$, and oxidises with a 30% increase in specific volume (see Fig. 3). In an oxidising environment at about 1000°C, this expansion opens up grain boundaries, exposes fresh surfaces for further attack, and so the oxidation becomes catastrophic (see Fig. 4).

Reaction-bonded silicon nitride (RBSN), made by nitriding silicon-powder shapes in molecular nitrogen at about 1400°C, is up to 25%

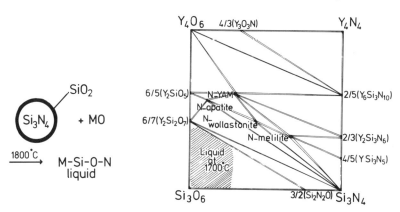

FIG. 1. Reaction of metal oxide additive with silicon nitride.

FIG. 2. Behaviour diagram of the Y–Si–O–N system.

Åkermanite

$Ca_2Mg[Si_2O_7]^{6-}$

$Y_2Si[Si_2O_3N_4]^{10-}$

$\equiv Si_3N_4 \cdot Y_2O_3$

$$\downarrow O_2 \quad 1000°C$$

$\Delta V = 30\%$

FIG. 3. Oxidation of N-melilite, $Y_2Si[Si_2O_3N_4]$.

porous and so its strength is low but it can be post-sintered (PSRBSN) to high density by heat-treatment at 1600–1800°C with MgO or Y_2O_3 sintering aids added in different ways: (i) mixed with the silicon powder before nitriding; (ii) by multiple impregnation of liquid containing the additives in solution; and (iii) by vapour infiltration (e.g. of MgO) from a powder bed. The processes are claimed[4,5] to combine the ease of fabrication of RBSN with the properties of HPSN. However, all densifying additives degrade the properties of silicon nitride to some extent. Hot-isostatic pressing (HIP) uses the least amount of additive but its cost seems likely to be uneconomic for the majority of applications. An alternative approach is to use the principles of ceramic alloying inherent in the production of 'sialons'— the acronym given[6] to phases in the Si–Al–O–N and related systems

FIG. 4. Silicon nitride bar hot-pressed with 15 wt % Y_2O_3 and oxidised at 1000°C for 120 h.

that were discovered independently in Japan[7] and at the University of Newcastle upon Tyne.[8]

2. THE Si–Al–O–N SYSTEM

2.1. β'-Sialon

As shown by Fig. 5, silicon nitride is built up of SiN_4 tetrahedra joined in a three-dimensional network by sharing corners in the same way that SiO_4 units are joined in chains, double chains, rings, sheets and networks to form the almost infinite variety of mineral silicates that make up the earth's lithosphere. Indeed, the atomic arrangement in β-Si_3N_4 is the same as that in beryllium and zinc silicates, Be_2SiO_4

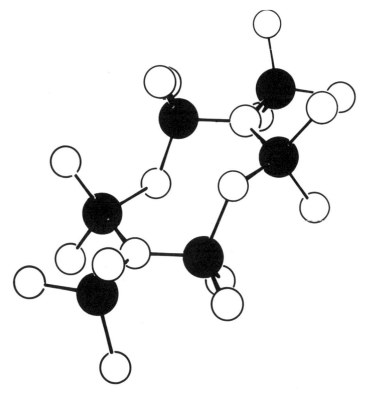

FIG. 5. The crystal structure of β-Si_3N_4 and β'-$(Si, Al)_3(O, N)_4$. ●, metal atom; ○, non-metal atom.

and Zn_2SiO_4, and the same simple principles of crystal chemistry apply. In particular, aluminium plays a special role in silicates because the AlO_4 tetrahedron is about the same size as SiO_4 and can replace it in the rings, chains and networks provided that valency compensation is made elsewhere in the structure. Thus, it is not too surprising that up to two-thirds of the silicon of β-Si_3N_4 can be substituted by aluminium without change of structure provided that an equivalent concentration of nitrogen is replaced by oxygen:[8]

$$Si^{4+}N^{3-} \rightleftharpoons Al^{3+}O^{2-} \tag{1}$$

In the Si–Al–O–N phase diagram at 1800°C shown in Fig. 6, aluminium content is plotted in equivalents on the x-axis (the balance being silicon) and oxygen content on the y-axis (balance, nitrogen). The β'-sialon phase, in which the β-Si_3N_4 crystal structure is retained, extends over a range of composition

$$Si_{6-z}^{24-4z}Al_z^{3z}O_z^{\overline{2z}}N_{8-z}^{\overline{24-3z}} \tag{2}$$

which represents the contents of the hexagonal unit cell with z varying between 0 and 4.

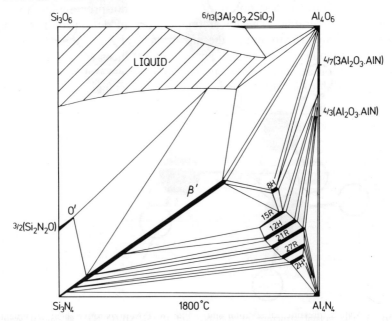

FIG. 6. The Si–Al–O–N behaviour diagram at 1800°C.

The relationship between sialon and silicon nitride is analogous to that between brass and copper. Pure copper is soft and weak, but up to 40% of the copper atoms can be replaced by zinc, without changing the structure, to give a harder, stronger alloy that melts at a lower temperature and so can be fabricated more easily than copper.

Because of its atomic arrangement, β'-sialon has mechanical and physical properties like silicon nitride, e.g. high strength and a small coefficient of thermal expansion; for β' with $z = 3$, $\alpha = 2\cdot7 \times 10^{-6}/°C$ compared with $3\cdot5 \times 10^{-6}/°C$ for β-Si_3N_4. Chemically, however, β'-sialon has some of the characteristics of aluminium oxide, but with important modifications. No matter how much Al and O is substituted for Si and N, the Al is four-coordinated by O, (AlO_4), and not six-coordinated (AlO_6) as in alumina. The Al–O interatomic bond strength in β' is therefore about 50% stronger than in Al_2O_3.

The β'-sialon is a solid solution and, like all solutions, its vapour pressure is lower than that of the pure solvent, i.e. silicon nitride (see Fig. 7). Thus, compared with silicon nitride, β'-sialon forms more liquid at a lower temperature with oxide additives such as magnesia and yttria. Control of the volume of liquid allows the material to be densified by pressureless sintering, i.e. without hot-pressing and like a conventional ceramic. The lower temperature of densification avoids excessive grain growth and so the fine-grain strength of the material is retained. Finally, the lower vapour pressure reduces volatilisation and decomposition at high temperatures; under normal preparative conditions, β'-sialon is thermodynamically more stable than β-silicon nitride.

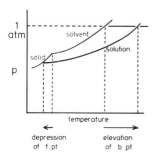

FIG. 7. Vapour pressures of solution and solvent.

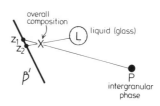

FIG. 8. Post-preparative heat treatment of β' + glass to give β' + YAG.

2.2. Densification with Yttria

After reaction sintering with yttria or yttria plus alumina, the mixed oxide and nitride powders corresponding to the required sialon composition give β' and the yttrium–sialon liquid necessary for densification. The latter cools to produce an intergranular glass. As shown later, this oxynitride glass is more refractory and stronger than an oxide glass but by post-preparative heat treatment at about 1400°C, or by controlled cooling, it can be reacted to give crystalline yttrium–aluminium–garnet ('YAG', $Y_3Al_5O_{12}$) in the grain boundaries together with a slightly changed β'-sialon composition:

$$Si_5AlON_7 + Y\text{-}Si\text{-}Al\text{-}O\text{-}N \rightarrow Si_{5+x}Al_{1-x}O_{1-x}N_{7+x} + Y_3Al_5O_{12} \quad (3)$$
$$\beta'\text{-sialon} \qquad \text{glass} \qquad\qquad \beta'\text{-sialon} \qquad\qquad \text{'YAG'}$$

These relationships are shown schematically by Fig. 8 from which it can be seen that the line of the β'-sialon phase and the position of the intergranular phase 'P' define a plane in the Jänecke prism (Fig. 9) representing the five component Y–Si–Al–O–N system. In order to effect the reaction given by eqn. (3), the overall composition 'X' and

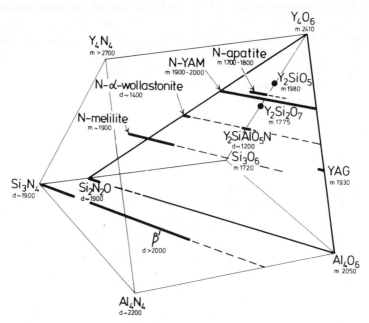

FIG. 9. Jänecke prism behaviour diagram showing some crystalline phases in the Y–Si–Al–O–N system.

the composition of the intergranular glass must also lie on this same plane. Figure 10 shows that the region of oxynitride glass composition formed by furnace cooling from 1700°C extends from the triangular oxide face (the Y–Si–Al–O glasses) of the Jänecke prism to a limit where about one in four oxygens are replaced by nitrogen. This glass region intersects with the β'-YAG plane as shown in Fig. 11, but comparison with the behaviour diagram of Fig. 12 at 1700°C shows that the oxynitride liquid formed at high temperature has not necessarily the same composition as the glass produced from it by cooling. Phase relationships and the products of glass devitrification change markedly with changes of heat-treatment time and temperature, and so it is not always easy to produce a two-phase ceramic of β' + YAG. Several of the crystalline phases shown in Fig. 9 and some of the polytypoids of Fig. 6 frequently occur in minor amounts due to non-uniform powder preparation, compositional variations caused by high-temperature volatilisation, incorrect compensation for surface oxides on the nitride particles, inadequate heat-treatment schedules and the incomplete control of many other processing variables. The aluminium-substituted N-α-wollastonite, Y_2SiAlO_5N (see Fig. 9), is unstable above about 1100°C but it commonly occurs in sialon

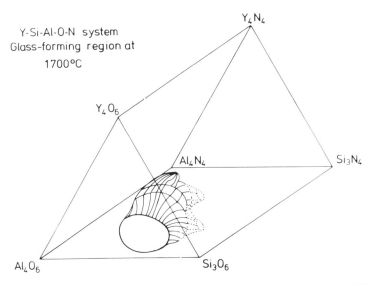

FIG. 10. Glass-forming region in the Y–Si–Al–O–N system on cooling from 1700°C.

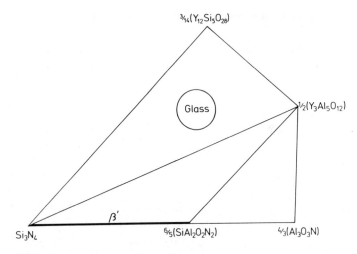

FIG. 11. Glass-forming region on the β'-YAG plane of the Y–Si–Al–O–N system after cooling from 1700°C.

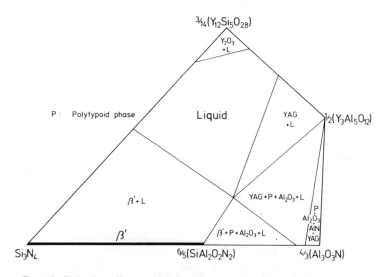

FIG. 12. Behaviour diagram showing phases on the β'-YAG plane at 1700°C.

reactions by glass devitrification. Its composition is within the Y-sialon glass region and so because its crystallisation requires only small atomic displacements it is formed readily at low temperatures.

3. COMMERCIAL SIALON CERAMICS

Based upon the above principles, extensive process development[9,10] has resulted in the present commercial production of β'-sialons. This is a family of materials consisting of two groups with different main microstructural constituents: (i) β'-sialon plus glass; and (ii) β'-sialon plus YAG. The strength of the glass-containing materials is high at room temperature, with three-point modulus of rupture approaching 1000 MPa, but it decreases as the intergranular glass softens above 1000°C. In type (ii) material, since the grain-boundary phase is an oxide, the oxidation resistance is excellent (Fig. 13), and because there is no intergranular glass it has good creep properties (Fig. 14). It has a somewhat lower room-temperature strength than type (i) but it retains this strength up to 1400°C.

The proportions of the vitreous and crystalline intergranular phases can be varied by changing the powder composition and heat treatment, and so the final mechanical and physical properties may be modified to suit particular applications. Properties of a type (i) β'-sialon are listed in Table 1; it is used for extrusion tooling, drawing dies and plugs, welding components, and industrial wear parts.

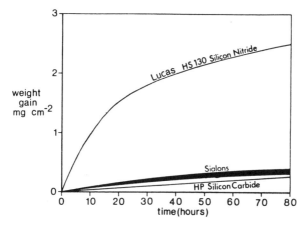

FIG. 13. Oxidation of β'-sialon at 1400°C in flowing dry air; after Arrol.[11]

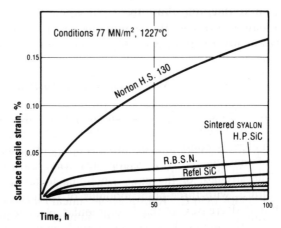

FIG. 14. Creep of β'-sialon at 1227°C and 77 MPa; after Lumby et al.[12]

TABLE 1
Typical Physical Properties of Type (i) β'-Sialon

Property	Value	Units
Three-point modulus of rupture, 20°C	945	MPa
Weibull modulus	15	—
Tensile strength, 20°C	450	MPa
Compressive strength, 20°C	3 500	MPa
Young's modulus of elasticity, 20°C	300	GPa
Hardness, 20°C	1 800	HV0·5
Fracture toughness (K_{1C})	7·7	MPa m$^{1/2}$
Poisson's ratio	0·23	—
Density	3 250	kg m^{-3}
Thermal expansion coefficient (0–1200°C)	$3·04 \times 10^{-6}$	K^{-1}
Thermal conductivity, 20°C	22	Wm^{-1} K^{-1}
Electrical resistivity, 20°C	10^{10}	ohm m
Permittivity at 10 GHz, 20°C	8·2	—
Loss tangent at 10 GHz, 20°C	0·002	—
Thermal shock resistance (quenched in cold water)	900	T°C
Coefficient of friction (sialon on sialon in 10 W 40 oil at 80°C)	0·04	—

3.1. Forming, Sintering and Machining

The appropriate mixes of nitride, oxide and oxynitride powders can be cold or warm formed prior to sintering by using all the shaping methods normally employed for oxide ceramics. These include: (a) isostatic pressing and, for large numbers of components, uniaxial pressing with and without binders and lubricants; (b) warm or cold extrusion with addition of plasticisers to produce continuous sections of pre-formed material; (c) injection moulding for intricate shapes; and (d) slip casting in aqueous media.

Pre-forms may be machined in the green state and, after debonding where necessary in air at up to 500°C, are then sprayed with a protective coating of refractory oxides before sintering in a nitrogen atmosphere at 1750–1850°C. Reproducible linear shrinkage occurs and, because this is allowed for, the final diamond machining of the hard, fully dense component to precise dimensional tolerances is reduced to a minimum.

3.2. Engineering Applications

Although the main motivation for the development of engineering ceramics was, and still is, the prospect of the ceramic gas turbine, its commercial realisation might well take another decade and, meanwhile, the unique properties of β'-sialons and their ease of fabrication are being applied in other directions.

So far, the most successful application of type (i) sialons has been as a cutting tool for machining metals. Figure 15 shows a selection of tool tips, and their performances in cutting cast iron, hardened steel and a nickel-based alloy are compared with those of cobalt-bonded tungsten carbide and of alumina in Table 2. The hot hardness of the sialon (see Fig. 16) is higher than that of the others and so it can be most effective at very high cutting speeds where the working temperature at the tool tip may exceed 1000°C. The 'lead time' for machining some of the RB211 aeroengine turbine discs was reduced to less than one quarter by using sialon inserts. They are being manufactured under licence by two major tool companies, one in Sweden and one in the USA.

Excellent thermal shock resistance, high-temperature mechanical strength and electrical insulation combine to make sialon unexcelled in welding operations. Figure 17 shows location pins for the resistance welding of captive nuts on vehicle chassis. The usual hardened steel pins in alumina insulating sleeves last for 7000 operations, i.e. a shift;

FIG. 15. Selection of sialon cutting tool tips.

TABLE 2
Cutting Performances of Co-Bonded WC, Al_2O_3, and a β'-Sialon

		Cast iron	Hardened steel EN31	Incoloy 901
WC	Cutting speed, m/min Depth of cut, mm Feed rate, mm/rev	250 6·5 0·50	5	20
Al_2O_3	Cutting speed, m/min Depth of cut, mm Feed rate, mm/rev	600 6·5 0·25	Impossible to cut	300 No second entry
β'-sialon	Cutting speed, m/min Depth of cut, mm Feed rate, mm/rev	1100 10·0 0·50	120 0·5 0·25	300 2·0 0·25

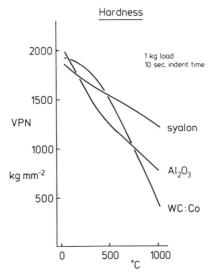

FIG. 16. Hot hardnesses of sialon, Al_2O_3 and WC:Co cutting tool tips.

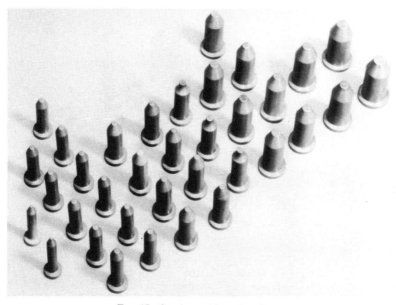

FIG. 17. Captive weld nut location pins.

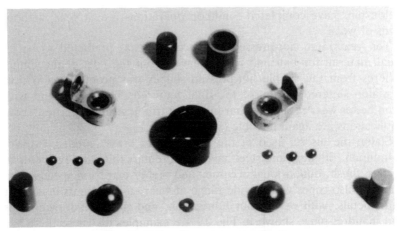

Fig. 18. Ball, roller and trunnion bearings.

Fig. 19. Metal tube-drawing plugs; coracle for pulling GaP single crystal from the melt.

sialon pins have completed 5 million operations, i.e. a year, without signs of wear.

Ten years ago, hot-pressed silicon nitride was predicted to be an ideal material for ball and roller bearings but the cost of machining spheres from the simple hot-pressed shapes was prohibitive. With a β'-sialon sintered to almost the final required dimensions, and with even better wear resistance, hardness and tribological properties, these applications are again possible (see Fig. 18).

Sialon die inserts used in the extrusion of brass, copper, bronze, aluminium, titanium and steel have given remarkable improvements in surface finish, dimensional accuracy and higher extrusion speeds. The material also copes with a wide range of wear environments in contact with metals, with or without lubrication, and the tube-drawing dies and mandrel plugs shown in Fig. 19 are examples of where there are marked increases in productivity compared with the conventional use of tungsten carbide.

Other applications depend on the resistance of sialons to most molten metals, including steel, although they are attacked by slag. It is

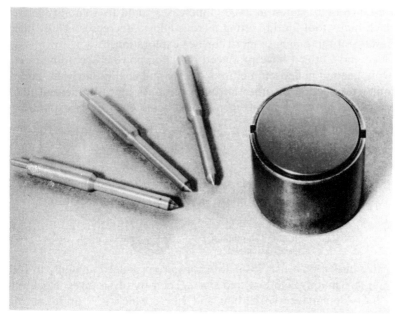

FIG. 20. Diesel fuel needle valves; diesel tappet shim.

being used in casting and metal spraying. The centre-piece of Fig. 19 is a coracle used in pulling single crystals from a gallium phosphide melt that attacks most other metals; it is a slip-cast β'-sialon. Examples of sialon automotive components are the fuel needle valves and the tappet shim of Fig. 20; after running for 60 000 km the tappet wear was less than 0·75 μm.

4. PROCESSING ROUTES AND RAW MATERIALS

Silicon and aluminium are the two most abundant metallic elements while oxygen and nitrogen make up the earth's atmosphere and so there is no possibility of a raw materials shortage for sialon manufacture. In what is now a large field, the β'-phase is the only sialon that has been explored in detail for its technological potential; others with at least equal promise are derived from α-silicon nitride and silicon oxynitride, Si_2N_2O. Moreover, the production of β'-sialon by different methods over its full range of composition has not yet been fully investigated. With increasing Al and O contents, the oxidation resistance of β'-sialon at 1400°C increases[13] and its solubility in iron, which limits tool life in cutting ferrous alloys, decreases. Thus, higher z-values of the β'-phase merit further exploration.

Lee and Cutler[14] have produced β'-sialon powder according to eqn. (4) by heating pelletised mixtures of clay and coal in nitrogen, and the similar production of other nitrogen ceramics by carbothermal reduction of mixed oxides in nitrogen is now being extensively explored.

$$3Al_2[Si_2O_5](OH)_4 + 15C + 5N_2 \rightarrow 2Si_3Al_3O_3N_5 + 6H_2O + 15CO$$
clay · · · · · · coal · · · nitrogen · · · β'-sialon, $z = 3$

(4)

Then, Umebayashi[15] in Japan obtained β'-sialon by reaction in nitrogen of volcanic ash (impure silica) and aluminium powder:

$$2SiO_2 + 4Al + 2N_2 \rightarrow Si_2Al_4O_4N_4 \qquad (5)$$
volcanic · aluminium · · · · · β'-sialon, $z = 4$
ash · · · · powder

Rice husks, of which 13 million tons are produced annually in India alone, are mainly cellulose and silica. Their pyrolysis gives 'black ash', an intimate mixture of carbon and silica, which is a useful starting material for silicon nitride and sialon production. Clay and coal, volcanic ash and rice husks are not going to produce pure β'-sialons for sophisticated applications like the ceramic turbine but they will provide useful refractory bricks, furnace linings and materials resistant to molten metals. Just as there are many grades of alumina ranging from (i) single-crystal sapphire, through (ii) transparent polycrystalline lucalox and (iii) 99·7% recrystallised alumina to (iv) the debased alumina refractories, so there will be a variety of β'-sialon grades, each of which will have its specific applications and its appropriate methods of manufacture. Unlike alumina, β'-sialon has excellent thermal shock properties and so can replace alumina, with advantage, in many of its uses.

5. α'-SIALONS

α'-structures based on the $Si_{12}N_{16}$ unit cell of α-silicon nitride occur in M–Si–Al–O–N systems where M includes Li, Ca, Y and all the rare earth elements from Nd to Lu. Appropriate mixtures of the nitrides, or of nitrides plus oxide, are heated without pressure at 1750°C in nitrogen or argon:

$$0·5Ca_3N_2 + 3Si_3N_4 + 3AlN \rightarrow Ca_{1·5}[Si_9Al_3N_{16}] \quad (6)$$

or

$$CaO + 3Si_3N_4 + 3AlN \rightarrow Ca[Si_9Al_3ON_{15}] \quad (7)$$

Unlike that of β, the α unit cell has two large interstitial sites that can accommodate additional atoms: α' is derived by partial replacement of Si^{4+} with Al^{3+} and valency compensation is effected by cations such as Ca^{2+} occupying these interstices in the Si–Al–N network. The structural principle is similar to that in the formation of 'stuffed' derivatives of quartz in which Al^{3+} replaces Si^{4+} and positive valency deficiencies are compensated by 'stuffing' Li^{1+} or Mg^{2+} into interstitial sites:

$$Si_2O_4 \rightarrow Li[SiAlO_4] \quad (8)$$
$$\text{quartz} \quad \beta\text{-eucryptite}$$
$$Si^{4+} \rightleftharpoons Li^{1+}Al^{3+} \quad (9)$$

When α' is synthesised entirely from nitrides the valency compensation is due solely to the additional modifier cations and the limiting compositions contain not more than two of these per unit cell, e.g. $Ca_2[Si_8Al_4N_{16}]$ or $Y_2[Si_6Al_6N_{16}]$ although these limits have not been

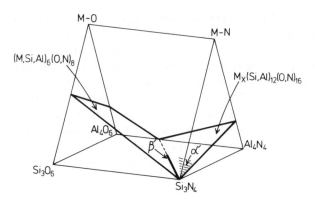

FIG. 21. Phase relationships between α' and β' sialons.

achieved experimentally. Where a modifier oxide is used, oxygen replaces nitrogen in the network but its limit is probably about one oxygen atom per unit cell. The α'-phases are stable in inert, nitriding and carburising atmospheres up to 1750°C. Like β'-sialons, they can be prepared by carbothermal reduction of mixed oxides in nitrogen. They have good oxidation resistance up to 1350°C and have coefficients of thermal expansion similar to that of silicon nitride ($\sim 3 \times 10^{-6}$/°C at 0–1250°C). Prospects for their technological application are as good as those for β'-sialon.[16]

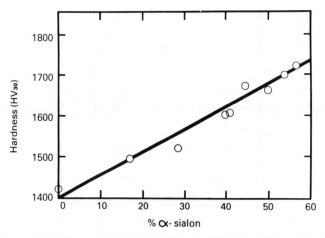

FIG. 22. Variation of hardness with α' content in $\alpha' : \beta'$ sialon composites (after Sandvik Hard Materials Limited).

Phase relationships between α' and β' are represented in Fig. 21. The phases are ideally compatible and composites of them can be prepared with different $\alpha':\beta'$ ratios from the appropriate oxide–nitride powder mixes by pressureless sintering in a single-stage process. The rates of α' and β' formation differ, and so by varying the composition and the reaction sintering conditions it is possible to vary and control the microstructure in a way that is impossible for a single-phase sialon. These composites are already being used in preference to β' in cutting tool applications because, as shown by Fig. 22, the hardness increases linearly with increasing α' content while strength remains unchanged. The hardness increment remains almost constant with increasing temperature even though the overall values decrease, and so at 1000°C the $50\alpha':50\beta'$ composite is about twice as hard as the pure β'-sialon (see Fig. 23).

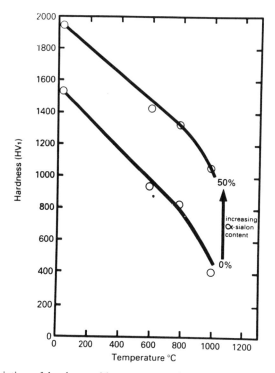

FIG. 23. Variation of hardness with temperature for $50\alpha':50\beta'$ composite compared with $100\beta'$ (after Sandvik Hard Materials Limited).

6. O'-SIALONS

There is a limited solubility (~10 mol % Al_2O_3 at 1800°C, see Fig. 6) of alumina in silicon oxynitride, Si_2N_2O, to give an O'-sialon solid solution without change of structure. A suitable additive such as Y_2O_3 lowers the solidus temperature and, by increasing the volume of liquid, allows densification by pressureless sintering in the same way as in the processing of β'-sialons.

If the overall compositions lie on the $Si_2N_2O-Al_2O_3-Y_2Si_2O_7$ plane of the Y–Si–Al–O–N system shown by Fig. 24, mixed powders of Si_3N_4, SiO_2, Al_2O_3 and Y_2O_3 react and densify at 1600–1800°C to produce O'-sialon ceramics with an intergranular glass phase that can be devitrified by post-preparative heat treatment to give $Y_2Si_2O_7$. Once again the products have low coefficients of thermal expansion ($\alpha \sim 2 \cdot 9 \times 10^{-6}$/°C) and good thermal shock properties. They are oxidation resistant to 1350°C and should be as useful as α' and β' in engineering applications.[17]

6.1. O'–β' Composites

Dense O'–β' ceramics have been obtained[18] by pressureless sintering the appropriate powder mixes and devitrifying the intergranular glass to give $Y_2Si_2O_7$ and YAG. It was hoped to combine the strength of β' with the oxidation resistance of O'. Microstructures have been changed by varying the proportions of the two compatible sialon phases, and with approximately 50O':50β' the material is oxidation resistant to above 1300°C. Much more research and development are required, however, before the potential of these and other composites can be assessed.

6.2. Electrical Properties

Silicon oxynitride is built up of SiN_3O tetrahedra joined by sharing corners and the orthorhombic crystal structure consists of irregular but parallel sheets of covalently bonded silicon and nitrogen atoms linked by Si–O–Si bonds. It can be regarded as a defect lithium–silicon nitride (see Fig. 25) in which one nitrogen is replaced by oxygen and lithium is removed for valency compensation:

$$LiSi_2N_3 \rightarrow \square Si_2N_2O \qquad (10)$$

Because of the structural similarities between the nitride and oxynitride there is some mutual solid solubility and up to

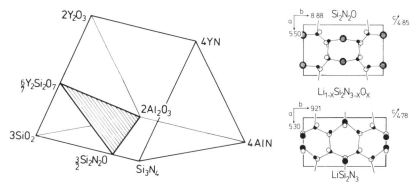

FIG. 24. The Si_2N_2O–Al_2O_3–$Y_2Si_2O_7$ plane of the Y–Si–Al–O–N system.

FIG. 25. Crystal structures of Si_2N_2O and $LiSi_2N_3$.

12 mol % $LiSi_2N_3$ dissolves in Si_2N_2O. Thus, in the O'-sialons the replacement

$$O^{2-} \rightleftharpoons Li^{1+}N^{3-} \qquad (11)$$

occurs separately from or simultaneously with those represented by eqns (1) and (9). The Li^{1+} ions, and perhaps other cations, are located between the parallel silicon–nitrogen sheets and it seems feasible that these materials might provide ceramics with fast cation transport and so become useful solid electrolytes similar to the β-aluminas.

7. NITROGEN GLASSES

Glasses occur in all the sialon systems so far studied and are important because the mechanical properties of the nitrogen ceramics, particularly their high-temperature strength and creep resistance, depend on the amount and characteristics of the grain-boundary glass. The glasses are also of interest in their own right, and a systematic study of the Mg, Ca, Y and Nd sialon systems[19] shows that by fusing powder mixtures of SiO_2, Al_2O_3, Si_3N_4 and AlN with the appropriate metal oxide at 1600–1700°C in a nitrogen atmosphere and then furnace cooling, glasses containing up to 15 atomic % N are obtained, i.e. one oxygen in four can be replaced by nitrogen.

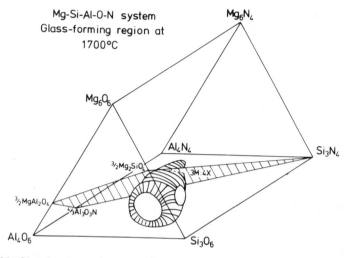

FIG. 26. Glass-forming region in the Mg–Si–Al–O–N system on cooling from 1700°C.

Figures 10 and 26 show the glass forming regions in the respective Y–Si–Al–O–N and Mg–Si–Al–O–N systems on cooling from 1700°C. Initially, as nitrogen replaces oxygen, glass formation is facilitated and the vitreous region expands. In oxide glasses Si and Al are 'network formers' because they are 4-coordinated by oxygen (MO_4). Other

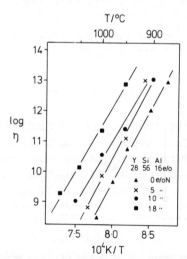

FIG. 27. Viscosities of Y–Si–Al–O–N glasses with increasing nitrogen concentration.

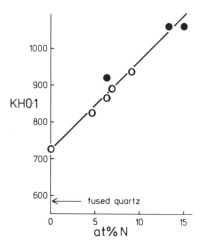

FIG. 28. The hardness of Y–Si–Al–O–N glasses. ○, Shillito et al.[20]; ●, Messier and Broz.[21]

cations are 'network modifiers' and are 6-coordinated (MO_6). However, in nitride systems such cations are 4-coordinated (e.g. MgN_4) and they become 'network formers'. With increasing replacement in the glass of oxygen by nitrogen, the increasing covalency of the bonding gives it more directional character and increases the tendency to crystallinity in the structure. The net effect of these opposing trends, as nitrogen replaces oxygen, gives first an expansion and then a contraction of the glass-forming region. Ultimately a limit is reached beyond which no glass is obtained by normal furnace cooling.

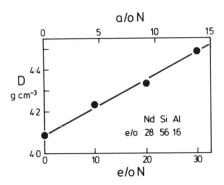

FIG. 29. The variation of density of Nd–Si–Al–O–N glasses with increasing nitrogen concentration.

Oxygen in the glass network is coordinated by only two metal atoms whereas nitrogen is bonded to three ligands. Thus, as might be expected, the viscosity of the glass (Fig. 27), the hardness (Fig. 28), and the density (Fig. 29) all increase with increasing nitrogen concentration. Where more refractory and erosion-resistant glasses are required, for example in radome applications, these nitrogen glasses are possible candidates.

8. CONCLUSIONS

Simple principles of crystal chemistry have been applied to produce new crystalline and vitreous metal–Si–Al–O–N materials by 'ceramic alloying'. The first of these to be commercially developed, β'-sialon,

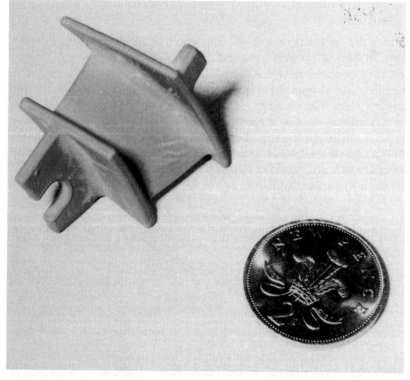

FIG. 30. β'-sialon turbine stator guide vane.

has applications as cutting tools, engine components and wear parts. It promises to be a leading contender for the ceramic gas turbine. Other sialons, particularly those based upon α-silicon nitride and O-silicon oxynitride, show equal promise. By varying the proportions of the compatible phases in composites of $\alpha'-\beta'$ and $O'-\beta'$, microstructures can be tailored to give desirable physical and mechanical properties more easily than with a single-phase sialon.

Figure 30 shows a sialon turbine stator guide vane made by injection moulding. It points the shape of things to come, but it also emphasises the difficulties to be surmounted in processing and in fabricating useful engineering components without sacrificing the properties and homogeneity that, at present, are obtained only with severe size limitations.

ACKNOWLEDGEMENTS

Presentation and publication of this paper has been assisted by the award of a Leverhulme Emeritus Fellowship. Figures 22 and 23 are reproduced by permission of Sandvik Hard Materials Limited.

REFERENCES

1. PARR, N. L., MARTIN, G. F. and MAY, E. R. W. *Study and Application of Silicon Nitride as a High Temperature Material,* Admiralty Materials Laboratory Report No. A/75(s), 1959.
2. WILD, S., GRIEVESON, P., JACK, K. H. and LATIMER, M. J. In *Special Ceramics, Vol. 5,* Popper, P. (Ed.), British Ceramic Research Association, Stoke-on-Trent, 377, 1972.
3. HAMPSHIRE, S. and JACK, K. H. In *Special Ceramics Vol. 7,* Taylor, D. E. and Popper, P. (Eds), *Proceedings of the British Ceramic Society, Vol. 31,* 37, 1981.
4. GIACHELLO, A. and POPPER, P. *Ceramurgia International,* **5** (1979) 110.
5. MANGELS, J. A. and TENNENHOUSE, G. J. *American Ceramic Society Bulletin,* **50** (1980) 1216.
6. JACK, K. H. *Transactions and Journal of the British Ceramic Society,* **72** (1973) 376.
7. OYAMA, Y. and KAMIGAITA, O. *Japanese Journal of Applied Physics,* **10** (1971) 1637.
8. JACK, K. H. and WILSON, W. I. *Nature,* **238** (1972) 28.
9. LEWIS, M. H., BHATTI, A. R., LUMBY, R. J. and NORTH, B. *Journal of Materials Science,* **15** (1980) 103.

10. LUMBY, R. J., BUTLER, E. and LEWIS, M. H. In *Progress in Nitrogen Ceramics*, Riley, F. L. (Ed.), Martinus Nijhoff, The Hague, 323, 1983.
11. ARROL, W. J. In Ceramics for high performance applications, *Proceedings of the Second Army Materials Technology Conference*, Hyannis, 1973, Burke, J. J., Gorum, A. E. and Katz, R. N. (Eds), Brook Hill Publishing Company, Chestnut Hill, U.S.A., 729, 1974.
12. LUMBY, R. J., NORTH, B. and TAYLOR, A. J. In Ceramics for high performance application II, *Proceedings of the Fifth Army Materials Technology Conference*, Newport, R.I., 1977, Burke, J. J., Lenoe, E. N. and Katz, R. N. (Eds), Brook Hill Publishing Company, Chestnut Hill, U.S.A., 839, 1978.
13. PETZOW, G. and GREIL, P. In *Proceedings of the First International Symposium on Ceramic Components for Engine*, Hakone, 1983, Somiya, S., Kanai, E. and Ando, K. (Eds), KTK Scientific Publishers, Tokyo, 177, 1984.
14. LEE, J. G. and CUTLER, I. B. *American Ceramic Society Bulletin*, **58** (1979), 869.
15. UMEBAYASHI, S. In *Nitrogen Ceramics*, Riley, F. L. (Ed.), Noordhoff, Leyden, 323, 1977.
16. HAMPSHIRE, S., PARK, H. K., THOMPSON, D. P. and JACK, K. H. *Nature*, **274** (1978) 880.
17. TRIGG, M. B. and JACK, K. H. *Proceedings of the First International Symposium on Ceramic Components for Engine*, Hakone, 1983, Somiya, S., Kanai, E. and Ando, K. (Eds), KTK Scientific Publishers, Tokyo, 199, 1984.
18. SUN, W. Y., THOMPSON, D. P. and JACK, K. H. In Tailoring multiphase and composite ceramics, *Proceedings of the Twenty-First University Conference on Ceramic Science*, Pennsylvania State University, 1985, to be published.
19. DREW, R. A. L., HAMPSHIRE, S. and JACK, K. H. In *Special Ceramics, Vol. 7*, Taylor, D. and Popper, P. (Eds), *Proceedings of the British Ceramic Society, Vol. 31*, 119, 1981.
20. SHILLITO, K. R., WILLS, R. R. and BENNETT, R. B. *Journal of the American Ceramic Society*, **61** (1978) 537.
21. MESSIER, D. R. and BROZ, A. *Journal of the American Ceramic Society*, **65** (1982) C-123.

COMMENTS AND DISCUSSION

Chairman: R. J. BROOK

H. Moeller: What material was used for indentation in your hot-hardness testing?

K. H. Jack: A diamond indentor, I believe. These results are not from Lucas Cookson Syalon. Dr Ekström could comment on the Sandvik technique.

T. Ekström: We use a diamond indentor which will work up to 1000°C without having to be replaced.

Jack: The figures presented for β'-sialon were from Kennametal. Sandvik show similar results but it is only these tool manufacturers who have measured hot-hardness. I have no results available from Lucas Cookson Syalon.

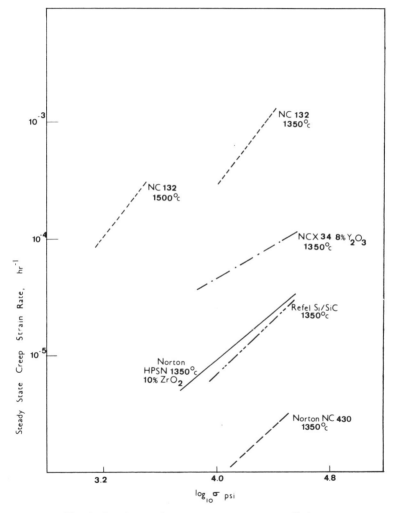

FIG. 1. Steady state flexural creep rate versus applied stress.

R. W. Davidge: If the hardness of the α'-sialon is better than that of the β'-sialon, why is there a 50–50 mix? What is the advantage of that compared with all α' for metal cutting?

Jack: Because it is a bit more difficult to make pure α' by normal processing.[1] It is easier to produce a 50–50 mixture. Also one of the potential advantages is in tailoring the microstructure. If you have two completely compatible phases then it is beneficial if you can get a dispersion of one in the other. Sandvik may have solved this. Have you?

Ekström: No. It will take another ten years because there are so many dimensions to play around with.

Jack: Yes. A lot more processing research is required—process development, but potentially I think this type of approach could be very good.

L. Brosnan: Is the use of sialon grain in abrasives a commercial proposition?

Jack: I don't know because I think you need other properties for good abrasives such as specific fracture surfaces.

Brosnan: There is also the question of bonding agents.

Jack: I don't know if anyone has looked at this. I suspect not, but it would be a good field to explore.

R. C. Gambie: The data on creep which you presented (see Fig. 14) shows a high surface tensile strain rate for Norton HS130 material compared with syalon, RBSN and Refel SiC. I would like to point out that the HS130 was material originally supplied by Lucas under the Norton/Lucas licence arrangement and must be at least 10 years out of date! The more recent data shown in Fig. 1 shows steady-state flexural creep rate versus applied stress for a variety of Norton hot-pressed silicon nitrides and SiC materials.

Editor's Note: Compare also with data of Lewis *et al.*, this volume, pp. 175–190.

Reference
1. SLASOR, S. and THOMPSON, D. P. This volume, pp. 223–230.

2

Discussion Session: Aspects of Commercialisation of Advanced Ceramics and Future Opportunities

Chairman: R. J. BROOK

CHAIRMAN'S OPENING REMARKS

We have the challenge thrown down by Professor Jack with respect to a material with a vast range of properties.

Dr Alan Rae (formerly of Anzon Ltd) was to have introduced this session and he is responsible for 'Rae's 1st rule':

'If you build a better mousetrap the world will beat a path to your door, . . . unless you're trying to sell ceramics!'

I would like to set the context for commercialisation of advanced ceramics since we have representatives from industry, universities and government from different countries.

The Ceramics industry can be broken down into four traditional sectors: (1) Refractories; (2) Heavy clayware; (3) Whitewares; (4) Cements and Glasses—which are developing classically at relatively slow rates and which are not the cause for the large excitement which has occurred in ceramics over the past 5–10 years. There were over 200 participants at the Oyez 1st European Engineering Ceramics meeting in London in February 1985.[1] They were not there to hear about tableware or cements. The reason they were there was because of the fine ceramics sector which is variously known by other names (special ceramics, technical ceramics, engineering ceramics). This sector breaks down on a market basis into three subsectors.

The first of these consists of 3·5 billion dollars per annum of electrical components where we use the special properties that ceramics give with respect to magnetic and electrical behaviour. Those properties are not available in other systems and, therefore, the marketing has been relatively straight forward. It is a healthy sector—relatively large growth rates of about 15% per annum—and I think those who are in electrical engineering have plotted the way forward there very carefully. It is a fast moving field but it is not the one that draws people to conferences of this kind.

The cutting tool sector, worth about $1 billion per annum (if you include tungsten carbide) has been mentioned previously with respect to the sialons[2] and this is a success story for engineering ceramics. It is an existing market which is growing and we can link it with the electrical ceramics market in following a successful development pattern.

The one sector with which the great uncertainty exists is the use of ceramics as structural engineering components, that is, in load bearing systems at high temperatures. It is a small market at the moment, $250 million per annum, but that is the one which is expected to grow rapidly and that is the one that people ask questions about.

There are difficulties with it. The engine market really breaks down into two further sectors; there is the gas turbine where there is the very clear advantage in going to high temperatures. If you can increase the temperature from 1100°C to 1300°C the economic advantage is clear and ceramic components could be used in the turbine blades, combustion chamber, bearings, etc., of such an engine. There are a number of programmes around the world involved with this. The economic advantage is clear but the reliability in use is critical. If you are using these materials in aircraft, for example, you have to be sure they are going to work, so the technical challenge is peculiarly difficult because we have to solve the reliability problem but the inducement in terms of financial advantage is very clear. So that is a slight paradox and a difficulty which suggests that it will be 15–20 years before we see ceramics in aircraft engines.

On the diesel side, it is more attractive because you can put components into the engine which may fail but, provided that not too many customers are irritated, then it is alright because you can always get out of a car and walk home! The economic advantage is not so clear. You can use ceramics as wear components, as thermal insulation, or as low inertia systems for turbocharger rotors, a luxury market. If you were to buy a car with a ceramic rotor, you would get away from the traffic lights one second before the person in a car with a metal turbo-charger! Now that may sell motor cars because if you bought a luxury car which cost a great deal and it was one second slower at the lights than some rival vehicle, then you would be irritated. It is, however, hardly the great promise that was expected from ceramics, satisfying a small luxury market rather than bringing the great savings in fuel which were expected.

I think engines, then, are a problem. With the gas turbine there is potentially a great advantage but the technical task is extremely difficult whereas in the diesel where the technical task is still difficult, but possible, the advantages are less apparent. Components have been made in certain materials, particularly partially stabilised zirconia, but from the marketing side it seems less clear. And then there are a number of smaller applications like wire-drawing dies, scissors, etc., which use ceramics.

The technical target has been well identified by the community who want to make dense materials. They should be free totally from faults which are bigger than say 10 μm. There should be no grain boundary glass if they are to be used at high temperature, as has been made clear,[2] and they should be tough.

The technical targets are known. If we look at the research underway, most people follow the classical ceramic processing route.

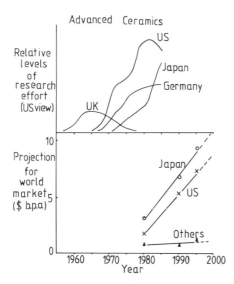

FIG. 1. US view of national efforts and projected world markets for advanced ceramics.

There is great emphasis on powders. Many people have recognised this and are working on it. There is great emphasis on homogeneity, that is, how do you fabricate systems where the powder packing is uniform throughout, where you do not have regions that are dense and others which are not? Finally there is much activity in terms of toughening and the zirconia toughening of alumina is an example of developments that have taken place. So I think the technical lines along which people are moving are clear.

Figure 1 shows the American predictions of how things are going.[1,3] The upper portion shows the national efforts on funding over the last few years and the present situation is that there is a large American programme, a substantial German interest and lately a significant Japanese involvement. The lower portion shows the predicted markets with Japan and the United States dominating the situation. It also shows clearly the position the Americans have assigned to the rest of us!

The link between investment and return in years to come is one which some believe in and one which we in Europe should take very seriously. This is the background for the discussion session.

REFERENCES

1. *Proceedings of the First European Symposium on Engineering Ceramics*, London, February 1985, Oyez Scientific and Technical Services Ltd.
2. JACK, K. H. Sialons: A Study in Materials Development. This volume, pp. 1–30.
3. LENOE, E. M. and MEGLEN, J. L. *American Ceramic Society Bulletin*, **64** (1985) 271.

COMMENTS AND DISCUSSION

R. J. Brook: The first thing I would ask is if anyone disagrees with the somewhat gloomy prognosis I have made which is rather a contrasting picture to the very bullish proposal that Professor Jack has made. Do people see that as an unfair contrast?

K. H. Jack: I am only 'bullish' because I think the developments will come from Japan. It depends on attitudes to markets. The big steel companies in Japan are diversifying into fine ceramics because they see in the not too distant future (and Japan has the highest profitability from steel compared with any other country) that the Koreans will have the edge over them. So, Nippon Steel and Kawasaki Steel are putting a tremendous amount of very sophisticated effort into the development of fine ceramics. Now the Irish steel industry couldn't do this, nor could the British steel industry, nor even could the European steel industry do it. Duncan Davies tried to stimulate the 'affluent' companies in Britain like Shell, ICI to invest in ceramics[1] but they are not interested because very few people in Europe will touch anything that is the slightest bit speculative. The Japanese, however, are prepared to wait a long time for the return on investment.

Brook: It is an important distinction and perhaps someone from industry here would like to comment upon this.

At the Oyez (Ref. 1 of paper) meeting the Japanese analyst that was there said that there was no nervousness in Japan about where the precise markets would come from. It was sufficient to invest in ceramic technology and learn how to use ceramics and then the markets would be there. Am I being naive?

K. Parkinson: No, they did it before in other areas.

S. Hampshire: So what is the explanation for all the feet-dragging in Europe compared with the Japanese and the US?

Parkinson: I don't know.

D. Broussaud: You said that ceramic manufacturers were making scissors and so on. I think it is very serious because all the high performance ceramics will be used in non-sophisticated areas before being introduced into engines and so on. I know that in France they are looking for those types of markets for surgical instruments, cutlery, industrial cutters and so on where ceramics have never been tried before and I think we shall see many applications in those non-sophisticated fields.

Brook: Can we follow that point through? Just some features of the market. I think it is the Japanese view that faith in your ability to manufacture ceramics is sufficient. How can you afford as an industrial nation in the future, able to manufacture metals and polymers, not to be able to make ceramics. The approach is 'let us do it and the markets will come'. I think that one thing with respect to cutting tools and scissors and so on is an element of secrecy. Many people and companies have recognised ceramics as a possibility, many see it as a way for steel companies, for example, to diversify and so if they identify a market they keep very quiet about it. Do you think that is true?

Broussaud: Yes.

Brook: So this view that markets are difficult to find may not be fair.
J. T. Smith: Let us try a different approach to markets. First of all we have to accept that the electronics ceramics market belongs to the Japanese. With regard to the cutting tool market, which you identified as $1 billion with tungsten carbide, I prefer to consider it as a $20 million market without tungsten carbide, worldwide. So we are looking at Kennametal, Lucas-Cookson, Sandvik, GTE, etc., and much of that $20 million is in alumina and as silicon nitride and sialon represent about 10% of that or $2 million, then it really is quite a small market at the moment. And the question is why?—it is probably twofold.

(1) Machines—you've got to have machines to take the silicon nitride tools. That is a slow process and an investment process.
(2) You've got to educate the people to use the tools.

Another difference, as I see it, between Japan on the one hand and the United States with the rest of Europe on the other is monetary. With money costing close to 15% at the moment, it is very difficult to compete with Japan where the cost of money is only half that plus the investment is defrayed by tax incentives, so it comes down to a business situation. Those of us in research can enjoy that research but the business man must convince the bank and that's the difference.

Brook: But if you have a share of some existing market in cutting tools, like, for example, tungsten carbide, then you can introduce ceramics—you have got an existing market. It is a big advantage when compared with someone just coming in to a new market like the chemical companies would have to do.

A. Roos: I think there is one basic thing that is lacking, not only in individual countries but in Europe as a whole—we have never learnt, and are still not prepared, to do things together, to join forces and as long as we are not prepared to do this we will be at a loss. That is the whole problem; there is money enough for investment.

Brook: Any comment from a European on that? I take it you (to Nosbusch) would agree?

H. Nosbusch: That gentleman would be very welcome in Brussels!

Brook: These are some features of the market we have discussed. I think the engines paradox is one which I've tried to identify. Needs—people have mentioned funding and cooperation. That is a problem certainly we have in Britain. Then there is the risk that individual countries will try to put their own act together. So this lack of coordination is a difficulty. The education of engineers was mentioned.

I think I'll come back to Professor Jack's remarks—it seems to me that the problems with ceramics are twofold, they are brittle and they are expensive whereas he would claim they are low-cost.
Jack: No, I am saying that there are low-cost refractory materials made from cheap raw materials but I do not think that you are going to get a cheap ceramic for a high performance application such as an aircraft engine.
P. Kennedy: I think the problem at the moment is demand. If the demand was there, these advanced ceramics could be made perhaps at sufficiently low cost.
Brook: If you have looked at Professor Timoney's engine in the exhibition there is a metal component at $35 but the ceramic equivalent is $7000. Is mass production going to bring it down that far?
W. G. Long: It will never get down to $35.
Brook: And does that matter? My experience of the motor industry is that it matters a lot. One company's attitude at the moment is that they know about ceramics, they've done some experiments with them, have characterised them and, the moment they become cost effective, they will use them.
Broussaud: There is the attachment cost also.
T. Murakami: I feel that I carry a big responsibility since there are very few Japanese people here! Thank you all for your kind remarks about Japan but I must say that I learned very much from the UK and Europe on how to use ceramics and we have still a lot to learn from you. With regard to the price problem, all Japanese ceramics manufacturers will be working closely with diesel engine manufacturers from the design point of view and will sell ceramic turbochargers cheaper than Inconel 743. I don't know if that is profitable but they will do it in order to get the market.
Brook: That is a very interesting comment. You are actually a user of ceramics? (Yes) How many other potential users are here rather than manufacturers? Yes, some.
R. Wallace: Could I make one comment on reliability. We are trying to push forward silicon carbide as a replacement for hard metals. The comment you get from customers is that the hard metal industries do not know what they are doing in promoting quality and the ceramics industry is even worse. There is that resistance; you've got to convince people of consistent quality.
Brook: Yes, I think that is recognised now.
R. T. Cundhill: But there is no real paradox with gas turbine engines

for aeroplanes. Just changing the specification of a metal alloy can take 5 years.

Brook: So a 20 year pay-off is not too serious a problem?

Cundhill: No, at the moment, we are working on the bearings for the next generation of engines, not the current ones.

A silicon nitride ball of approximately 25 mm diameter, made in 1984, cost $100 a ball. Well that is not too bad because a carbide ball of approximately the same size would cost $35. And going into net-shape manufacturing can reduce that.

That is not really a problem though. The other thing is that a big bearing in one of these engines may take 26 of these balls, so that's $2600, or $1300 next year. But the bearings themselves will sell for between $5000 and $10 000 each. So the cost is not too important and I think we can educate our customers in turn. The cost will come down and will not be prohibitive.

Brook: That is very encouraging and fits with the view that there is more commercial progress than immediately meets the eye.

Jack: Dr Shaw can probably tell us how the Dyson Group do things. They have a good export business to Japan in ceramic materials. How does this happen?

B. Shaw: From a small beginning. There is an application and, in that particular case (not the type of material we are discussing now), there was a market, which we didn't immediately identify, which grew, maybe by a small amount of luck. I think it comes back to this act of faith all the time. We're a bit of a bottom-line brigade in this part of the world always looking for ready markets, now.

Brook: That is partly scale though. Recently, I was in China and was surprised to find that they are spending four times as much on engineering ceramics as the British government is. Perhaps I shouldn't be. It seems that they have great advantages of scale and of coordination. Where you have a target that is technically difficult and perhaps an act of faith is required, then using the market to lead you there is perhaps expecting too much?

Broussaud: Yes, in many countries there is much emphasis on and significant funding for research on new materials. In the States they are building up a large programme on ceramic fibre composites and so what are the prospects for these types of materials?

Brook: I think it comes under one of the three main lines of investigation that I mentioned: powders, homogeneity and toughening. Fibres are one of the key ways to toughening and I get the impression that is one of the richest areas for exploration—things are very quiet, it is difficult to get clear information on what people are doing, it is a sensitive area and it is technologically difficult as well. This is one of the cases where the US information embargo is most clear. That is always a good sign that something is happening!

Finally, let me draw this together a little. It has been rather more encouraging that I was expecting at the outset. One feature is that people seem to have found this sufficiency of faith a reasonable thing. That is certainly a view which is held elsewhere and perhaps we shall have to build up confidence in waiting for the larger term. I think the coordination issue is one which has been made in a heartfelt way, which Europe is attempting to do something about and to which meetings like the present one have some contribution to make.

Reference

1. *Symposium on New Developments and Investments in Ceramics,* London, April, 1985, The Fellowship of Engineering.

3

Analysis of Coarsening and Densification Kinetics During the Heat Treatment of Nitrogen Ceramics

H. PICKUP, U. EISELE, E. GILBART AND R. J. BROOK
Department of Ceramics, University of Leeds, LS2 9JT, UK

ABSTRACT

In view of the recognised difficulties associated with the identification of atomic mechanisms during the microstructural development of ceramics, a technique has been developed which allows estimation of the relative rates of densification and coarsening (grain growth) in a sintering compact. The results of applying the method to two silicon nitride powders are given and the possible benefits of the method discussed.

1. INTRODUCTION

Despite the very many studies that have been made of the subject, it is now recognised that attempts to identify the mechanisms of microstructural change during sintering can be unrewarding. The range of possible mechanisms, the limited spans of experimental conditions over which a given mechanism may be dominant, the many departures from geometric ideality encountered in a typical sintering compact, and the complex interaction between the two general processes (densification and grain growth), whereby a compact may lower its total interfacial energy, make it difficult first to identify unambiguously the controlling mechanism in any given experiment and secondly to be confident that a similar mechanism will apply in a parallel study of the same chemical system in which for example slightly different conditions of temperature, pressure, atmospheric environment or even powder source are used. This difficulty in drawing conclusions of general validity from mechanism studies under a particular set of conditions can seriously limit the applicability of such work.

A possible response to this problem is to recognise the value as a figure of merit of the relative rates of densification and of grain growth.[1] For many applications of ceramics where high strength is required, materials of high density are sought. Since grain growth acts to reduce the rate of densification by increasing such terms[2] as the path length for atom diffusion from the grain boundaries to the pores, there is value in using processing conditions both where the densification rate at a given grain size is high and where the grain growth rate at a given density is low; the need for large grains, if it exists, can be met by annealing the fully dense material subsequent to densification.

A clear target in processing, therefore, is to enhance the ratio of densification rate to grain growth rate for the system under study. The purpose of this paper is to apply a simple experimental procedure for measuring this ratio to two silicon nitride powders and to assess its merit as a means of selecting and refining processing conditions.

2. EXPERIMENTAL METHOD

The simplest form of the procedure involves the use of hot-pressing kinetics to calculate the ratio $(\dot{G}/G)(\rho/\dot{\rho})$ where G is grain size and ρ density. Extensions of the method[3] allow its use both for pressureless sintering studies and for the estimation of variants on the ratio, notably the resulting change in pore size, r, as a consequence of the combined effect of coalescence ($\dot{r}$ positive) and densification ($\dot{r}$ negative); these will not be discussed further here.

The assumption made at the outset is that the densification rate depends upon the grain size. With the form of dependence commonly supposed,[4]

$$\dot{\rho} = KG^{-m} \tag{1}$$

where K is a constant and m an exponent. While the method is not specific to a given mechanism, the requirement for a grain size dependence makes it inapplicable to systems which densify by power law creep (dislocation processes).[5] It is thus important to measure the stress dependence of the densification rate in a preliminary experiment to ensure that such processes are not dominant, i.e. that $n < 3$ in

$$\dot{\rho} = K'\sigma^n \qquad (2)$$

where σ is the stress applied in hot-pressing.[4]
By solving[6] for G in eqn. (1), taking the time derivative and simplifying

$$\frac{\dot{G}}{G}\frac{\rho}{\dot{\rho}} = -\frac{1}{m}\frac{\ddot{\rho}\rho}{\dot{\rho}^2} = \Gamma^* \qquad (3)$$

The desired ratio (the inverse of the figure of merit is used for historical reasons[1]) is therefore readily estimated if one has a continuous record of the sample density as a function of time during hot-pressing.

3. RESULTS

The two silicon nitride powders examined were prepared in different ways: the first (Starck LC12) is made by the direct nitridation of silicon; the second (Toshiba 'a') is made by the carbothermal reduction of silica in a nitriding atmosphere. Scanning electron micrographs of the two powders are shown in Fig. 1.

The powders were hot-pressed[7] in graphite dies at 1650°C using an applied pressure of 10 MPa. The same sintering aid was used in the two cases, namely, 10 wt % of an equimolar mixture of MgO/Y_2O_3. Continuous readout of the sample dimension parallel to the hot-pressing direction coupled with measurement of final sample size and density, allowed continuous access to sample density as a function of time.

A polynomial function was fitted to the data in the density versus log time format; this is shown for the two powders in Fig. 2. As noted earlier, a preliminary to the evaluation of the ratio is the verification that power law creep processes are not responsible for the observed microstructural changes. This is shown in Fig. 3 where data for the pressure dependence of the densification rate are given for a Starck silicon nitride powder; the close to linear dependence is similar to that found in other studies of silicon nitride densification[8–10] in the presence of sintering aids and is indicative that power law creep processes are not active.[4]

To develop the ratio, the first and second derivatives of the

FIG. 1. Scanning electron micrographs of the two examined powders: (a) Starck LC12; (b) Toshiba 'a'.

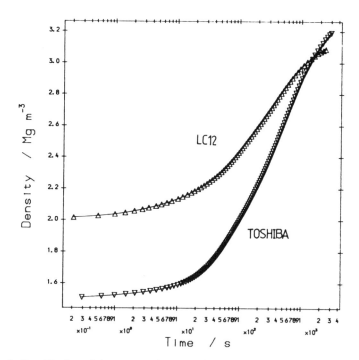

FIG. 2. Densification of the two powders at 1650°C with 10 MPa applied pressure. The functional form (lines) for the data (points) is fitted in this representation.

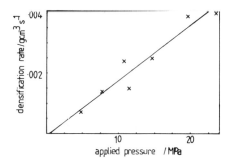

FIG. 3. The densification rate for silicon nitride hot-pressed with sintering aids commonly shows the approximately linear dependence on applied pressure as shown here for Starck powder.

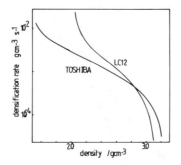 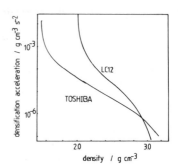

FIG. 4. The dependence of densification rate on density for the two powders.

FIG. 5. The dependence of the second derivative, $\ddot{\rho}$, on density.

density–time data are evaluated at different densities (Figs. 4 and 5); the expression Γ^* as given in eqn. (3) can then be developed as shown in Fig. 6.

4. DISCUSSION

A feature apparent from the micrographs of the two powders is the difference they show in particle size distribution. With few exceptions, the particles in the silica derived powder (Fig. 1(b)) show a uniform size which is in good agreement with the mean size as measured by sedimentation (1·1 μm).

This feature is reflected in the densification plots. Thus the density at the beginning of hot-pressing (Fig. 2) is greater for the wider

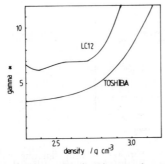 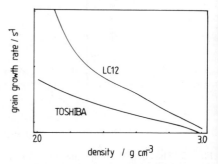

FIG. 6. The ratio of grain growth rate to densification rate for the two powders.

FIG. 7. The relative grain growth rate for the two powders.

distribution powder as expected from packing efficiency arguments.[11] This initial density is not identical to the green density (the powder has already been brought to temperature in the former instance) but good correlation between the two has been found both for this and other systems.

The size distribution aspect is also the basis for a first explanation of the behaviour in Fig. 4. The wide distribution with the presence of coarse and fine powders is more prone to coarsening processes as in Ostwald ripening;[12] the resulting particle growth causes a strong fall in densification rate as observed. The narrow size distribution in contrast is initially resistant to ripening and the retention of a relatively high densification rate, even to high densities ($\rho_{th} \sim 3{\cdot}2\,\text{g cm}^{-3}$), is observed.

The second derivative plots can also be interpreted in this light. Noting from eqn. (3) that

$$\frac{\dot{G}}{G} = -\frac{1}{m}\frac{\ddot{\rho}}{\dot{\rho}} \qquad (4)$$

Figs. 4 and 5 combine to yield a plot of the relative grain growth rate (Fig. 7) where the relative stability of the narrow distribution powder against coarsening can be seen. The actual ratio plot (Fig. 6) suggests something close to a factor of two difference in the overall ratio between the two powders. The plot also shows a weakness of the Γ^* format in that the simple use of the ratio (eqn. (3)) yields a function which rises to high value as higher densities are reached; the use of the method which monitors the overall change in pore size is more stable in this regard.[3]

The use of the method is therefore demonstrated by this application to the two silicon nitride powders, the pattern of behaviour expected on the basis of differences in particle size distribution being reflected in the evaluated ratio. More generally the method can be used to explore the ratio when processing differences are more complex, i.e. in selecting favourable processing conditions (temperature, particle size, additive type) by monitoring changes in the ratio as the conditions are changed. An advantage is that no assumption is made in respect to mechanisms beyond the need for a grain size dependent densification rate. The fact that the information is provided relatively quickly (a matter of hours for the full Γ^* versus ρ evaluation for a given set of conditions) also makes it attractive by comparison with full microstructural study. There is no doubt, however, that where detailed study of a material is attempted, the combination of both kinetic and microstructural approaches offers greater protection against error in[6] the interpretation of results.

ACKNOWLEDGEMENTS

Support of this work by the Commission of the European Communities is gratefully acknowledged.

REFERENCES

1. HANDWERKER, C. A., CANNON, R. M. and COBLE, R. L. Advances in Ceramics, **10** (1984) 619.
2. HERRING, C. Journal of Applied Physics, **21** (1950) 301.
3. EISELE, U. In preparation.
4. CANNON, W. R. and LANGDON, T. G. Journal of Materials Science, **18** (1983) 1.
5. VIEIRA, J. M. and BROOK, R. J. Journal of the American Ceramic Society, **67** (1984) 450.
6. BROOK, R. J., GILBART, E., SHAW, N. J. and EISELE, U. Powder Metallurgy, **28** (1985) 105.
7. WESTON, R. J. and CARRUTHERS, T. G. Proceedings of the British Ceramic Society, **22** (1973) 197.
8. BOWEN, L. J., WESTON, R. J., CARRUTHERS, T. G. and BROOK, R. J. Journal of Materials Science, **13** (1978) 341.
9. BOWEN, L. J., CARRUTHERS, T. G. and BROOK, R. J. Journal of the American Ceramic Society, **61** (1978) 335.
10. RAHAMAN, M. N., RILEY, F. L. and BROOK, R. J. Journal of the American Ceramic Society, **63** (1980) 648.
11. FROST, H. J. and RAJ, R. Journal of the American Ceramic Society, **65** (1982) C-19.
12. FISCHMEISTER, H. and GRIMVALL, G. Materials Science Research, **6** (1973) 119.

COMMENTS AND DISCUSSION

Chairman: R. W. DAVIDGE

A. Briggs: You do seem to be quite definite about there being no dislocation mechanisms involved in sintering. On what basis do you make that statement?

R. J. Brook: It is a long standing and furious debate that has taken place in the sintering community over many years—why do I make it?—as I see it, the yield stresses for ceramic systems which we believe to be of interest are much greater than the stresses that are engendered by surface energy forces and therefore on those grounds you may expect that dislocations may not be so important in sintering itself. But I think that such theoretical approaches are risky. What you should do is to measure it, and when we measure it we find that for the majority of systems we look at you get a linear or square law dependence of the densification rate on the pressure. There are exceptions—alkali halides will give you power of 3; magnesium oxide, if it is very pure, in a hot-pressing experiment will give you a power of 3—so that is a dislocation creep mechanism for densification and you cannot use the present analysis. But I think that silicon nitride does not have a reputation for being plastic and our experience supports that.

G. Wotting: As far as I understood, your mechanism concerning the grain growth rate is based on solid state sintering mechanisms—is this right?

Brook: No, one of the initial slides which showed grain growth and separation of boundaries from pores is for oxides in the solid state. I think there is no assumption in the analysis about the mechanism by which the grains grow and this could apply to Ostwald ripening in liquid phase systems for example. Any method which reduces the amount of surface area in the sample will give coarsening and will make the driving force for densification less.

Wotting: If you use silicon nitride as an example you need solution and reprecipitation of grains. Can you calculate these processes as simply as you can, for example, with solid state sintering processes for oxides or silicon carbide?

Brook: The model is more difficult and Ostwald ripening is a more complex one. Is it the reaction at the surface between the grains and the fluid or is it diffusion in the fluid which is controlling? We do not understand these things as mechanisms, but in this analysis there is no assumption about that. All I am saying is that I can measure, by this analysis, the ratio between the densification rates and grain growth. You can change the quantity of liquid phase additive, for example, and you can change the chemistry of the liquid phase additive and you can see, experimentally, how this ratio changes. I agree that the mechanisms are difficult to argue about and the complexity is so great that if you solve it for one system, then it may be different if you change to a different powder, for example. So I think looking for mechanisms is hard work and difficult to relate to other systems which are rather close. This is an analysis which does not depend on mechanisms.

D. P. Thompson: I was interested in your final graph which shows the function increasing for both powders used which I presume is more to do with densification taking place rather than the actual coarsening. From your results, can you say that, in general, there is no relationship between the grain growth and densification? You are obviously implying that it is either one or the other. Is it completely random between the two, or can you say that for a given powder, generally, it would be either densification or grain growth which causes your ratio to increase?

Brook: This is the difficulty of the subject. The densification rate depends upon the grain size. Therefore it depends in turn on the rate of grain growth, and the grain growth rate depends upon such things as the size of the pores, and therefore it depends on the density, so these two rates are interlinked. If you change a parameter like the quantity of additive which affects one of them it will then affect the other, and so in any real system you have a balance between these two. They will both be taking place and what we try to do in processing is to shift that balance so that it puts more emphasis on densification as a way to reduce this surface energy than on coarsening or grain growth. So you tune your processing parameters: the composition, the quantity, the temperature, the particle size, the particle size distribution, and so on, so that the ratio becomes more favourable. It is important to emphasise that they both occur, they interact with one another and therefore separating them is a difficult thing to do and that is why you see the ratio rather than the single parameter by itself.

K. H. Jack: Although you do not want to discuss mechanisms, the fact is that you have an additive there and one would expect, therefore, that liquid is involved in whatever mechanism that you do not want to discuss!

Brook: You are talking to the wrong person. I very much like discussing mechanisms. I have found by bitter experience that audiences do not like it when I talk about mechanisms; they do not believe me! Therefore I now do not bother talking about mechanisms.

Jack: It seems however that you have a very valuable analytical tool.

Brook: I think the difficulty with mechanisms is that you can spend a lifetime as with alumina with magnesia. For thirty years people have been trying to understand sintering of alumina using Linde 'A' powders and magnesia added as nitrate. Now I think we are close to the answer for that system. When we switch to a different alumina, we get a different mechanism. It really is extraordinarily complicated. We learn a lot from mechanism studies and I think it is a terrible mistake to withdraw from trying to understand mechanisms. If you stop thinking about what the atoms are doing, I think a lot of your understanding and chances to develop things are lost, but I find audiences sceptical if I cannot say 'this is definitely the mechanism and anybody who does not agree is an idiot!'

Jack: But in trying to identify mechanisms you must have found differences when the quantity of additive is varied and major differences between additives. You have not said anything about that.

Brook: As you showed in the work with magnesia and yttria additives[1] the behaviour is very different and the balance between the rates at which processes happen is very different. What we are trying to find is a method that is quite quick but will allow you to rank order experiments. We try to see some picture emerging.

P. Kennedy: If you vary the temperature and perhaps a plastic flow mechanism came your way, would you be able to identify this or might you misinterpret some of the results on account of this?

Brook: I think it is true that of all the variables, although temperature is practically very important, in terms of our understanding of what is going on, it is one of the least useful. Particle size dependence of rates can be used with the scale laws to try to find a mechanism. The additive dependence is capable of giving simple rules. Varying the temperature gives an activation energy, and you can compare it with other activation energies in the literature. Now if you look at the silicon nitride literature you have a very wide choice of activation energies. So somewhere in the literature you will find the same activation energy as the one you have! But so many things are temperature dependent: the quantity of liquid, the viscosity of the liquid, the diffusion rates in the liquid and the solubility in the liquid, so the activation energy is a melange of all sorts of different things—it is not a simple parameter—but, in terms of practical processing, the temperature is very important.

R. W. Davidge: Empiricism rules, OK!

Reference

1. HAMPSHIRE, S. and JACK, K. H. In *Special Ceramics, Vol. 7,* Taylor, D. E. and Popper, P. (Eds), *Proceedings of the British Ceramic Society,* **31** (1981) 37.

4
Sinterability of Silicon Nitride Powders and Characterisation of Sintered Materials

V. Vandeneede, A. Leriche and F. Cambier

Centre Recherches de l'Industrie Belge de la Céramiques, 4 Avenue Gouverneur Cornez, B-7000 Mons, Belgium

AND

H. Pickup and R. J. Brook

Department of Ceramics, University of Leeds, LS2 9JT, UK

ABSTRACT

Seventeen different silicon nitride powders which are commercially available are characterised from the point of view of chemical and physical properties. Their densification behaviour with MgY_2O_4 as sintering aid is compared in the case of pressureless sintering and hot pressing.

The influence of the physicochemical characteristics of the powders on their sinterability and on the mechanical properties of the sintered materials is discussed on the basis of microstructural examination.

1. INTRODUCTION

Silicon nitride materials were first produced by the reaction bonding technique which showed their great potential for use as structural components at high temperatures in various environments.[1] However, this technique does not produce fully dense materials, the remaining porosity (about 20%) leading to poor mechanical properties (modulus of rupture of the order of 200 MPa). On the other hand, dense silicon nitride ceramics with strength up to 800 MPa were prepared by hot pressing using mainly magnesia[2] or yttria[3] as sintering aids. The shaping of complex components involves the use of diamond machining and therefore HPSN components are very expensive impeding

their mass production. Fine silicon nitride powders which have been available for a few years allow us to consider the possibility of pressureless sintering giving dense silicon nitride materials.
The silicon nitride powders are mainly produced by four different routes. Three different raw materials can be used: metallic silicon, oxide and halide, for instance silicon chloride. The first route, the most used currently, is the direct nitridation of silicon following:

$$3Si + 2N_2 \rightarrow Si_3N_4 \qquad (1)$$

The second method is the carbothermal reduction of silica in a nitrogen atmosphere:

$$3SiO_2 + 6C + 2N_2 \rightarrow Si_3N_4 + 6CO \qquad (2)$$

Silicon nitride powders can also be synthesised by a vapour phase reaction between silicon chloride and ammonia:

$$3SiCl_4 + 16NH_3 \rightarrow Si_3N_4 + 12NH_4Cl \qquad (3)$$

The last route follows the same reaction as the previous route, it starts from the same raw materials but proceeds by formation of imides, then decomposition:

$$SiCl_4 + 6NH_3 \rightarrow Si(NH)_2 + 4NH_4Cl \qquad (4)$$

$$3Si(NH)_2 \rightarrow Si_3N_4 + 2NH_3 \qquad (5)$$

It is well known that the sinterability of the powder depends strongly on the physical and chemical characteristics. The purpose of this paper concerns the physical and chemical characterisation of most of the powders available on the market, comparative sinterability evaluation and microstructural and mechanical properties determination on the sintered materials.

2. EXPERIMENTAL

Seventeen different powders were collected from European, American and Japanese producers. The main impurities (free Si, Fe, C, Cl_2, Al, Ca) were specified by the suppliers; oxygen was the only element to be analysed. This was carried out through CO_2 conversion and IR absorption measurement.

The tap density was determined using a method described by Berrin *et al.*,[4] the size of agglomerates using the Sedigraph, and the surface area by the BET method.

$\alpha-\beta$ phase contents were obtained for the powders and for the densified materials following the method of Gazzara and Messier.[5] SEM and TEM examination of the powders permitted determination of the average crystal size and the shape of agglomerates, and SEM photographs were also obtained on polished, chemically etched surfaces of sintered materials.

The powders were mixed with 10 wt % MgY_2O_4 and hot pressed (1650°C, 10 MPa) until densification became very slow. The same mixtures were also isostatically pressed to obtain cylinders (7 mm diameter, 25 mm length), which were sintered in a powder bed consisting of boron nitride and silicon nitride mixed with the same quantity of magnesia as in the sample. These pellets were fired at 1700°C for 2 h. After firing the specimens were cut to prepare $30 \times 4 \times 3$ mm³ bars ($20 \times 4 \times 3$ mm³ for hot pressed specimens). The bars were used to determine the Young's modulus (Grindosonic) and the modulus of rupture (3 point flexure test, span 15 mm, cross head speed 0.05 mm min^{-1}).

3. RESULTS AND DISCUSSION

The chemical characteristics of the different powders are presented in Table 1. For each type of powder, the nature of the precursor materials: silicon, silica or silicon chloride, the overall purity and the percentage of main impurities are indicated. These impurities are directly related to the powder synthesis route. The powders 1–14 are synthesised by direct nitridation resulting in the presence of residual free silicon, iron, which is a catalyst for the reaction, and carbon. Note that powders 10 and 11 contain a higher impurity level (oxygen, iron, aluminium, calcium). These powders are in fact obtained by grinding of reaction bonded pieces. The carbothermal reduction of silica produces powders such as no. 15, characterised by high oxygen (2·65%) and carbon (0·9%) contents. Chlorine (0·01–0·1) is present as an impurity in the powders synthesised from silicon chloride; this is the case for powders 16 and 17. The powders obtained by this route are generally purer than those coming from other routes with the exception of powder 12.

The physical and crystallographic characteristics of the powders are

TABLE 1
Chemical Characteristics of the Powders (%)

Powder no.	Raw material	Purity	O_2	Free Si	Fe	C	Cl_2	Al	Ca
1		97·0	0·86	<0·1	0·04	0·4	—	<0·1	<0·06
2		96·5	0·60	<0·6	0·12	0·4	—	<0·2	<0·08
3		97·0	1·00	<0·1	0·04	0·1	—	<0·2	<0·06
4		96·5	1·34	<0·1	0·04	0·1	—	<0·15	<0·06
5		97·0	1·47	<0·1	0·01	0·13	—	0·06	0·01
6		98·8	1·28	0·3	0·08	0·2	—	0·03	0·01
7		—	1·55	<1·0	0·5	0·2	—	0·5	0·3
8	Si	96·5	0·34	<1·0	0·25	—	—	0·25	0·01
9		96·9	1·05	2·0	0·5	0·2	—	0·5	0·3
10		—	2·62	—	<1·5	—	—	<1·5	<1·5
11		—	1·74	—	<1·5	—	—	<1·5	<1·5
12		99·9	1·45	<0·072	0·04	—	—	0·09	0·001
13		—	1·00	—	—	—	—	—	—
14		95·8	2·57	—	0·19	0·22	—	0·26	0·26
15	SiO_2	96·5	2·65	—	0·01	0·9	—	0·2	0·01
16	$SiCl_4$	99·9	1·04	—	0·005	0·1	0·1	<0·001	<0·001
17		98·2	1·90	—	<0·01	—	0·01	<0·005	<0·005

listed in Table 2. This includes the β phase content, the tap density, the size of crystals and of agglomerates (Sedigraph method) and the surface area. SEM examination showed that Si_3N_4 particles are agglomerated in different shapes (chains, clusters, etc.) and in different sizes and that elongated particles (length L, width l) are present in some grades. The results of these examinations are also reported in Table 2. The lowest β-Si_3N_4 contents were measured for powders 14, 15 and 17, that is, one for each synthesis route. The highest β level is that of the two ex-RBSN powders (10 and 11). Generally higher tap densities can be reached with the 'direct nitridation method' powders (except 3 and 8), and the lowest result corresponds to powder 15 (carbothermal reduction of silica).

The values of agglomerate size and of surface area show a small degree of agglomeration for powders prepared by reaction with silicon chloride (16 and 17). On the other hand, average crystal size values show that very fine particles can be obtained by all 3 methods. Highest surface area is demonstrated by powder 5 ($26 \cdot 5 \, m^2 \, g^{-1}$); however some other Si_3N_4 powders also have high values for this parameter: powders 17 ($14 \cdot 8 \, m^2 \, g^{-1}$), 16 ($9 \cdot 9 \, m^2 \, g^{-1}$), 15 and 1 ($8 \cdot 7 \, m^2 \, g^{-1}$). Elongated

TABLE 2
Physical Characteristics of the Powders

Powder no.	β-Si$_3$N$_4$ (%)	Tap density (Mg m^{-3})	Agglomerate size (μm) (Sedigraph)	Crystal size (μm)	Surface area (m^2 g^{-1})	Shape and size of agglomerates (μm)	Elongated particles (L/l)
1	5.0	0.91	1.0	2.0	8.7	chains >50	5.3 a few
2	6.0	0.85	6.0	0.8	2.8	chains >50	10.0 many
3	6.0	0.57	0.9	1.0	8.5	clusters ~15	6.1 many
4	6.0	0.67	0.7	0.5	11.0	clusters ~5	8.7 a few
5	5.5	0.96	0.8	0.5	26.5	chains >50	15.0 a few
6	5.0	0.82	2.8	0.3	6.5	clusters ~2	10.0 many
7	22.0	1.09	12.0	1.0	3.2	chains ~13	10.0 many
8	23.0	0.69	3.8	1.0	2.1	chains ~13	10.0 many
9	14.0	0.89	1.7	1.0	2.8	clusters ~5	7.5 a few
10	34.0	1.06	8.0	1.0	3.3	balls ~7	– none
11	35.0	0.97	8.0	1.0	3.5	{chains ~100, large grains ~60}	20.0 a few
12	6.5	0.91	1.4	0.2	6.9	clusters	20.0 many
13	28.0	1.00	11.0	0.2	2.5	clusters	7.6 many
14	1.0	1.10	0.7	0.2	6.3	clusters ~2	7–8 none
15	1.0	0.52	1.1	0.3	8.7	{chains >100, balls ~20}	7–8 none
16	15.0	0.70	0.4	0.5	9.9	clusters ~2	7.2 a few
17	2.0	0.74	0.3	0.1	14.8	chains 10–20	20.0 a few

particles are sometimes present in the observed powders (mainly from direct nitridation). An example is given in Fig. 1 (SEM 30 000 X, powder 6). The occurrence of these elongated particles is also different according to the powder (see Table 2). Generally the silicon nitride agglomerates as chains or as clusters. Comparing the size of the agglomerates measured by the Sedigraph after ultrasonic dispersion, and by direct observation, most of the powders are weakly agglomerated with the smallest agglomerates observed in powders 14 and 16.

The parameters describing the densification of pressureless sintered and hot-pressed silicon nitride are reported in Table 3. The first and fourth columns present the values of green densities (ρ_0) which confirm the tap densities (Table 2), showing that the most easily compacted powders are references 5, 10 and 11 and that the worst ones are 6, 15, 16 and 17. The second and third columns give the bulk densities after pressureless thermal treatment and the corresponding open porosity. Under these conditions, for pressureless sintering, the theoretical density of $3 \cdot 2 \, \text{g cm}^{-3}$ has not been reached. The best

FIG. 1. Scanning electron micrograph of powder 6.

TABLE 3
Densification Parameters of Pressureless Sintered and Hot-pressed Materials

Powder no.	Pressureless sintering			Hot pressing		
(*)	ρ_0	ρ	P	ρ_0	ρ	P
1	1·73	—	—	1·85	2·79	13
2	1·60	—	—	1·75	2·45	23
3	—	—	—	1·77	2·73	15
4	1·74	2·74	17	1·91	2·95	2
5	1·90	2·79	15	2·02	3·09	0
6	1·44	2·25	31	1·61	2·86	—
7	1·70	1·96	39	1·64	2·57	14
8	1·60	1·92	41	1·54	2·63	20
9	1·75	1·91	41	1·81	2·74	17
10	1·95	2·53	23	2·14	3·23	0
11	1·94	2·52	23	2·17	3·17	0
12	1·62	2·47	25	1·78	2·72	—
13	1·82	2·22	32	2·04	2·61	14
14	—	2·88	6	—	—	—
15	1·55	2·93	9	1·51	3·20	0
16	1·59	2·87	10	1·55	3·12	0
17	1·63	2·91	11	1·52	3·18	0

* ρ_0: green density (Mg m^{-3}); ρ: bulk density (Mg m^{-3}); P: open porosity (volume %).

densities (~2·9 or about 90% of theoretical) are obtained for the 4 powders 14–17. The easily compacted powder, reference 5, reaches about 85% of the theoretical density under these experimental conditions. The fifth and sixth columns correspond to bulk densities and remaining open porosities of hot-pressed materials. No open porosity was found for materials 5, 10, 11, 15, 16, and 17. On the other hand full density is reached for powders 10 and 11 which are easily compacted and also for powders 15 and 17 showing their high sinterability. Powders 5 and 16 also achieve bulk densities higher than 3·0.

Computer treatment of hot pressing data produces the variation of densification rate as a function of the density. The different powders can be separated into 4 classes according to their sintering behaviour. The first class (Fig. 2) includes the powders which present a very low starting density (~1·6) but which reach, after hot-pressing, densities

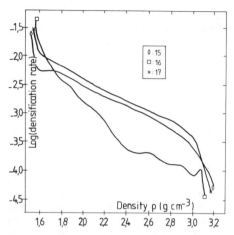

FIG. 2. Densification rate versus density: hot pressing of powders 15, 16 and 17.

close to the theoretical (powders 15–17). Note that powders 15 and 17 show a very similar variation of densification rate, that is to say a sudden slowing of the rate at the start of the sintering, then a relatively fast densification rate until the theoretical density is reached. Powder 16 presents a slower densification rate than 15 and 17. The second class (Fig. 3) puts together the badly compacted powders which reach after firing very low densities (~2·6–2·9). This class includes the

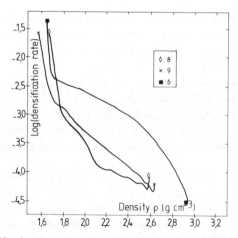

FIG. 3. Densification rate versus density: hot pressing of powders 6, 8 and 9.

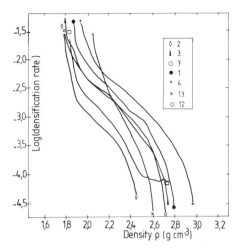

FIG. 4. Densification rate versus density: hot pressing of powders 1, 2, 3, 4, 7, 12 and 13.

powders 8, 9 and 6. Powder 6 presents the fastest densification rate for this class. The third type of powder (Fig. 4) forms pellets of average density ~1·9, which after firing have bulk densities less than 3·0. This is the case for powders 1 to 4, 7, 12 and 13. The fourth class (Fig. 5) concerns the powders which present a high green density (~2·1) and which reach densities close to theoretical. Such behaviour is observed for the ex-RBSN powders 10 and 11, and for powder 5.

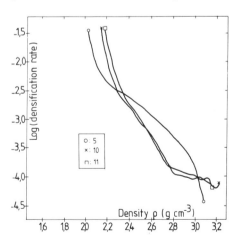

FIG. 5. Densification rate versus density: hot pressing of powders 5, 10 and 11.

Pickup et al.[6] have described a method to compare the firing behaviour of powders in terms of the densification rate and the grain coarsening rate. Densification rate, $\dot{\rho}$, is linked to the grain size, G, as follows:

$$\dot{\rho} = KG^{-m}$$

where K and m are constants, m being characteristic of the densification mechanism. From this equation one can obtain the value of Γ^*,

$$\Gamma^* = \frac{\dot{G}\rho}{G\dot{\rho}} = -\frac{1}{m}\frac{\rho\ddot{\rho}}{\dot{\rho}^2}$$

Γ^* is therefore proportional to the ratio of grain coarsening rate over densification rate and has to be minimised. Table 4 reports the Γ^* values for each powder at various densities (2·0, 2·4, 2·8). The lowest values are obtained for powders 15, 16, 17 and 6. However powder 6 has a larger value of Γ^* at higher density and does not reach a density higher than 2·9. Figure 6 represents the Γ^* variation of those powders achieving dense HPSN versus the bulk density. It appears that powders obtained from $SiCl_4$ and SiO_2 (respectively 16, 17 and 15) behave better than the other ones. Moreover the Γ^* values for

TABLE 4
Γ^* Values of Si_3N_4 Powders

Powder no.	$\rho = 2·0$	$\rho = 2·4$	$\rho = 2·8$
1	>9·5	7·4	—
2	8·8	>9·5	—
3	9·3	7·5	—
4	>9·5	6·9	9·5
5	—	6·2	7·0
6	3·5	4·5	6·0
7	11·0	4·0	—
8	6·0	5·5	—
9	9·8	13·0	—
10–11	—	12·0	6·5
12	10·0	7·5	—
13	—	13·0	—
15	4·0	3·5	5·0
16	5·8	6·9	4·0
17	3·4	3·5	4·0

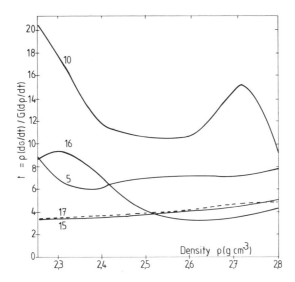

FIG. 6. Γ* variation versus density for powders leading to fully dense hot-pressed ceramics.

powders 15 and 17 remain constant as far as the densification proceeds. Powder 5 is also interesting but the Γ* value is higher. Powders 10 and 11 are characterised by high Γ* values. Generally, Γ* is higher for agglomerated powders (see Table 2—Sedigraph data). This allows us to think that densification rate is reduced due to the high degree of coordination of the grains discouraging the removal of porosity.

The β-Si_3N_4 content and mechanical properties (E, σ_f) data for pressureless sintered and hot-pressed materials are presented in Table 5. As expected, the most sinterable powders give ceramics characterised by the best mechanical properties. For sintered materials, powders 16, 5 and 14 have the higher moduli of rupture. It can be noticed that strength values of 555 MPa (powder 16) and 437 MPa (powder 5) are very high in view of the residual porosity, respectively 10 or 15 vol. %. SEM examination of the two materials shows that powder 5 has coarser grains (~5 μm length) than powder 16 (~3 μm, Fig. 7); however aspect ratios seem to be in the same range (~5). Highly sinterable powders, 15 and 17, produce relatively poor strength values (278, 259 MPa). Chemical treatment of these materials shows overetched surfaces (Fig. 8 No. 17) when compared with others

TABLE 5
β-Si$_3$N$_4$ Content and Mechanical Properties of Pressureless Sintered and Hot-pressed Materials*

Powder no.	Pressureless sintering			Hot pressing		
	β	E	σ_F	β	E	σ_F
1	—	—	—	70	211	432
2	—	—	—	60	148	246
3	—	—	—	76	205	398
4	89	157	219	69	255	549
5	86	184	437	62	280	539
6	80	79	92	—	—	—
7	76	18	21	94	183	224
8	68	56	45	94	154	218
9	78	31	23	47	161	176
10	88	132	118	100	276	525
11	69	113	91	100	275	396
12	92	129	157	—	—	—
13	87	71	71	89	155	267
14	94	183	354	—	—	—
15	63	198	278	14	294	310
16	90	195	555	100	300	692
17	86	201	259	51	253	473

*β: percentage of β-Si$_3$N$_4$; E: Young's modulus (GPa); σ_F: flexural strength (MPa).

prepared in the same way (Fig. 7). The oxygen content (Table 1) of these powders suggests the presence of an amorphous phase, which explains the lower flexural strength. Moreover, β phase conversion is less (only 63%) during the sintering of powder 15, consequently the microstructure shows more equiaxed grains (Fig. 9). For hot pressed materials the best modulus of rupture is also obtained for reference 16, but values of or exceeding 500 MPa can be reached by powders 4, 5, 10 and 17. Powder 17 contains only 51% of β-phase and has a microstructure consisting of very fine particles ($\sim 0.3\ \mu$m) and elongated grains ($\geqslant 1\ \mu$m). The small crystal size confirms the low value of Γ^*. As for pressureless sintering the chemical etching of powder 17 reveals a significant amorphous phase volume. Although sintered up to theoretical density, materials from powder 15 do not have high fracture strengths. In fact the microstructural examination (Fig. 10)

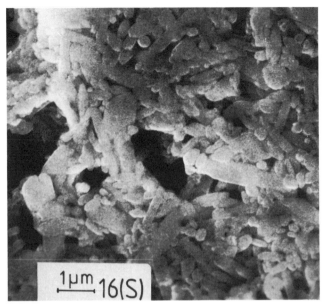

FIG. 7. Chemically etched surface of pressureless sintered material, powder 16.

FIG. 8. Overetched surface: pressureless sintered material, powder 17.

FIG. 9. Chemically etched surface of pressureless sintered material, powder 15.

FIG. 10. Chemically etched surface of hot pressed material, powder 15.

reveals the absence of cross-linked elongated grains, confirmed by the low β content (14%). However the grain size is small which is compatible with the low Γ^* value.

4. CONCLUSIONS

Seventeen different silicon nitride powders which are candidates for the production of SSN materials have been compared. Wide differences in behaviour are, apparently, related to the various characteristics of the powder. High sinterable raw materials require ultrafine crystal size leading to high densification rate; low agglomeration in order to favour the removal of porosity; high α/β phase ratio to obtain a high aspect ratio in the sintered material. Such raw materials can be obtained by the different production routes. However mechanical properties seem to be closely dependent on the nature and the volume of liquid phase appearing during the sintering and on the degree of conversion into β-Si_3N_4.

Better mechanical properties are thus obtained from high purity raw materials which have reasonably low levels of oxygen. These results suggest the merit of production routes using halides as starting materials.

ACKNOWLEDGEMENTS

This work was performed under CEC contracts SUT-117B and SUT-121 UK. The authors thank AERE Harwell for oxygen content determinations.

REFERENCES

1. JACK, K. H. Sialons: a Study in Materials Development. This volume, pp. 1–30.
2. WESTON, R. J. and CARRUTHERS, T. G. *Proceedings of the British Ceramic Society*, **22** (1973) 197.
3. KATZ, R. N. and GAZZA, G. In *Nitrogen Ceramics*, Riley, F. L. (Ed.), Noordhoff, Leyden, p. 417, 1977.
4. BERRIN, L., JOHNSON, D. W. and NITTI, D. J. *American Ceramic Society Bulletin*, **51** (1972) 840.

5. GAZZARA, C. P. and MESSIER, D. R. *American Ceramic Society Bulletin*, **56** (1977) 777.
6. PICKUP, H., EISELE, U., GILBART, E. and BROOK, R. J. Analysis of Coarsening and Densification Kinetics During the Heat Treatment of Nitrogen Ceramics, this volume, pp. 41–51.

COMMENTS AND DISCUSSION

Chairman: R. W. DAVIDGE

R. J. Brook: (addressed to Smith). How many silicon nitride powder manufacturers can the world afford? We have about twenty, I think, at the moment, and there are more and more as time goes on. Will it be like the motor industry where after a few years they will sort themselves out and there will only be one left?

J. T. Smith: Ultimately, nitrided silicon will be economically attractive for a certain quality of product with a certain level of performance, but for more demanding performances the chemically derived products will be favoured. In the future the chemical people will produce a product which will probably force the nitrided people out of business.

5

Microstructural Development and Secondary Phases in Silicon Nitride Sintered with Mixed Neodymia/Magnesia Additions

B. SARUHAN, M. J. POMEROY AND S. HAMPSHIRE

Materials Research Centre, National Institute for Higher Education, Limerick, Plassey Technological Park, Limerick, Ireland

ABSTRACT

The pressureless sintering of silicon nitride powder with mixed neodymia/magnesia additives is reported. Of the sixteen additive combinations studied the majority give rise to rapid densification after firing in 1 atm nitrogen for 30 min at a temperature of 1650°C and densities of 3050 kg m^{-3} are achieved. Such rapid densification is attributable to a large volume of liquid formed during densification. The α–β transformation is controlled by the additive level and combination ratio and, in certain compositions, the optimum requirement of high density and virtually complete transformation are obtained simultaneously. The resulting β-grain morphologies are acicular and interlocking suggesting that ceramic compositions within this easily densified silicon nitride/additive system should have good mechanical properties. Preliminary work shows that the intergranular glass can be devitrified to give more beneficial crystalline phases.

1. INTRODUCTION

Silicon nitride-based ceramics have achieved considerable commercial success because of their high strengths, wear resistance and resistance to thermal shock. A significant feature of the fabrication process for β'-sialons, for example, is the ease with which they can be densified with low levels of additives by pressureless sintering.[1] Furthermore, an interlocking fibrous grain morphology, which gives rise to their high strengths, is observed[2] simultaneously with completion of densification. Rapid densification is obtained with β'-sialons because of the large volume of liquid formed during sintering but because aluminium and oxygen are absorbed from this liquid into precipitating β'-sialon grains, the amount of liquid remaining after complete densification is small.[3] Moreover, since any remaining grain-boundary glass can be converted into a beneficial refractory phase, such as yttrium aluminium garnet, high strengths can be maintained to temperatures of up to 1300°C.[2]

Wötting and Ziegler[4] have demonstrated the importance of the shape and aspect (length to diameter) ratio of β silicon nitride grains in determining strength and resistance to fracture of material pressureless sintered with mixed yttria–alumina additives. These authors have also shown that grain morphology and aspect ratio can be controlled by the amount of additive used during sintering. Thus there is a clear need to identify additive systems which promote a fibrous grain morphology.

Magnesia as a single additive to silicon nitride promotes good densification but an equi-axed low β phase material, whereas yttria additions result in a fibrous high β phase material but densification is poor.[5] The object of studies being carried out at NIHE Limerick is to examine different additive systems for silicon nitride which do not involve the use of alumina and previous work[6] concentrated on mixed magnesia–yttria additions. Results obtained showed that although a fibrous high β-phase grain morphology could be easily developed, densification was difficult. More recently, mixed magnesia–neodymia additions have been investigated and this paper reports the initial results obtained.

2. EXPERIMENTAL

The silicon nitride powder used in this work was supplied by H. C. Starck (Berlin) (LC 12 grade). Analar grade magnesia and 99·99% pure neodymia were used as mixed additives.

The powders were blended in iso-propanol for 5 h after which the alcohol was evaporated off and the powders dry mixed for a further 30 min. Quantities of powder were then die-pressed and embedded in a mixture of 50 wt % boron nitride + 50 wt % silicon nitride and relevant additive combination. The compacted pellets were fired for 30 min at a temperature of 1650°C in 1 atm nitrogen.

Green and fired densities and volume shrinkages were determined using a mercury displacement balance. The phases present in the fired pellets were determined by X-ray powder diffraction using a Hägg–Guinnier camera and Cu $K\alpha$ radiation. Relative amounts of α and β phases were estimated by comparing the intensities of the X-ray reflections from the α (102) and (210) planes with those of the β (101) and (210) planes.

The microstructures of the fired pellets were examined using scanning electron microscopy. Both fracture surfaces and polished and etched surfaces were examined.

Post preparative heat treatments were carried out at a temperature of 1300°C for 36 h in 1 atm nitrogen.

The heat treated pellets were subjected to phase analysis by X-ray diffraction in order to assess whether any glass formed during firing was converted into a crystalline refractory phase.

3. RESULTS

3.1. Effect of Additive on Densification and Transformation

3.1.1. Effect of different magnesia/neodymia mole ratios and total number of moles of additive

Table 1 gives the compositions investigated where both the total number of moles of additive and the magnesia to neodymia ratio were

TABLE 1
Compositions for Samples Where Both the Additive Level and Magnesia to Neodymia Mole Ratio are Varied

| Weight % | Additive | Magnesia | Total % | Green | Fired |
| | (wt %) | to neodymia | addition | density | density |
MgO	Nd_2O_3	mole ratio	(mole %)	(kg m^{-3})	(kg m^{-3})
1·5	12·7	1	10·4	2 110	2 820
3·0	10·4	2·4	14·1	2 090	2 890
4·0	8·9	4·0	16·3	2 060	3 050
5·0	7·4	5·6	18·5	2 090	2 950
7·0	4·5	13·0	22·5	2 060	2 830
8·5	2·2	32·0	24·5	2 050	2 800

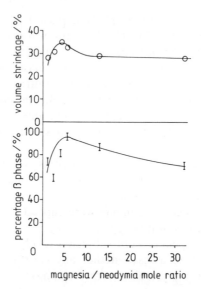

FIG. 1. Effect of magnesia/neodymia mole ratio (additive level varied) on shrinkage and β formation after sintering with silicon nitride at 1650°C for 30 min.

varied. Green and fired bulk densities are also given. Figure 1 shows the effect of mole ratio on volume shrinkage and amount of β phase formed after 30 min at 1650°C. A maximum in the volume shrinkage occurs at a magnesia to neodymia mole ratio of four and corresponds to a density of 3050 kg m^{-3}. A similar effect is observed for the formation of β phase where levels of 95% are observed for a magnesia to neodymia mole ratio of 5·6 and a total additive level of 18·5%. Accordingly this series of additive combinations gives rise to the simultaneous achievement of high density and complete transformation of α to β silicon nitride. This is required in order to prevent grain coarsening which may occur if complete conversion is achieved before full densification.[6]

3.1.2. *Effect of varying magnesia to neodymia mole ratio with constant total additive level*

Four additive combinations were investigated in order to assess the effect of mole ratio on volume shrinkage and transformation for a constant level of additive. These combinations are given in Table 2 together with green and fired densities. Figure 2 shows the results

TABLE 2
Compositions for Samples Where the Additive Level is Held Constant but the Magnesia to Neodymia Mole Ratio is Varied

Weight % MgO	Additive (wt %) Nd$_2$O$_3$	Magnesia to neodymia mole ratio	Total % addition (mole %)	Green density (kg m^{-3})	Fired density (kg m^{-3})
1·7	14·4	1	12·5	2 130	3 050
3·5	6·1	4·8	13·9	2 030	2 870
3·9	2·6	12·5	13·5	1 980	2 800
4·0	1·9	18·5	13·7	2 000	2 880

obtained. In contrast to the results obtained when both mole ratio and additive level are varied, the volume shrinkage is little affected by varying the magnesia to neodymia mole ratio alone. The fired density for a mole ratio of 1:1 is 3050 kg m^{-3}. However, in contrast to the additive series referred to in Section 3.1.1 only some 65% α-β conversion occurs. The extent of transformation declines and becomes constant at a level of 50–55% as the mole ratio is increased above a value of 5.

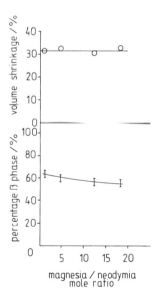

FIG. 2. Effect of magnesia/neodymia mole ratio (additive level constant) on shrinkage and β formation after sintering with silicon nitride at 1650°C for 30 min.

3.1.3. Effect of varying amount of additive for fixed magnesia to neodymia mole ratio

When using a combined magnesia–yttria additive, Giachello et al.[7] observed that subsequent heat treatments at a temperature of 1400°C for 15 h gave rise to formation of a crystalline mixed silicate $Mg_5Y_6Si_5O_{24}$. The presence of this phase in grain boundary regions enables strengths to be maintained to higher temperatures than for non-annealed samples. Accordingly, compositions based on a magnesia to neodymia ratio of 5:3 were investigated in order to determine whether a similar phase could be formed in a system already shown to exhibit excellent densification behaviour.

Table 3 shows the additive levels employed, together with green and fired densities. Figure 3 shows the effect on densification and transformation of varying the total molar percentage of additive. Volume shrinkage is dependent upon additive concentration and levels in excess of 10 mole percent give rise to good densification behaviour. With respect to transformation, Fig. 3 shows that the amount of β silicon nitride formed is very high for all of the compositions investigated.

3.2. Secondary Phases

In some fired specimens, phases other than α or β silicon nitride were observed. For those compositions containing more than 12 wt % neodymia an unidentified phase was observed which is most probably either $Nd_2O_3:Si_6N_8$ or a melilite phase. Results for compositions with less neodymia show that apatite may form as a secondary phase. Work is continuing to confirm the exact nature of these secondary phases.

TABLE 3
Compositions for Samples Where Magnesia to Neodymia Mole Ratio is Held Constant but the Additive Level is Varied

Weight % Additive (wt %)		Magnesia to neodymia mole ratio	Total % addition (mole %)	Green density (kg m^{-3})	Fired density (kg m^{-3})
MgO	Nd_2O_3				
4·0	19·8	1·67	22·6	2 130	2 950
3·2	16·1	1·67	18·1	2 100	3 050
2·6	13·2	1·67	14·8	2 070	2 980
2·2	10·8	1·67	12·3	2 050	2 800
1·8	8·8	1·72	10·0	2 030	2 750
1·4	7·7	1·67	7·9	2 020	2 600

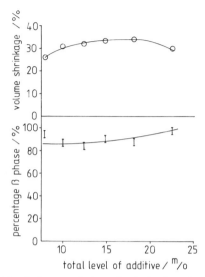

FIG. 3. Effect of total level of magnesia + neodymia (mole ratio constant) on shrinkage and β formation after sintering with silicon nitride at 1650°C for 30 min.

3.3. Microstructural Features

For all of the compositions investigated the microstructures of the fired material were typified by an interlocking acicular β silicon nitride grain morphology. A typical fracture surface grain structure is shown in Fig. 4. Figure 5 shows a polished and etched section for the composition containing 16 mole % additive with a magnesia to neodymia mole ratio of four. This microstructure is typical of others observed for combined magnesia–neodymia additions and the interlocking nature of the β silicon nitride grains is more clearly visible.

Although the aspect ratios have not been determined in a statistical manner, it appears that additive level and combination ratio do affect this parameter. In almost all cases the aspect ratio is greater than six with a maximum observed ratio of nine.

3.4. Post-preparative Heat Treatments

Selected compositions which demonstrated good densification behaviour were subjected to post-preparative heat treatments in order to devitrify the glass formed after sintering. In almost all cases the major phase produced had an apatite structure. For the magnesia–neodymia–silica system the composition of this phase is $MgNd_4Si_3O_{13}$.

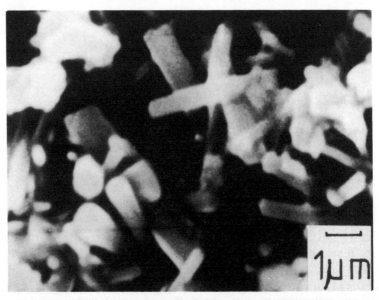

FIG. 4. Scanning electron micrograph of fracture surface of silicon nitride sintered at 1650°C for 30 min with 1·5 wt % MgO/12·7 wt % Nd_2O_3.

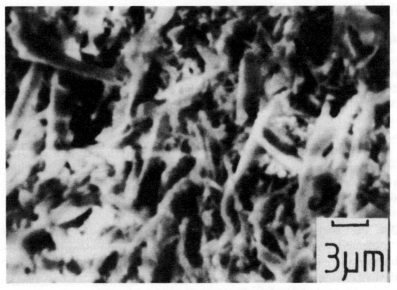

FIG. 5. Scanning electron micrograph of polished section of silicon nitride sintered at 1650°C for 30 min with 4·0 wt % MgO/8·9 wt % Nd_2O_3.

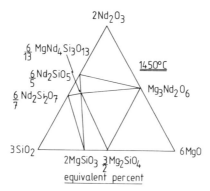

FIG. 6. Phase relationships at 1450°C in the MgO-Nd$_2$O$_3$-SiO$_2$ system.

However, X-ray analyses of the products of heat treatments demonstrate that the apatite contains a small amount of nitrogen since the characteristic d spacings observed for the oxide are displaced slightly.

In order to assess possible products of devitrification the magnesia–neodymia–silica system was investigated at a temperature of 1450°C. Figure 6 shows the phase relationships observed. It is seen that the apatite phase is in equilibrium with five other phases which accounts for its frequent occurrence as a product of post-preparative treatments.

A significant feature of the magnesia–neodymia–silica system, when compared to the magnesia–yttria–silica system, is that the phase Mg$_5$Nd$_6$Si$_5$O$_{24}$ (the neodymium analogue of the Mg$_5$Y$_6$Si$_5$O$_{24}$ phase reported by Giachello et al.[7]) cannot be prepared at 1450°C.

4. DISCUSSION

The results show that, without question, the mixed magnesia–neodymia additive system allows rapid densification of Starck LC 12 silicon nitride by pressureless sintering. According to Hampshire and Jack[8] and Wötting and Ziegler[4] rapid densification occurs when a large volume of low viscosity liquid is formed during sintering. The latter authors attribute the development of elongated β silicon nitride grains to slow diffusion of material from dissolving α particles to growing β grains. High viscosity liquids are characterised by slow diffusion rates. In a comparison of kinetics and mechanism for the single additives, magnesia and yttria, Hampshire and Jack[8] observed that magnesia causes a large volume of low viscosity liquid to be formed from which equi-axed β grains develop, whereas with yttria a smaller volume of high viscosity liquid arises from which higher aspect ratio grains grow.

Since an elongated, fibrous grain morphology is developed with combined magnesia–neodymia additives, it is clear that the liquid formed is of high viscosity. Therefore the rapid densification rates observed must be due to the formation of a large volume of liquid.

The large number of different additive levels and ratios which give rise to high densities suggest that the region of liquid formation within the five component Mg–Nd–Si–O–N Jänecke prism must be extensive. Drew et al.[9] have shown that in the magnesium and yttrium sialon systems the glass formation region is large and, in these systems, rapid densification rates are observed. Accordingly, it appears that, in terms of glass formation, there are parallels between the Y–Si–Al–O–N system and the Mg–Nd–Si–O–N system, partly because of similar eutectic temperatures. The complete Mg–Nd–Si–O system is not reported in the literature but a pseudo-binary $MgSiO_3$–Nd_2SiO_5 phase diagram[10] shows a eutectic at 1470°C. Hampshire and Jack[5] report that when nitrogen is added to silicate systems a reduction in the eutectic temperature is observed and therefore the lowest liquidus temperature for the Mg–Nd–Si–O–N system will be well below 1470°C. Thus, it is expected that there will be similarities in the densification rates for the Y–Si–Al–O–N and Mg–Nd–Si–O–N systems.

It is demonstrated that the choice of level of additive and the mole ratio of magnesia to neodymia for good densification and good microstructural development is wide. However, it must be noted that this choice is constrained by the occurrence of quaternary oxynitride phases when high levels of neodymia are used. It has been shown that such phases in the Ce–Si–O–N[11] and Y–Si–O–N[12] systems oxidise with a considerable increase in volume causing the disintegration of the ceramic component.

With respect to microstructure, preliminary results show that the aspect ratios of β silicon nitride grains vary between 6 and 9 depending on additive level. Wötting and Ziegler[4] show that high aspect ratios contribute to high strength and resistance to fracture when using mixed yttria–alumina additives. Clearly, until these parameters are assessed for mixed magnesia–neodymia additions then few conclusions can be drawn as to the quality of silicon nitride-based materials made using these additives. However, on the basis of the initial observations of microstructures and the high densities achieved, promising mechanical properties may be expected.

Whilst the densification rates of sialons and silicon nitride plus magnesia–neodymia additives may be similar it must be noted that, for the latter, the large volume of liquid formed during sintering remains as a glass after firing. This is in contrast to the β'-sialons where the liquid volume is reduced as aluminium and oxygen are incorporated into the β' grains. In order to produce a high strength ceramic material capable of maintaining its strength to high temperatures it is necessary to devitrify this integranular glass. The results obtained show that for the combined magnesia–neodymia additives, this glass may be devitrified to form an apatite phase as the major component which, however, may contain a small amount of nitrogen. It is necessary to firmly establish whether this refractory phase undergoes a volume expansion on oxidation, causing rupture of fabricated components. Results from limited experiments suggest that a N-apatite of composition $MgNd_9Si_6O_{25}N$ can be safely maintained at 1000°C in air for 48 h with no degradation and negligible weight gain.

5. CONCLUSIONS

The results show that combined magnesia–neodymia additions are promising densification aids for the pressureless sintering of silicon nitride. A wide choice of additive level and relative amounts of magnesia and neodymia exists within which a high density product with an interlocking fibrous grain structure can be easily and rapidly obtained. Work is continuing on further characterisation of these materials.

ACKNOWLEDGEMENTS

This work is supported by the Commission of the European Communities—DGXII—under contract SUT-126-EIR(H) as part of the Raw Materials (Technical Ceramics) research programme.

We wish to acknowledge the work of Ms D. Bourke in the preparation of samples for the determination of phase relationships.

REFERENCES

1. JACK, K. H. Sialons: a Study in Materials Development, this volume, pp. 1–30.
2. LUMBY, R. J., BUTLER, E. and LEWIS, M. H. In *Progress in Nitrogen Ceramics*, Riley, F. L. (Ed.), NATO ASI Series E65, Martinus Nijhoff, The Hague, p. 683, 1983.
3. LEWIS, M. H. and LUMBY, R. J. *Powder Metallurgy*, **26** (1983) 73.
4. WÖTTING, G. and ZIEGLER, G. *Ceramics International*, **10** (1984) 18.
5. HAMPSHIRE, S. and JACK, K. H. *Special Ceramics* Vol. 7, Taylor, D. E. and Popper, P. (Eds), *Proceedings of the British Ceramic Society*, **31** (1981) 37.
6. HAMPSHIRE, S. and POMEROY, M. J. *Annales de Chimie, Science des Materiaux*, **10** (1985) 65.
7. GIACHELLO, A., MARTINENGO, P. C., TOMASSINI, G. and POPPER, P. *American Ceramic Society Bulletin*, **59** (1980) 1212.
8. HAMPSHIRE, S. and JACK, K. H. In *Progress in Nitrogen Ceramics*, Riley, F. L. (Ed.), Martinus Nijhoff, The Hague, p. 225, 1983.
9. DREW, R. A. L., HAMPSHIRE, S. and JACK, K. H. In *Progress in Nitrogen Ceramics*, Riley, F. L. (Ed.), Martinus Nijhoff, The Hague, p. 323, 1983.
10. FEDOROV, N. F., ANDEEV, I. F. and TUNIK, T. A. In *Phase Diagrams for Ceramicists*, American Ceramic Society Inc., Columbus, Ohio, 1978.
11. QUACKENBUSH, C. L. and SMITH, J. T. *American Ceramic Society Bulletin*, **59** (1980) 533.
12. HAMPSHIRE, S. and JACK, K. H. In *Proceedings of the First International Symposium on Ceramic Components for Engine, Hakone, 1983*, Somiya, S., Kanai, E. and Ando, K. (Eds), KTK Scientific Publishers, Tokyo and Riedel Pub. Co., Dordrecht, p. 350, 1984.

COMMENTS AND DISCUSSION

Chairman: D. P. THOMPSON

D. P. Thompson: Firstly, I wish to comment that if you get the apatite phase formed, then you can get an oxide apatite in that system as well so that under oxidation it will remain as an apatite structure. Secondly, have you determined the eutetic temperature in the system between silicon nitride and the apatite?
M. J. Pomeroy: No, not directly.
S. Hampshire: We believe it could be less than 1450°C.
R. W. Davidge: The compounds that you have been identifying are things like $MgNd_4$, $MgNd_9$ whereas the magnesia/neodymia ratios you have used are quite different values. Is there any merit in looking at ratios with high neodymia to low magnesia?

Pomeroy: Yes. I would suggest that the limiting value you could look at under these conditions (1650°C/30 min) is 1·6. If you start to go much lower than that, then sintering becomes quite difficult, but if you look at the phase diagram then a composition which will give a grain boundary composition somewhere between forsterite and apatite is more beneficial.

Davidge: So if you have an optimum ratio for sintering, you have not necessarily got the right ratio for making the beneficial compounds, so far?

Pomeroy: No, but we have a very wide choice of compositions giving good densification and the possibility is that you may be able to produce the right grain boundary phases also.

P. Popper: When we did the work on sintering of silicon nitride with magnesia plus yttria[1] we found that we had to include magnesia in the powder bed, otherwise we lost it from the sample. Have you had any experience of this and have you checked this?

Pomeroy: The powder bed is the same composition as the sample so includes magnesia and neodymia and suppresses weight losses which at 1650°C are of the order of 1–2%.

A. Hendry: Would you like to comment on why the fibrous microstructure develops?

Pomeroy: I think it depends on the type of additive present and the previous work[2] on identifying the rate-controlling process (either diffusion through the liquid or solution/precipitation) supports this. With magnesia, the α–β transformation is retarded and a more equiaxed grain structure results and that is either controlled by the rate of solution or precipitation. Wötting and Ziegler[3] suggest that surface tension forces have some effect. With yttria, where there is a very high viscosity liquid, silicon nitride dissolves more rapidly and the process is diffusion controlled so it appears that in this system the diffusion distances are shorter and there is time for selectivity in terms of where the atoms fit onto the precipitating grains.

G. Wötting: Do you think there is any nitrogen in the grain-boundary glass?

Pomeroy: We have not analysed the grain-boundary glasses but we have produced Mg–Nd–Si–O–N glasses containing up to 20 equiv. % N.

Wötting: This is a very interesting point because I believe that Nd favours dissolution of nitrogen into glasses and this will affect the properties of any grain boundary phases.

References

1. GIACHELLO, A., MARTINENGO, P., TOMMASINI, G. and POPPER, P. *American Ceramic Society Bulletin*, **59** (1980) 1212.
2. HAMPSHIRE, S. and JACK, K. H. In *Special Ceramics*, Vol. 7, Taylor, D. E. and Popper, P. (Eds), *Proceedings of the British Ceramic Society*, **31** (1981) 37.
3. WÖTTING, G. and ZIEGLER, G. *Ceramics International*, **10** (1984) 18.

6

Microstructural Development, Microstructural Characterization and Relation to Mechanical Properties of Dense Silicon Nitride

G. WÖTTING, B. KANKA AND G. ZIEGLER

Deutsche Forschungs- und Versuchsanstalt für Luft- und Raumfahrt (DFVLR) eV, Institut für Werkstoff-Forschung, D-5000 Köln 90, West Germany

ABSTRACT

Microstructural development during liquid-phase assisted densification of Si_3N_4 powder compacts is a function of powder properties, type and amount of additives as well as of processing conditions (time, temperature, pressure). Varying these parameters results in—besides different densities and phase compositions—a different mean grain size and especially grain morphology of the β-grains. These characteristics mainly determine the mechanical properties at temperatures $>1000°C$. In order to optimize dense Si_3N_4 materials by correlating microstructural characteristics with mechanical properties, quantitative analysis of the grain structure is necessary. Thus, a method has been developed to quantify microstructures consisting of elongated, rod-like β-grains. By means of these characteristics it is possible to discuss the interdependence between starting composition, processing conditions, microstructure and resulting mechanical properties of dense Si_3N_4 materials. This leads to a model for the microstructural development during liquid-phase assisted sintering of Si_3N_4 and the relation between microstructural characteristics and mechanical properties.

1. INTRODUCTION

It is well known that, dependent on the properties of Si_3N_4 starting powders, the type and amount of additives as well as the processing parameters temperature, time and pressure, different microstructures and materials with different properties are obtained after sintering.[1-7] Especially for Si_3N_4 it was found by analyzing various sets of samples that mainly the elongated grain morphology of the β-grains and simultaneously small grain sizes result in good mechanical properties. For example, determination of the fracture energy γ of hot-pressed Si_3N_4 (α-Si_3N_4 + 5 wt % MgO) as a function of hot-pressing time resulted in a maximum which coincides with complete α- to β- conversion. After prolonged soaking time, γ (as well as fracture strength and fracture toughness) decreased mainly due to globularization which reduces the aspect ratio of the β-grains;[8] additionally, grain growth has a certain effect. Thus, in order to optimize mechanical properties of dense Si_3N_4 materials it is necessary to know which factors affect the grain morphology and the grain size during liquid phase-assisted sintering. In order to find these parameters, a quantitative microstructural characterization method has to be performed for materials which are preferably composed of rod-like grains, like dense Si_3N_4. This report presents a method to characterize such microstructures as well as the correlation between additive composition and processing parameters on the one side and the microstructure and mechanical properties of sintered Si_3N_4 on the other side.

2. MICROSTRUCTURAL DEVELOPMENT DURING SINTERING OF Si_3N_4

Up to now, dense Si_3N_4 can only be obtained by a liquid phase-assisted sintering process. The densification mechanisms seem to be rearrangement, solution–diffusion–reprecipitation, and coalescence. For the microstructural development, solution–reprecipitation processes and coalescence are of greatest importance. The resulting microstructure and mechanical properties are strongly influenced by various parameters. One significant parameter is the type of the Si_3N_4-starting powder:

— Starting with thermodynamically stable β-Si_3N_4-phase, during sintering a microstructure is developed consisting of large equiaxed grains, resulting in relatively low mechanical properties.[1,3]
— Starting, however, with thermodynamically unstable α-Si_3N_4-

phase, the resulting microstructure is composed of elongated prismatic grains which interlock and therefore give rise to significantly improved mechanical properties.

Therefore, one prerequisite for obtaining prismatic β-grains after densification is the use of Si_3N_4-powders consisting mainly of the α-phase. As an approximation it was demonstrated[1] that the resulting aspect ratio $\bar{a}$ of the β-grains (= ratio length to thickness of prismatic grains) depends on the phase composition of the starting powder as follows:

$$\bar{a} = 1 + \alpha/\beta$$

Due to the limited knowledge of physical and chemical conditions in the immediate surrounding of the growing crystal, the reason for this idiomorphic crystallization of prismatic grains during liquid phase sintering of α-Si_3N_4 may not be given unequivocally. Nevertheless, crystallographic investigations permit some conclusions to be drawn:[9,10]

— Idiomorphic crystallization is often observed if no nuclei of the phase to be precipitated are present. In this case, crystallization may not occur until a certain supersaturation is reached which leads to a spontaneous nucleation, often accompanied by a rapid crystallization. Within certain limits, the crystallization rate shows a power-law dependence on the degree of supersaturation.

— Reasons for the idiomorphic crystallization may be anisotropic diffusion or thermal conductivity, resulting in local differences in the supersaturation and thus in directed crystallization (outer effects), or stacking faults, resulting in preferred adsorption of further atomic units (inner effects). More disturbed surfaces, therefore, show higher growth rates than less disturbed ones.

Thus, idiomorphic crystallization occurs far from the thermodynamic equilibrium and leads to crystal morphologies with higher energy. In order to reduce the interfacial energy after supersaturation has slowed down, globularization takes place with prolonged soaking time.

In order to find some relations between the parameters starting composition, process conditions and microstructural characteristics, especially grain morphology and grain size, a quantitative microstructural analysis has to be performed.

3. MICROSTRUCTURAL CHARACTERIZATION

All conventional methods to characterize microstructures are based on the assumption of nearly equiaxed grain shapes.[11] Some corrections are possible by using shape factors. However, there have been only a few attempts to quantify microstructures consisting of markedly anisotropic grains, e.g. grains which may be described as rotational ellipsoids.[12]

As mentioned above, high-strength dense Si_3N_4 preferably consists of small grains with a high aspect ratio. The presence of such grains in the microstructure may be demonstrated by overetching polished microsections (Fig. 1(b)).[13] In normally etched microsections (Fig. 1(a)), however, only the thickness of the visible grains are true values, as for prismatic grains the thickness is not dependent on the grain orientation. On the other hand, the determination of the visible length from microsections yields a random value, depending on the orientation of the section. With the assumption that all grains within one sample have approximately the same aspect ratio, a relatively simple method may be used to characterize such microstructures (Fig. 2). This assumption seems reasonable as nucleation should occur at about the same degree of supersaturation, and crystallization takes place in a nearly constant compositional environment within the whole specimen because no solid solution is to take place. This 'true'-aspect ratio is

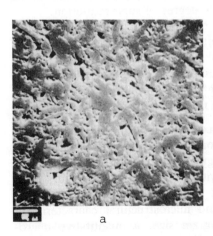
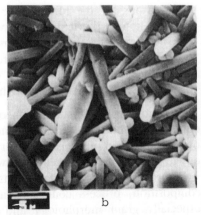

a b

FIG. 1. SEM-micrographs of a normally etched (a) and an overetched (b) polished microsection of SSN.

Microstructural Development and Characterization of Dense Silicon Nitride

Assumptions:
— All grains are prismatic
— The aspect ratio of all grains of one sample is ~ constant

Analysis:

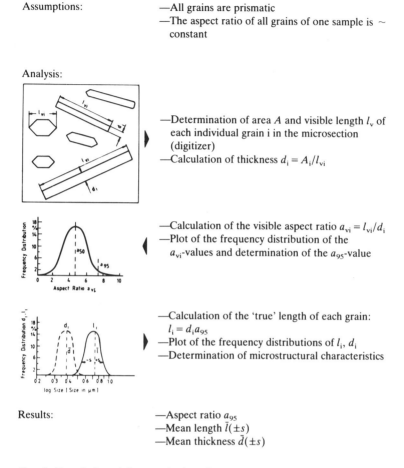

— Determination of area A and visible length l_v of each individual grain i in the microsection (digitizer)
— Calculation of thickness $d_i = A_i/l_{vi}$

— Calculation of the visible aspect ratio $a_{vi} = l_{vi}/d_i$
— Plot of the frequency distribution of the a_{vi}-values and determination of the a_{95}-value

— Calculation of the 'true' length of each grain:
$l_i = d_i a_{95}$
— Plot of the frequency distributions of l_i, d_i
— Determination of microstructural characteristics

Results:
— Aspect ratio a_{95}
— Mean length $\bar{l}(\pm s)$
— Mean thickness $\bar{d}(\pm s)$

FIG. 2. Description of the quantitative microstructural characterization method.

obtained from the frequency distribution of the aspect ratios of all grains visible in the microsection (e.g. Fig. 1(a)), choosing the value $\bar{a}_{95}$ (5% of all visible grains show higher, 95% lower $\bar{a}$-values). This choice primarily depends on the statistic consideration that about 10% of the grains are cut parallel to their crystallographic c-axis and therefore show the realistic $\bar{a}$-value. The value $\bar{a}_{95}$ is the mean value of 10% of these highest $\bar{a}$-values. This measure was verified experimentally by decomposing samples completely by strong chemical solution and determining the aspect ratio of the remaining single grains.[14]

4. EXPERIMENTAL RESULTS

In order to find some relations between the starting composition, processing parameters and microstructural characteristics, different sets of samples were analyzed.[5–7,13]

Set I

The first set concerns the sintering behavior and microstructural development as a function of the Al_2O_3-addition to a starting powder composition of $Si_3N_4 + 15$ wt % Y_2O_3 (Fig. 3). Under atmospheric N_2-pressure complete densification by sintering of compositions containing only Y_2O_3 is difficult, but simultaneous addition of Al_2O_3 improves densification. Constant density is reached with more than 3 wt % Al_2O_3-addition. As can be seen from Fig. 3, a continuous decrease of the aspect ratio $\bar{a}$ of the β-grains occurs with increasing Al_2O_3-addition. As $\bar{l}$, the mean grain length, remains nearly constant after reaching a high degree of densification, this means a continuous increase of the mean grain thickness (not shown in Fig. 3).

Set II

The second example shows the effect of the amount of Y_2O_3- and Al_2O_3-addition at constant Y_2O_3 to Al_2O_3 ratio on density and microstructural characteristics of SSN (Fig. 4). In this case, at nearly constant density an increase of $\bar{a}$ at nearly constant $\bar{l}$-values is found with rising additive amount, indicating a simultaneous decrease of $\bar{d}$ (not shown in Fig. 4).

Set III

In this case, the composition $Si_3N_4 + 15$ wt % $Y_2O_3 +$

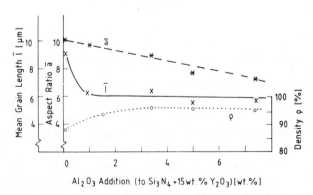

FIG. 3. Density and microstructural characteristics of SSN + 15 wt % Y_2O_3 as a function of Al_2O_3-addition (sintering conditions: 1800°C, 2 h + 1820°C, 2 h).

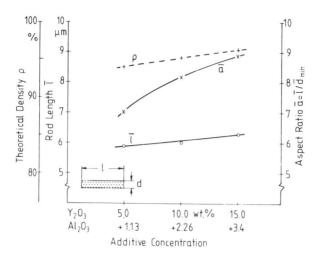

FIG. 4. Density and microstructural characteristics of SSN as a function of $Y_2O_3 + Al_2O_3$—addition at constant ratio Y_2O_3/Al_2O_3 (sintering conditions: 1800°C, 2 h + 1820°C, 2 h).

3·4 wt % Al_2O_3 was sintered under different conditions (Fig. 5). Already after 2 h, about 96% theoretical density was achieved, prolonged soaking times only providing small changes in density. Longer soaking times, however, result in a continuous decrease of the microstructural characteristics $\bar{a}$ and $\bar{l}$, indicating nearly constant

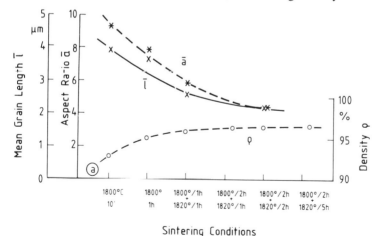

FIG. 5. Density and microstructural characteristics of SSN + 15 wt % Y_2O_3 + 3·4 wt % Al_2O_3 as a function of the sintering conditions.

$\bar{d}$-values. This means, after short soaking times, grains with very high aspect ratios have crystallized. With increasing soaking times, globularization of these grains occurs.

5. DISCUSSION

Sets I and II show variations of the final aspect ratio with the chemical composition of the additives. By use of the phase diagram Y_2O_3–Al_2O_3–SiO_2 (Fig. 6) which represents the theoretical composition of the pure oxide secondary phase during sintering (including the SiO_2 of the Si_3N_4-powder) it can be seen that in the first set (marked E1/4–E1/0) with increasing Al_2O_3 addition to Si_3N_4 + 15 wt % Y_2O_3,

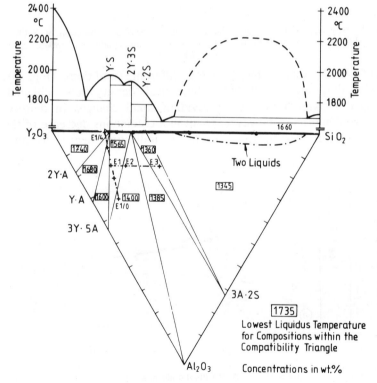

FIG. 6. Phase relationships and liquidus temperatures in the Y_2O_3–Al_2O_3–SiO_2 system.

the liquidus temperature decreases from about 1740°C to 1400°C. In the second set (marked E1–E3), with decreasing amount of Y_2O_3 + Al_2O_3 addition (at constant ratio Y_2O_3/Al_2O_3), the liquidus temperature is reduced from about 1600°C to about 1350°C. This allows the conclusion to be drawn that the viscosity of the liquid phase at the sintering temperature is higher for compositions with high liquidus temperatures. Lower viscosities, however, result in a faster material transport by diffusion. Thus, supersaturation after spontaneous nucleation (see Section 2) decreases faster and reduces the rate of the idiomorphic crystallization. Additionally, globularization processes after slowing down of the supersaturation are accelerated by high diffusion rates and low viscosities. Both effects lead to a decrease in the resulting aspect ratio of the grains in the sintered material. This, however, means that an unequivocal relation exists between the viscosity of the liquid phase and the aspect ratio of the precipitated crystals: the higher the viscosity at the sintering temperature, the higher the resulting aspect ratio.

Set III demonstrates the globularization processes taking place with increasing sintering time. The highest $\bar{a}$-value is developed after a short time. With slowing down of supersaturation, globularization starts. The change in microstructure is demonstrated by the following SEM micrographs (Fig. 7), including samples sintered under extreme conditions.

1800°C, 10 MIN   1800°C, 2H + 1820°C, 2H   50 SINTERING CYCLES

FIG. 7. SEM micrographs of SSN (15 wt % Y_2O_3 + 3·4 wt % Al_2O_3) sintered under different conditions.

6. CONCLUSIONS

Based on theoretical considerations, the experimental results may be summarized by a model which describes the microstructural development during liquid phase-assisted sintering of α-Si_3N_4, and the relation between microstructural characteristics and mechanical properties (Fig. 8).

After a short sintering time, grains with high aspect ratios develop due to idiomorphic crystallization from local supersaturations. After slowing down of supersaturation, these thermodynamically unstable grain structures reform to a more globular morphology in order to reduce their energy, associated with further densification of the sample. The aspect ratio formed during precipitation as well as the globularization rate seem to depend primarily on the viscosity of

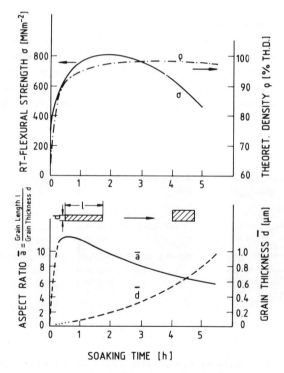

FIG. 8. Density, microstructural characteristics and resulting flexural strength of SSN as a function of sintering time (schematic plot).

the liquid phase and the diffusion rate, respectively: a high viscosity of the liquid phase leads to a slow reduction of the supersaturation and a low globularization rate, and finally results in a higher aspect ratio than in the opposite case. The resulting mechanical properties (up to the softening temperature of the mainly amorphous grain boundary phase) are mainly controlled by a superposition of the effects of residual porosity and aspect ratio, and only to a lower degree by grain growth of the materials. The mechanical properties of Sets I–III have not been discussed here, but they generally show a strong correlation to the aspect ratio at constant density.[5–7,13] High strength values are obtained with high aspect ratios, though the samples are not yet completely densified. This demonstrates the strong influence of the interlocking of elongated grains, providing crack deflection associated with fracture energy consumption. Prolonged soaking times at nearly constant density lead to a decrease in fracture strength as well as fracture toughness, as globularization occurs. In practice, a compromise must be made between the amount and viscosity of the liquid phase to achieve complete densification and to retain a high aspect ratio. Of importance, however, is that the aspect ratio may be influenced by the starting composition and the processing parameters.

The relation between aspect ratio and mechanical properties is valid up to the transformation temperature of the secondary phase which means up to about 1000°C. Nevertheless, efforts to improve strength below 1000°C by increasing the viscosity also affects high-temperature strength, as the softening of the material is shifted to higher temperatures.

REFERENCES

1. LANGE, F. F. *Journal of the American Ceramic Society,* **62** (1979) 428.
2. HAMPSHIRE, S. and JACK, K. H. In *Special Ceramics,* Vol. 7, Taylor, D. E. and Popper, P. (Eds), *Proceedings of the British Ceramic Society,* **31** (1981) 37.
3. KNOCH, H. and GAZZA, G. E. *Ceramurgia International,* **6** (1980) 51.
4. KNOCH, H. and ZIEGLER, G. *Science of Ceramics,* **9** (1977) 494.
5. WÖTTING, G. and ZIEGLER, G. *Ceramics International,* **10** (1984) 18.
6. WÖTTING, G. and ZIEGLER, G. In *Proceedings of the First International Symposium on Ceramic Components for Engine, Hakone, 1983,* Somiya, S., Kanai, E. and Ando, K. (Eds), KTK Scientific Publishers, Tokyo, p. 412, 1984.

7. WÖTTING, G. and ZIEGLER, G. *CFI-Fortschrittsber. der DKG*, **1** (1985) 33.
8. HIMSOLT, G., KNOCH, H., HUBNER, H. and KLEINLEIN, F. W. *Journal of the American Ceramic Society*, **62** (1979) 29.
9. WILKE, K. T. *Kristallzuchtung*, VEB Deutscher Verlag der Wissenschaften, Berlin, 1973.
10. CARMICHAEL, I. S. E., TURNER, F. J. and VERHOOGEN, J. *Igneous Petrology*, McGraw-Hill, New York, 1974.
11. UNDERWOOD, E. E., *Quantitative Stereology*, Addison-Wesley, Menlo Park, CA, 1970.
12. DE HOFF, R. T. *Transactions of the American Institute of Mechanical Engineers*, **224** (1962) 474.
13. WÖTTING, G. and ZIEGLER, G. *Science of Ceramics*, **12** (1983) 361.
14. KANKA, B., WÖTTING, G. and ZIEGLER, G. to be published in *CFI-Fortschrittsber. der DKG*.

COMMENTS AND DISCUSSION

Chairman: D. P. THOMPSON

R. J. Brook: Is the reason why you get the high aspect ratios at the beginning because of the different nucleation rates on the different surfaces of the crystal? Your explanation suggests that the viscosity erodes the aspect ratio which you have initially.

G. Wötting: We think that the reason is that the crystallization rate depends on the degree of supersaturation and high viscosity leads locally to a higher supersaturation and as the diffusion is slow, there is no reduction of the supersaturation. Thus, higher supersaturation leads to higher crystallization rate and it is known from the crystallographic literature that the crystallization shows a power law dependence on the degree of supersaturation. That is the reason for the high aspect ratios after short sintering times.

D. P. Thompson: You are assuming that the process all occurs on nuclei of silicon nitride?

Wötting: No, we assumed spontaneous, homogeneous nucleation. Of course, there may be heterogeneous nucleation on existing grains too but this should be homogeneous nucleation within the liquid phase.

K. H. Jack: I think that Professor Brook is worried about the same things that I am. You have to have differential nucleation and growth on different faces of the crystal, even accepting that you have a high supersaturation. Have you considered this?

Wötting: We found preferred growth in the *c*-direction and this can

occur because of anisotropic effects such as heat transfer as well as preferred adsorption of stacking faults relative to the c-axis.
Brook: You have to consider the rates at which growth occurs along each of the axes. If you say that the viscosity is changing the aspect ratio, then changing the viscosity changes the ratio at which those two growth processes occur. So the more viscous, the more you favour a-axis growth rather than c-axis growth.
Wötting: Well, why do whiskers grow? In my opinion, it is because of supersaturation which is so high that growth occurs only in one direction.
Jack: That accounts for the effect of supersaturation but not the effect of viscosity.
Wötting: Viscosity is related to diffusion and the higher the viscosity, the lower the diffusion rate.Thus you have locally supersaturation for longer times. For higher diffusion rates, supersaturation is slowed down and you have equilibrium growth.
A. Hendry: I think that there are two aspects that you are looking at. There is the nucleation of the crystals which is critically dependent on supersaturation—the higher the supersaturation, the greater the number of nucleation sites—but having started the process you then have to consider the growth rate and growth is not so critically dependent on supersaturation but is more dependent on the relative surface energies, the interfacial tensions, between the solid surfaces and the phase from which the surfaces are developing. What you are saying is that there is a correlation between viscosity and supersaturation but also presumably between viscosity and interfacial tension on the different faces of the growing crystal.
Wötting: I should mention that the results of the model presented are deduced from the system $Y_2O_3-Al_2O_3-SiO_2$ and we feel that, unless we enter a field of liquid immiscibility, then the properties of the liquids within the system are similar. If you change the system to e.g. $MgO-Nd_2O_3-SiO_2$ then the situation may be quite different. There will be differences in interfacial energy, wetting and solubility as well as viscosity and diffusion rates.

We have tried to explain our experimental results rather than develop a theory and I think the model presented is adequate.
Hendry: Do you think that it is necessary to have a phase transformation?
Wötting: Yes, absolutely, because the phase transformation leads to supersaturation and idiomorphic growth. If you start with the more

stable β phase, then with equilibrium growth the grain structure is determined finally by the interfacial energy.

Hendry: The supersaturation cannot be very high because the difference in ΔG between α and β at 1650°C is not great.

Wötting: But the sintering mechanisms are quite different. In one case you have complete solution of α and reprecipitation of β whereas if you start with large β grains, these do not dissolve in the liquid and only small grains dissolve and precipitate on the larger ones to minimize surface energy.

Thompson: What benefits in strength are achieved by developing high aspect ratios?

Wötting: For $Si_3N_4 + 15\% \ Y_2O_3 + 3\% \ Al_2O_3$ which is the optimum composition this has a room temperature strength of $750 \ MN \ m^{-2}$ and on increasing the Al_2O_3 to $7 \ wt \%$, the strength falls below $500 \ MN \ m^{-2}$. Fracture toughness and strength are correlated and also correlated to aspect ratio.

7

Sialon and Syalon Powders: Production, Properties and Applications

P. FERGUSON AND A. W. J. M. RAE[†]

Anzon Limited[‡], *Wallsend, NE28 6UQ, UK*

ABSTRACT

The role of sialon-based ceramics as engineering and refractory materials is reviewed. The commercialisation of β'-sialons incorporating yttrium oxide (Syalons) as a sintering aid is also described.

1. INTRODUCTION

In the 1970s sialon ceramics containing yttrium oxide, a liquid-phase sintering aid, were developed for high stress and high temperature applications. Methods patented worldwide by Lucas Industries plc are currently employed to manufacture these β'-yttrium sialons (Syalons). Syalon is a registered trade name of Lucas Cookson Syalon Limited. Based on β'-$Si_{6-z}Al_zO_zN_{8-z}$ ($z = 2 \cdot 0$–$4 \cdot 2$),[1] a silicon nitride derivative, Syalons combine the desirable properties of silicon nitride such as strength, hardness and heat resistance with improved properties such as toughness, chemical resistance and sinterability. Commercial production of high-purity silicon nitride, sialon and Syalon ceramic powders commenced in 1983–4 at Anzon Limited, a Cookson Group plc company. In 1982 a joint venture company, Lucas Cookson Syalon Limited, between Lucas Industries plc and Cookson Group plc was formed. The activities of Lucas Cookson Syalon Limited include the

[†] Now at TAM Ceramics Inc., Niagara Falls, New York, USA.
[‡] From 1st January 1986 Cookson Ceramics and Antimony Limited.

manufacture of Syalon components and the licensing of other companies to manufacture industrial ceramics, wear parts and automotive components from Syalon powder. At present, Lucas Cookson Syalon Limited have granted licences to seven companies in Japan and the USA. In keeping with their leading role in sialon ceramics Lucas Cookson Syalon Limited continue to support a substantial research and development programme. The present paper outlines the types and applications of Syalon powders manufactured by Anzon Limited.

2. POWDER PRODUCTS

Powder products presently available from Anzon Limited are listed in Table 1. They include silicon nitride, 21R 'AlN' polytype (Syalon 404), β'-sialon (Syalon 401) and β'-yttrium sialons (Syalon 102 and Syalon 201).

The Lucas patented method to manufacture Syalon ceramics involves sintering, at about 1800°C, a mixture of silicon nitride, alumina, 21R 'AlN' polytype and yttrium oxide. A schematic representation of the powder processing method employed by Anzon is shown in Fig. 1. Silicon nitride powder (>90% α), manufactured by nitriding high purity silicon powder, is ball milled with other constituents (alumina, etc.) to give a homogeneous powder of the correct particle size and purity. The resulting material is in the necessary form for the fabrication of Syalon components. Development of the powder processing procedure including the use of clean room facilities has enabled the modulus of rupture values for Syalon 102 to be improved from 826 MPa in 1981 to 960 MPa in 1984.

TABLE 1
Powder Products

A Silicon nitride
B Syalons
   (i) Engineering grades (Y–Si–Al–O–N)
      Syalon 102—High toughness
        (β' + glass)
      Syalon 201—High temperature strength
        (β' + YAG)
   (ii) Refractory grades (Si–Al–O–N)
      Syalon 401—β'-sialon, $z = 3$
      Syalon 404—21R 'AlN' polytype

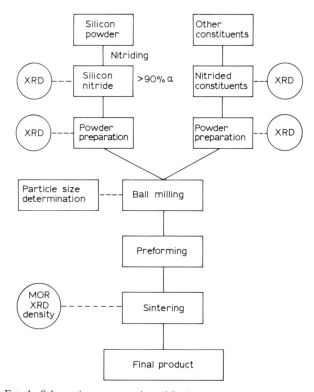

FIG. 1. Schematic representation of Syalon ceramic production route.

2.1. Engineering Syalon Grades

Currently, two types of β'-yttrium sialon powder are manufactured for engineering applications. These materials, designated Syalon 102 and Syalon 201 are available only to licensees of Lucas Cookson Syalon Limited. Sintered Syalon 102 powder produces a tough microstructure of β'-Syalon grains interconnected by a continuous glassy second phase which limits the upper operating temperature for such materials to around 800°C. Syalon 201 on the other hand is formulated to produce β' grains bonded by a crystalline YAG (yttrium aluminium garnet) phase. This intergranular phase gives good creep resistance and strength retention up to 1350°C. The basic properties of Syalon ceramics can be summarised as:

(i) High mechanical strength at ambient and elevated temperatures.

(ii) High specific strength resulting in weight savings over metallic systems.
(iii) High hardness, high toughness and low coefficient of friction. This combination of properties results in excellent wear, abrasion and erosion resistance.
(iv) Low coefficient of thermal expansion leading to good thermal shock resistance.

Components can be fabricated from Syalon ceramic powders using the normal forming techniques for oxide ceramics. These include isostatic pressing, uniaxial pressing, slip casting, extrusion and injection moulding. Pre-forms utilise organic binders which are removed in air at temperatures up to 500°C prior to final sintering at about 1800°C in a protective atmosphere. Final machining if required is performed using diamond grinding or ultrasonic drilling.

Areas in which Syalon ceramics are being evaluated include gas turbines (aero and auto), reciprocating engines (petrol and diesel) and industrial wear parts. Syalon ceramics are established materials for indexible cutting tools and metal forming dies and in both applications out-perform the conventional materials, cobalt-bonded tungsten carbide and alumina.[1,2]

2.2. Refractory Syalon Grades

In the refractories industry there is a growing demand for materials with very high dimensional accuracy and stability at high temperatures in corrosive environments. To meet this demand two Syalon powders, Syalon 401 (β'-$Si_3Al_3O_3N_5$) and Syalon 404 (21R 'AlN' polytype) have been developed and are commercially available. The potential usefulness of Syalon ceramics as refractories is due to their excellent thermal shock resistance ($\Delta T \sim 900°C$) compared to oxide ceramics ($\Delta T \sim 300°C$) and their good compatibility with molten ferrous and non-ferrous metals.[3,4] A recent commercial development in silicon carbide refractories for blast furnace and kiln furniture applications is the use of sialon as a bonding agent. Sialon bonded silicon carbide shows improved strength and alkali resistance compared to conventional silicon nitride/silicon oxynitride bonded silicon carbide. Currently Syalon ceramics are under consideration as continuous casting dies and crucibles for vacuum investment casting.

A comprehensive study, supported by the EEC, to develop Syalon refractories is underway at Anzon Limited. The initial phase of this work has been to assess the effect of adding Syalon powders to common refractory materials, e.g. alumina, magnesia, etc. An example of this work is illustrated by the hot-pressing (1700°C 1 h) of controlled additions of Syalon 102 with refractory grade lime stabilised (3·5 wt %) zirconia (see Figs. 2 and 3). Increasing the Syalon content to 5 wt % produces a dense, strong microstructure as indicated by the decrease in porosity and the increase in hardness of the material (see Fig. 2). X-ray diffraction powder photographs of the hot-pressed samples show that a destabilisation of the cubic zirconia occurs for the control sample (no Syalon addition), with the cubic content decreasing to 50% (see Fig. 3(a)). This change is shown by the intensity variation of the X-ray reflections for the cubic (c) and monoclinic (m) phases. Destabilisation of cubic zirconia is rectified by Syalon addition, a gradual restabilisation of the cubic zirconia occurs with increasing additive until at 5 wt % Syalon the cubic:monoclinic ratio is restored to the value (70:30) of the as-received material. Claussen et al.[5] have reported similar effects when adding reactive nitrides, e.g. Mg_3N_2, AlN, Si_3N_4 or ZrN to zirconia. However, in the present investigation a cubic Zr(O, N) phase was also observed to accompany the stabilisation of the cubic zirconia. Similar results to this were found with Syalon 401 and Syalon 404 additives. This work is still in progress and, on conclusion, will be reported elsewhere.

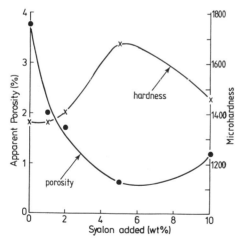

FIG. 2. Variation of apparent porosity and microhardness of hot-pressed lime stabilised zirconia with Syalon 102 additions.

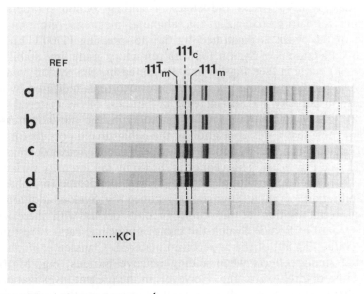

FIG. 3. X-ray diffraction powder photographs (a, b, c, d and e) corresponding to hot-pressed samples containing 0, 1, 2, 5 and 10 wt % Syalon 102.

3. SUMMARY

The combination of excellent chemical, physical and mechanical properties of sialon-based ceramics is currently being exploited in engineering and refractory applications. The improved sinterability of Syalon powders compared to silicon nitride powders allows the production of fully dense components. An initial investigation has demonstrated that Syalon powder additions improve the properties of a number of refractory materials.

ACKNOWLEDGEMENTS

The authors gratefully acknowledge the Board of Anzon Limited for permission to publish this paper. The financial support for this work was provided for by the EEC.

REFERENCES

1. JACK, K. H. In *The Chemical Industry,* Ellis Horwood Limited, Chichester, p. 271, 1982.
2. COTHER, N. E. and WILSON, F. Paper presented at *Scanwire '85 Conference,* Copenhagen, May 1985.
3. JACK, K. H. *Transactions and Journal of the British Ceramic Society,* **72** (1973) 376.
4. LUMBY, R. J., NORTH, B. and TAYLOR, A. J. In *Ceramics for High Performance Applications II,* Burke, J. J., Lense, E. N. and Katz, R. N. (Eds), Brook Hill Publishing Co., Chestnut Hill, MA, p. 893, 1978.
5. CLAUSSEN, N., WAGNER, R., GAUCKLER, L. J. and PETZOW, G. *American Ceramic Society Bulletin,* **56** (1977) 301.

COMMENTS AND DISCUSSION

Chairman: D. BROUSSAUD

(Paper read by K. H. Jack for the authors)

P. Popper: When you add the sialon to partly stabilise zirconia, what atmosphere do you have?

K. H. Jack: This is a hot-pressing rig with an atmosphere which is probably a mixture of carbon monoxide, nitrogen, etc. With the zirconia I'm sure that it is the nitrogen which is the additional stabiliser and presumably you could add that in a variety of ways, not necessarily using sialon.

K. Notter: Are these powders used currently as abrasives or, for example, for grinding, lapping, polishing, and if so what sort of applications?

Jack: No, they are not and this was asked previously. I think there is great potential in exploring this with sialon.

G. Wötting: Is there a loss in the sintering activity using these powders instead of starting from the pure constituents because energy of formation of solid solutions will be lost?

Jack: That will only be true for Sialon 401. The other powders are just mixtures of components, that is the difference. There is the 404 which is 21R only and there is the 401 which is the β'. They are already reacted. These are the 'magic' ingredients which we hope will improve refractories and you do not need much of them.

A. Kennedy: You said that there are some kinds of powder that the man in the street can buy and some that are available only to the licensees. I don't understand this, why?

Jack: The reacted powders which the man in the street can buy probably involve no commercial confidentiality but if you buy the unmixed powders to make your own sialon then you have to have a bit of know-how which would be supplied to licensees by Lucas Cookson Syalon.

M. Ahmed: You mentioned the use in metal cutting; what sort of applications?

Jack: This is well established and being developed by Sandvik and Kennametal. Different materials are required for different applications but the advantages of using sialon are clear for cast iron and nickel-based alloys. High-speed machining using very high cutting speeds requires special cutting tools. You cannot use them with any advantage in ordinary small lathes because you then lose the advantage of very high speed.

Ahmed: Can you quantify what you mean by high speed?

T. Ekström: Up to $200–300$ m min^{-1}.

8

Reaction Sequences in the Preparation of Sialon Ceramics

W. Y. SUN, P. A. WALLS AND D. P. THOMPSON

Wolfson Laboratory, Department of Metallurgy and Engineering Materials, University of Newcastle upon Tyne, NE1 7RU, UK

ABSTRACT

Sialon ceramics are produced by a reaction sintering process whereby an appropriate mix of binary oxide and nitride powders is heated to temperatures at which eutectic liquids are formed which then initiate a combined process of densification and solution–reprecipitation of the final phase assemblage. At temperatures below the eutectic, side reactions occur which lead to intermediate crystalline phases being produced and after the liquid phase has appeared a continuous equilibration process takes place as the tie lines relating solid and liquid compositions in equilibrium continuously vary with temperature. Loss of gaseous species (particularly from liquid phases) is another factor which affects the composition and quality of the final product. It is clearly desirable to move from the physically and chemically inhomogeneous starting condition to the dense homogeneous finished product in a single firing operation provided that good quality material can be obtained.

Reaction sequences in the preparation of α', β' and O' ceramics (and binary mixtures of these) are described, and requirements for avoiding unwanted second phases are discussed.

1. INTRODUCTION

Nitrogen ceramics are being developed for engineering applications because of their good mechanical properties at high temperatures. The best properties are exhibited by fully dense materials, prepared by adding small amounts of oxide and nitride powders to the starting silicon nitride powder so that at firing temperatures an oxynitride liquid phase is formed which facilitates densification by a liquid phase mechanism. The glassy phase which forms when this liquid is cooled to room temperature is detrimental to the high-temperature strength and must therefore either be minimised or converted by subsequent heat treatment into a more refractory crystalline phase. In either case, the development of materials which simultaneously combine good room-temperature and high-temperature strength requires careful control of the type and amount of additive(s) used.

In early hot-pressing work on silicon nitride[1] it was noted that maximum strength was achieved after the $\alpha \to \beta$ transformation had taken place and the term 'grain refinement' was used to describe this. Since that time considerable attention has been paid to the quality of starting powders, powder processing techniques and the variables associated with densification and $\alpha \to \beta$ transformation to the extent that now fully dense materials with the desired phase composition can be made reproducibly by pressureless sintering techniques. Less attention has however been paid to the achievement of good microstructures. Obviously it is impossible to carry out a detailed assessment of the effects of microstructural variables unless each specimen is fully dense and has the same phase composition. This second consideration is more difficult to achieve in sialon systems where the chemical complexity results in a number of competing reactions taking place and where liquid and gaseous species play a significant part in the reaction. The simultaneous achievement of complete reaction, full density and a good microstructure in a single firing operation is therefore very demanding and it is perhaps not surprising that reproducibility is so difficult to achieve in large scale production.

The present paper describes reaction sequences which occur during the formation of some single phase and two-phase sialon materials and discusses the limitations that this places on the development of a good microstructure in the final material.

2. PREVIOUS WORK

2.1. Hot-pressed Silicon Nitride

Wild et al.[1] investigated the role of magnesium oxide in the hot-pressing of silicon nitride and established the reaction sequence shown in Fig. 1. The magnesium oxide, present initially as hydroxide, reacts with surface silica on the silicon nitride powder to form forsterite, Mg_2SiO_4, in the temperature range 1000–1400°C. Above this temperature α-Si_3N_4 starts to transform to β-Si_3N_4 by solution–precipitation in a liquid phase, and at the same time, this liquid starts to redissolve the forsterite. $\alpha \rightarrow \beta$ transformation is complete by 1700°C and on cooling, the liquid solidifies as a magnesium silicon oxynitride glass. Densification, which starts as soon as the liquid phase forms, is complete by 1600°C but the maximum strength occurs in samples hot-pressed at higher temperatures. The overall reaction sequence, consisting of the initial formation of a crystalline oxide phase followed by the appearance of liquid phase, $\alpha \rightarrow \beta$ transformation and final cooling to give β-Si_3N_4 plus glass, has provided the basic pattern for all subsequent development work on the sintering of silicon nitride with additives. Further examples are described in later sections.

2.2. β′-Sialon

Oyama and Kamigaito[2] in their initial development work on sialons referred to 'line-splitting' in the X-ray pattern of materials prepared by reacting silicon nitride/alumina powder mixes at 1700°C. This was due to the simultaneous occurrence of β-Si_3N_4 and a highly-substituted β'-sialon in the product. The explanation lies in the much slower reactivity of silicon nitride compared with alumina which results in all the alumina reacting with some silicon nitride in the presence of liquid to give a high-z β'-sialon and leaving the remaining silicon nitride unreacted but transformed to β by the liquid phase.

Prior to the work of Lumby et al.[3] the composition of β'-sialon was believed to lie along the Si_3N_4–Al_2O_3 join and as a result sialon

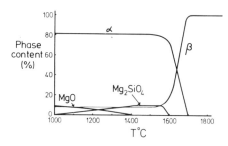

FIG. 1. Reaction sequence for silicon nitride hot-pressed with 10% MgO at 1000–1800°C.

X-phase[4] was always observed in the products. Gugel et al.[5] in an attempt to characterise this phase reported that large amounts of X-phase were produced at the start of the reaction when SiO_2/AlN powders were used whereas Si_3N_4/Al_2O_3 powders reacted in the same way gave X-phase only at the end of the reaction. These observations can be explained in terms of reaction kinetics (see Fig. 2) because silica reacts quickly with some aluminium nitride to give a liquid phase near the X-phase composition which readily crystallises X-phase by solution–precipitation and this X + AlN + liquid mix reacts further to give β'-sialon plus residual X-phase. The silicon nitride–alumina reaction on the other hand proceeds as described above with the formation of less liquid phase and hence less crystallisation of X-phase.

When the correct β' formula was established,[3] β'-sialons could be prepared with very little X-phase in the final product but trace amounts of polytypoid phases[6] sometimes occurred. A systematic study of reaction sequences involved in β' formation using four different reaction routes was described by North.[7] He showed that:

(i) X-phase occurs transitorily in amounts related to silica content of the starting mix;
(ii) the z-values of β' sialons are always higher at the start of a reaction corresponding to the slowest step being the reaction between α-Si_3N_4 and the first-formed high-z β' phase;

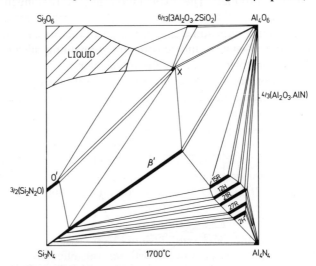

FIG. 2. Phase relationships in the Si_3N_4–SiO_2–Al_2O_3–AlN system at 1700°C.

(iii) the slowest step is the conversion of β-Si_3N_4 in the starting powder to β'-sialon;
(iv) direct routes, e.g. preparing $z = 2$ β' sialon from Si_2N_2O–AlN mixtures, lead to fewer side reactions and hence a higher yield of β' initially;
(v) after sufficient time and a high enough temperature, the fired product is the same whichever route is used and is the only reliable indication that equilibrium has been achieved.

These general principles are true for most sialon reactions and the next sections describe their application to the production of mixed-phase sialon materials.

3. RESULTS AND DISCUSSION

3.1. Preparation of O'–β' Sialons

Trigg and Jack[8] showed that approximately 7·5 mole % of alumina could be incorporated into the structure of silicon oxynitride to form an O' sialon. Between O' and β' sialon there exists a two-phase O'–β' region extending to a β' z-value of approximately 0·8. Preparing O'–β' materials without an oxide additive results in poor densification and unreacted α-silicon nitride in the product. In the present work yttrium oxide was added to the mix and the starting constituents of Y_2O_3 (Rare Earth Products Ltd, 99·9%), Al_2O_3 (Alcoa, grade A17), Si_3N_4 (Starck–Berlin, grade LC10) and SiO_2 (Thermal Syndicate plc, crushed quartz crystal) were mixed together to give a composition halfway between O' and β' at the aluminium-rich end of the two-phase region and containing in addition 15 wt % of $Y_2Si_2O_7$ to aid densification. The mixed powders were compacted into pellets and fired in a nitrogen atmosphere in a carbon resistance furnace at 1400–1800°C. The progress of the reaction is shown in Fig. 3.

A liquid phase forms from the starting constituents above 1300°C and α-Si_3N_4 dissolves in this liquid above 1400°C with precipitation of O'. This reaction proceeds more rapidly as the temperature is increased and at 1600°C, O' is the major phase along with α-silicon nitride and liquid. Above this temperature, solution of α-silicon nitride results in the precipitation of β' and some O' also redissolves so that at 1800°C the final equilibrium O'–β' assemblage has been achieved. This reaction sequence illustrates the effect of a changing liquid composition on the course of the reaction. At the lower

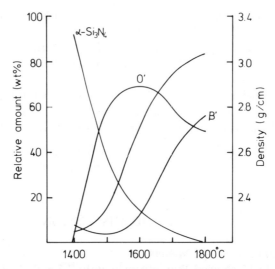

FIG. 3. Reaction sequence for the preparation of O'–β' ceramics from Si_3N_4, Al_2O_3, Y_2O_3, AlN mixtures at 1400–1800°C.

temperatures, the liquid is more oxygen rich and precipitates the more oxygen-rich product (O'); with increasing temperature the liquid becomes more nitrogen rich and precipitates the more nitrogen-rich product (β'). The selective precipitation of the two end-members allows the possibility of tailoring the microstructure by carrying out heat treatments at intermediate temperatures. Indeed, a distribution of high aspect ratio needles of O' in a β' matrix should produce a material with good mechanical properties.

3.2. Preparation of α'–β' Sialons

Cutting tools prepared from compositions in the α'–β'-glass region of the yttrium sialon system have a higher hardness than equivalent β'-glass materials and this results in increased wear resistance and a longer tool life.[9]

Reaction sequences in the formation of α'–β' materials densified with either lime or yttria have been studied by sintering mixed Si_3N_4, Al_2O_3, AlN powder compacts in a carbon element furnace in nitrogen at 1200–1800°C. The results for a calcium-containing mix designed to give equal amounts of α' and β' with no residual glassy phase are shown in Fig. 4. Initially the lime reacts with alumina and silica (on the surface of the silicon nitride powder) probably by a liquid phase

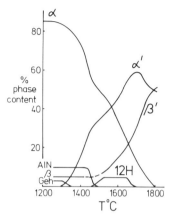

FIG. 4. Reaction sequence for the preparation of $\alpha'-\beta'$ ceramics from CaO, Al_2O_3, AlN, Si_3N_4 mixtures at 1200–1800°C.

mechanism to produce crystalline gehlenite ($Ca_2Al_2SiO_7$). With increasing temperature, the amount of liquid in the specimen increases, redissolving the gehlenite which finally disappears at 1350°C. Solution of α-Si_3N_4 in the liquid starts at 1300°C and an α' phase with high unit cell dimensions is precipitated. This reaction continues as the temperature increases but no further changes occur until at 1450–1500°C several reactions take place simultaneously. The aluminium nitride, which has so far played no part in the reaction, dissolves and 12H is precipitated. The 12H phase is only transient and disappears again by 1700°C. The unit cell dimensions of α' decrease at 1500°C to lower values which are then retained up to 1800°C (see Fig. 5). Finally at 1500°C, β'-sialon with a z-value of 1·0 starts to precipitate and increases in amount until by 1800°C, the final traces of α-Si_3N_4 have disappeared and α' and β' are present in equal amounts. The β' z-value remains constant throughout. The change in α' unit cell dimensions may be due to different α'-liquid tie lines at different temperatures but is more likely to be due to the more reactive oxide constituents in the starting mix promoting an oxygen-rich α' which at higher temperatures reacts with more α-silicon nitride to give the final equilibrium α' composition.

A similar reaction sequence is observed when yttrium oxide is used as the densifying additive (Fig. 6). In this case YAM ($Y_4Al_2O_9$) and YAG ($Y_3Al_5O_{12}$) are the crystalline oxides formed at low temperatures and the liquid phase appears at a temperature approximately

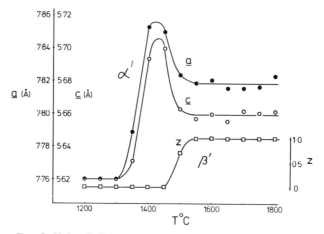

FIG. 5. Unit cell dimensions of α' and β' sialons shown in Fig. 4.

100°C higher than in the calcium case. The α' unit cell dimensions are again high initially and decrease as the reaction proceeds (see Fig. 7).

3.3. α'-Alumina Reaction Couples

An alternative method of preparing $\alpha'-\beta'$-glass materials is by reacting α'-sialon with alumina.[10] Obviously this reaction could be studied by intimately mixing powders of the two starting materials and

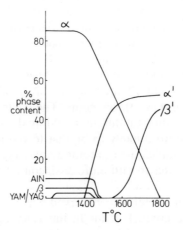

FIG. 6. Reaction sequence for the preparation of $\alpha'-\beta'$ ceramics from Y_2O_3, Al_2O_3, AlN, Si_3N_4 mixtures at 1200–1800°C.

Reaction Sequences in the Preparation of Sialon Ceramics 113

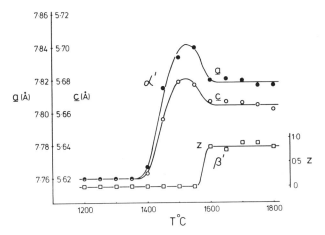

FIG. 7. Unit cell dimensions of α' and β' sialons shown in Fig. 6.

sintering or hot-pressing at 1700–1800°C, but in the present work an initial exploration was carried out using a reaction couple (see Fig. 8) because previous work had shown that this was a particularly effective method of studying all possible reactions between the two materials.

A preprepared dense sample of calcium α' sialon of composition $Ca_{0.48}Si_{9.42}Al_{2.58}O_{1.61}N_{14.39}$ containing 10% of additional calcium sialon glass was placed in contact with a short length of alumina rod and immersed in boron nitride powder in a hot-pressing apparatus as shown in Fig. 8. The temperature was raised to 1600°C at a pressure of 30 MPa and left for 30 min to allow the calcium sialon glass in the α' to melt and flow into the interface between the two materials. The pressure was then released and the temperature raised to 1700°C for 30 min to allow the reaction to proceed. At the end of the reaction the bond at the interface was extremely weak and the compact fractured easily into two halves. The break occurred at the alumina side of the interface leaving approximately 250 μm of interfacial region attached to the original α' sample. X-ray diffraction studies (Fig. 8) showed a region of 15R on the α' side of the break separated from the unreacted α' by a layer of β'. Figure 9 shows scanning electron micrographs across the interface at the break (a) and at the 15R-β'-α' region (b). The 15R consists of large thin plates aligned perpendicularly to the the interface and fanning out towards the β' region. The original interface is in the middle of this 15R region and can only be

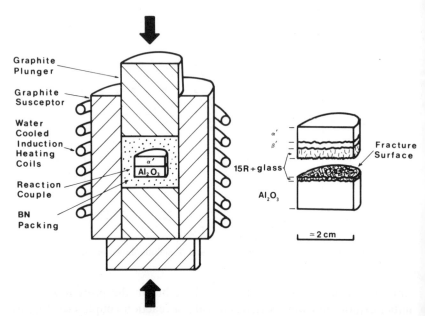

FIG. 8. Experimental arrangement for preparing $\alpha'-\beta'$-ceramics using an α'-Al_2O_3 reaction couple.

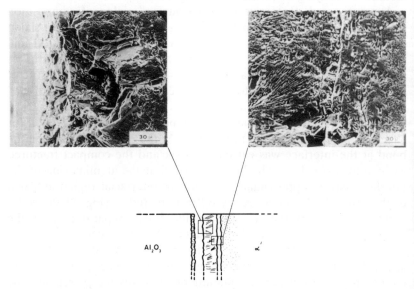

FIG. 9. Scanning electron micrographs of the fracture surface and the $\alpha'-\beta'$ region of the α'-Al_2O_3 reaction couple.

identified as the source point from which most of the plates fan out. Most of the reaction has occurred on the α' side of the interface due to diffusion of aluminium and oxygen through the calcium sialon liquid.

Confirmation of this result is shown by the X-ray digimaps in Fig. 10. In these micrographs black corresponds to low concentrations of an element and white to a high concentration, with shades of grey in between. The aluminium scan in (d) shows a gradual decrease in aluminium from 15R to β' and then to α' with a corresponding increase in silicon (see (c)). The calcium scan (b) shows some concentration of calcium in the α' region, none in the β' region but

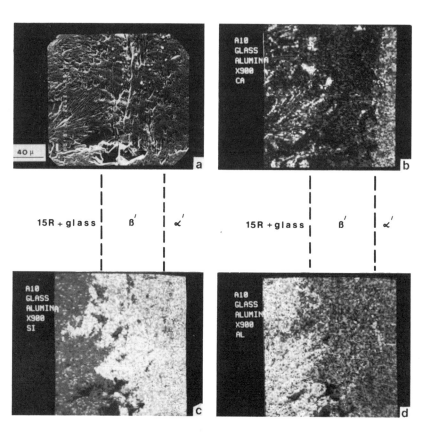

FIG. 10. The $\alpha'-\beta'$ region (a) and X-ray digimap scans for calcium (b), silicon (c) and aluminium (d) in the reaction zone of the α'-Al_2O_3 reaction couple.

scattered areas of high concentration in the 15R region corresponding to intergranular glass. It is perhaps surprising that the β' layer does not show similar glassy regions.

The results show that small additions of alumina to calcium α' sialon result in transformation to β' plus glass and also that reaction with more alumina moves the overall composition into a region of 15R plus liquid. The occurrence of transient polytypoid phases in α', β' or mixed sialon composites may therefore occur if there is inadequate mixing of the starting powders.

4. CONCLUSIONS

During the preparation of sialon ceramics, competing reactions occur which, depending on the nature of the starting materials and the firing cycle, can result in different amounts of product phases. In general, if the specimen is held for long enough at a high enough temperature (i.e. high enough to allow sufficient liquid phase to be produced but low enough to prevent weight loss of volatile species) then the desired final equilibrium product will be achieved.

In a starting mix of binary nitrides and oxides, the oxides react first to give metal silicates or aluminates which then subsequently dissolve in the grain boundary liquid phase as the temperature is increased. α-silicon nitride dissolves in this liquid and precipitates different sialon phases at different temperatures; calcium α'-sialon forms at 1300°C, yttrium α'-sialon at 1400°C, O$'$ sialon at 1400°C and β' sialon above 1500°C.

In the preparation of single-phase sialons, the single step firing of compositions in a two-phase sialon–liquid region makes it difficult to develop optimum microstructures. However, there is more flexibility in two-phase sialon compositions because the two phases form at different temperatures and selective heat-treatment cycles can be employed.

ACKNOWLEDGEMENTS

Two of us (W.Y.S. and P.A.W.) thank Lucas Cookson Syalon for financial support during the period in which the work was carried out.

REFERENCES

1. WILD, S., GRIEVESON, P., JACK, K. H. and LATIMER, M. J. In *Special Ceramics,* Vol. 5, Popper, P. (Ed.), British Ceramic Research Association, p. 377, 1972.
2. OYAMA, Y. and KAMIGAITO, O. *Yogyo-Kyokai Shi,* **80** (1972) 327.
3. LUMBY, R. J., NORTH, B. and TAYLOR, A. J. In *Special Ceramics,* Vol. 6, Popper, P. (Ed.), British Ceramic Research Association, p. 283, 1975.
4. THOMPSON, D. P. and KORGUL, P. In *Progress in Nitrogen Ceramics,* Riley, F. L. (Ed.), Martinus Nijhoff, The Hague, p. 375, 1983.
5. GUGEL, E., PETZENHAUSER, I. and FICKEL, A. *Powder Metallurgy International,* **7** (1975) 66.
6. THOMPSON, D. P., KORGUL, P. and HENDRY, A. In *Progress in Nitrogen Ceramics,* Riley, F. L. (Ed.), Martinus Nijhoff, The Hague, p. 61, 1983.
7. NORTH, B. M.Sc. thesis, University of Birmingham, 1976.
8. TRIGG, M. B. and JACK, K. H. In *Proceedings of the First International Symposium on Ceramic Components for Engine,* Hakone, 1983, Somiya, S., Kanai, E. and Ando, K. (Eds), KTK Scientific Publishers, Tokyo, p. 343, 1984.
9. EKSTROM, T. and INGELSTROM, N. Characterisation and Properties of Sialon Materials, this volume, pp. 231–254.
10. HAMPSHIRE, S., PARK, H. K., THOMPSON, D. P. and JACK, K. H. *Nature,* **274** (1978) 880.

9

Preparation and Densification of Nitrogen Ceramics from Oxides

S. A. SIDDIQI,† I. HIGGINS‡ AND A. HENDRY‡

Wolfson Laboratory, Department of Metallurgy and Engineering Materials, University of Newcastle upon Tyne, NE1 7RU, UK

ABSTRACT

A method of preparation of silicon nitride from silica by carbothermal reduction in nitrogen gas was first published in 1925 but received little further attention until the upsurge of interest in nitrogen ceramics in the 1970s. Several methods of formation of silicon nitride and sialons from oxides have now been patented and the carbothermal reduction process is a viable commercial alternative for the manufacture of a wide range of powders of different grades of purity. It is particularly suited to the production of low-grade sialons from clays and other naturally occurring raw materials.

The present paper describes the principles of the reduction and nitriding reactions and the thermodynamic basis for prediction of phase stabilities. Examples are given of the role of impurities on the reactions to form nitrides from oxides and of their effect on carbide formation. Methods of preparation of β'-sialon and of silicon oxynitride are described and the assessment of extent of reaction by measurement of carbon monoxide evolution is discussed.

Densification of nitrogen ceramics by reaction-sintering of the component compounds is well established but the densification of sialon powders prepared by carbothermal reduction has not been investigated. An example is given of the sintering of silicon oxynitride and the role of impurities inherited from the powder preparation is described.

† Now at Centre for Solid State Physics, University of the Punjab, Pakistan.
‡ Department of Metallurgy, University of Strathclyde, Glasgow, UK.

1. INTRODUCTION

A method of preparation of silicon nitride from silica by carbothermal reduction in nitrogen gas was first published in 1925 but received little further attention until the upsurge of interest in nitrogen ceramics in the 1970s. Several methods of formation of silicon nitride and sialons from oxides have now been patented and carbothermal reduction is rapidly becoming a viable commercial process for the manufacture of a wide range of powders of different grades of purity. It is particularly suited to the production of low-grade sialons from clays and other naturally occurring raw materials in an efficient and economic process.

1.1. The Si–C–O–N System

The preparation of nitrogen ceramics from oxides by carbothermal reduction is currently of interest[1–3] because of the availability of several different cheap raw materials and because of the versatility of the process. It is possible to prepare several different nitrogen-containing phases from oxides by varying the starting compositions and reaction conditions.

A simple example of the process is the formation of silicon nitride from silica according to the reaction

$$3SiO_2 + 6C + 2N_2 \rightarrow Si_3N_4 + 6CO \qquad (1)$$

Examples of potential silica sources include volcanic ash,[4] fumed silica[5] and powdered quartz.[6] Carbon sources include carbon black[1,5] and coal.[6] Alternatively the product of partially combusted rice husks, known as black ash, is a finely divided mixture of silica and carbon which can be used for silicon nitride, oxynitride or carbide production.[2,3] In the reaction to form silicon nitride, the α-structure forms via a mixed solid–solid and gas–solid reaction where silicon monoxide is generated as an intermediate. β-silicon nitride is considered to form from a liquid phase and hence the nature and abundance of any potential liquid-forming impurities in the starting materials is important in dictating the ratio of $\alpha:\beta$ in the product.

Existing thermodynamic data can be used to predict phase stability in the Si–C–O–N system. By considering all of the relevant chemical reactions, diagrams such as Fig. 1 can be drawn showing phase stability as a function of temperature and oxygen partial pressure. It should be noted that variations in the partial pressures of nitrogen and carbon monoxide in the range of interest in nitride formation and over

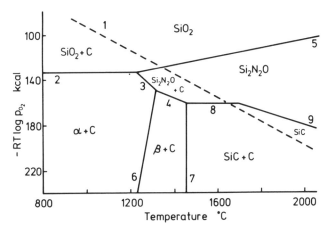

FIG. 1. Phase stability in the Si–C–O–N system.

the temperature range shown in Fig. 1 make very little difference to the positions of the phase boundaries. The partial pressures of nitrogen and carbon monoxide can effectively therefore be taken as constant when constructing such diagrams.[7] Care must be taken however when using this diagram. For example, at 1400°C and low partial pressure of oxygen β-silicon nitride is the predicted stable phase but in practice α-silicon nitride invariably forms under these conditions. This contradiction occurs because the diagram does not take account of other significant gaseous partial pressures in the system (notably silicon monoxide), relative reaction rates or the presence of liquid phases when assessing phase stability.

2. PRODUCTION OF β'-SIALON FROM KAOLIN

The production of phases in the Si–Al–O–N system from aluminosilicates is analogous to the production of silicon nitride from silica. For example, kaolinite which has a fixed Al:Si ratio of 1:1 can be used to prepare β'-sialon;

$$3(2SiO_2 \cdot Al_2O_3 \cdot 2H_2O) + 15C + 5N_2 \rightarrow 2Si_3Al_3O_3N_5 + 15CO + 2H_2O$$

(2)

This reaction has been studied in detail using a china clay (English

China Clays Ltd., Grade D) and a Welsh anthracite coal (Coedbach Colliery, NCB) as kaolinite and carbon sources. As in the formation of Si_3N_4 discussed above, Reaction 2 is thermodynamically favourable but the reaction mechanism depends on other factors.

The chemical analyses of the clay and coal starting materials is given in Table 1. The clay and coal were dry mixed in stoichiometric amounts according to Reaction 2 and either pressed into individual 1 g pellets or the powder was wetted, tumble-rolled and screened to give 1–2 mm spherical pellets which were reacted in a small packed bed. A vertical furnace of 25 mm inner diameter was used for the reactions with a nitrogen flow rate of 100 ml min^{-1} equivalent to a linear flow of 200 mm min^{-1}. All of the results reported here were obtained at 1400°C.

The behaviour of kaolinite on heating is well documented.[8] At about 500°C dehydration occurs giving an amorphous material (meta-kaolin) and at 1200°C nucleation of mullite occurs. At reaction temperature, 1400°C, therefore, the kaolinite has already broken down to a mixture of mullite and free silica;

$$3(2SiO_2.Al_2O_3.2H_2O) \rightarrow 3Al_2O_3.2SiO_2 + 4SiO_2 \qquad (3)$$

X-ray diffraction was used to identify the products of carbothermal reduction and Fig. 2(a) shows the yield of crystalline products as a function of time at 1400°C. Initially the only crystalline phase is mullite and the first reduced phase to appear is β-silicon carbide followed by

TABLE 1
Composition of the Raw Materials

	ECC (Grade D) clay (wt %)	Coedbach coal (wt %)	
SiO_2	46·88	Volatiles	7·8
Al_2O_3	37·65	Total carbon	97·0
Fe_2O_3	0·88	Ash	3·0
TiO_2	0·09		
CaO	0·03		
MgO	0·13		
K_2O	1·60		
Na_2O	0·21		
LOI	12·45		

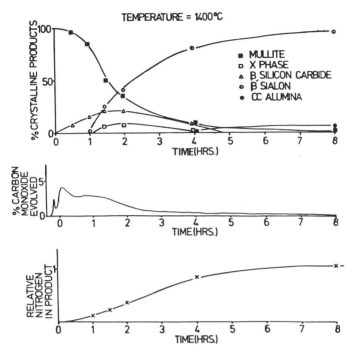

FIG. 2. Variation of the concentration of (top) crystalline phases, (centre) carbon monoxide in the outlet gas, and (bottom) nitrogen in the pellet, with time of reaction at 1400°C.

the low-nitrogen X-phase and then by β'-sialon. Under these conditions maximum β' yield occurs after 8 h and is accompanied by about 5% residual alumina. These two phases coexist on prolonged heating at this temperature.

The progress of reaction was also followed by monitoring the carbon monoxide concentration in the exhaust gases and Fig. 2(b) shows how the percentage of CO varies with time. A small peak at 500°C is associated with reactions between volatiles in the coal and water from the clay. A sharp peak in CO concentration immediately on reaching reaction temperature is followed by a broader peak which gradually decays as the carbon is removed from the sample by reaction. The area under the carbon monoxide curve is a measure of the amount of oxygen removed from the system and by comparison of Figs. 2(a) and 2(b) it can be seen that after 1 h of reaction a large amount of oxygen has been removed but only a small amount of nitrogen-containing crystalline phases have formed. Bulk nitrogen analysis of specimens was carried out and the results are shown in Fig. 2(c). The steady increase in nitrogen content of the samples with reaction time is consistent with the distribution of crystalline phases obtained by X-ray diffraction (Fig. 2(a)) and hence the rapid loss of oxygen from the system in the early stages of reaction is not associated with an equivalent gain in nitrogen content.

In order to determine if the early stages of reaction to produce sialon are independent of the composition of the reaction gas (nitrogen), clay:coal pellets were reacted initially in argon and then in nitrogen for different times. The results of these experiments are summarised in Fig. 3. Heating for 0·5 h in argon at 1400°C gives mullite and silicon carbide. A longer time in argon results in formation

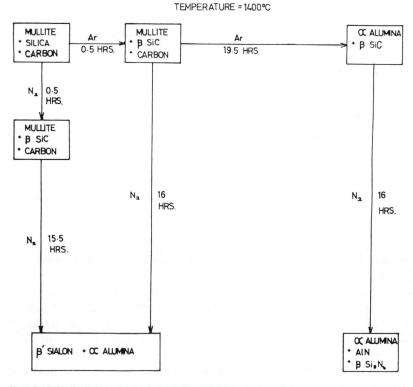

FIG. 3. Schematic representation of the effect of heating clay-coal pellets in argon and nitrogen gas.

of α-Al_2O_3, more SiC and less mullite until after 20 h the only crystalline phases observed by X-ray diffraction are α-Al_2O_3 and β-SiC. Hence the initial step in the carbothermal reduction process is reduction of silica to silicon carbide followed, in the absence of nitrogen, by reduction of the silica in mullite leaving alumina.

Figure 3 shows that a sample held for 0·5 h at 1400°C in argon has the same crystalline products as are obtained after 0·5 h in nitrogen and that identical products form in a sample which has been directly nitrided for 16 h and in one which has been heated for 0·5 h in argon prior to nitriding for 15·5 h. The early stages of reaction are therefore independent of the nitrogen gas atmosphere. However, when the products of a 20 h reaction in argon are subsequently nitrided, a mixture of AlN, α-Al_2O_3 and β-Si_3N_4 is formed and no β'-sialon is detected. Clearly, therefore, it is only the early stages of reaction which are independent of the presence of nitrogen gas.

The mechanism of transformation of mullite and silica to mullite and silicon carbide and hence to β'-sialon is currently being investigated and preliminary results show that the presence of impurities is extremely important. Sol–gel mixing[9] was used to prepare a silica–alumina–carbon mix from high-purity powders and corresponding to a β'-sialon, $z = 3$, composition after mixing. Treatment of this mix in nitrogen under the same conditions as the clay:coal pellets discussed above resulted in slow reactions and inhomogeneous products. However, when this mix was doped with 1–2 wt % of each of a number of metal oxides, the reaction rate increased dramatically and the resulting sialon was homogeneous. It is believed that these metal oxides result in the formation of a small amount of liquid phase at reaction temperature which acts as a reaction site for dissolution of reactants and formation of products. The mobility of reacting species is also increased via liquid diffusion thus giving a homogeneous β'-sialon. Native impurities in the clay and coal behave in a similar manner.

3. DENSIFICATION OF IMPURE SILICON OXYNITRIDE

It has been demonstrated above that impurities in the starting materials play an important role in the formation of nitride and oxynitride phases from oxides. However, these impurities are retained in the powder products and are also therefore important when considering sintering as they will again become liquid at densification temperature. An example of this effect in the densification of silicon oxynitride prepared from rice-husk ash will be given.

Rice-husk ash is a finely divided mixture of silica and carbon obtained by pyrolysis of rice husks in a limited supply of air[2] and contains alkali, alkaline earth and transition-metal oxides as impurities, the concentrations of which are dependent on the geographical origin of the rice husks. The major impurities in the ash used in the present work are calcium and potassium and an additional 2 wt % CaO was added to aid formation of oxynitride in the powder preparation stage.[2] The carbothermally produced silicon oxynitride powder with these inherited impurities was densified by hot-pressing and the sample pellet shrinkage was measured by the linear displacement of the press plattens monitored by a position transducer. The experimental technique is described in detail elsewhere.[2,10] From the observed behaviour it may be shown that densification proceeds via a liquid phase sintering mechanism as proposed by Kingery.[11]

3.1. The Kingery Model

Three stages of densification are recognised in the Kingery model and are shown schematically in Fig. 4 and graphically in Fig. 5.[10] Initially a liquid forms as the sample approaches densification temperature thus drawing the particles together by capillary pressure and forming a neck (Fig. 4). In the second stage, material is dissolved from the neck into the liquid, transported away from the neck and reprecipitated elsewhere in areas of lower stress. The equation governing this important second stage of densification is

$$\Delta V/V_0 = kt^{1/n} \quad (4)$$

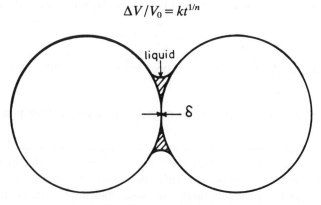

FIG. 4. The Kingery model of densification.

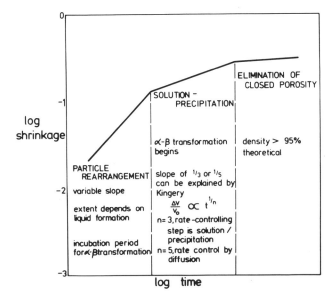

FIG. 5. Graphical representation[10] of the three stages of densification in the Kingery model.

where ΔV is the change in volume at time t, V_0 is the original volume, k is a constant and the exponent n has a value of 3 if solution or reprecipitation is the rate-controlling step and of 5 if diffusion through the liquid is rate controlling (Fig. 5). Thus a graph of $\log(\Delta V/V_0)$ against $\log t$ is a straight line for this stage of densification with a slope of either 1/3 or 1/5 if the Kingery model is valid. The third stage of densification is the elimination of closed porosity but is small in extent and does not significantly increase the final bulk density.

4. RESULTS AND DISCUSSION

Figure 6 shows bulk density plotted against densification temperature for 1 h hot-pressings of silicon oxynitride powder prepared at 1400°C by carbothermal reduction of rice-husk ash. The presence of liquid phases at powder preparation temperature is confirmed by the increase in bulk density which becomes apparent at 1400°C. However temperatures of 1600°C and above are required before the viscosity of the liquid is sufficiently low to allow full densification to occur. Figure 7

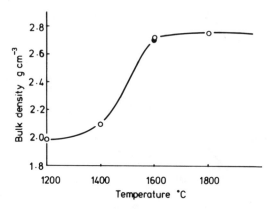

FIG. 6. Bulk density as a function of temperature for hot-pressed silicon oxynitride.

shows a plot of $\log(\Delta V/V_0)$ against $\log t$ for samples hot-pressed at 1600°C and its similarity to Fig. 5 is evident. The applicability of a liquid-phase sintering model may therefore be assumed and the data give a slope of approximately 1/5 ($n = 5$) in the second stage of densification indicating, after Kingery,[11] that diffusion of material from the inter-particle necks through the liquid is the rate-controlling step in sintering.

Scanning electron microscopy (SEM) was also used to investigate the densification process and Figs. 8 and 9 illustrate successive stages of densification. The initial spherical particles produced by carbothermal

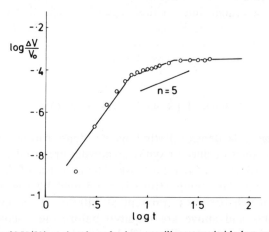

FIG. 7. $\log(\Delta V/V_0)$ against $\log t$ for impure silicon oxynitride hot-pressed at 1600°C.

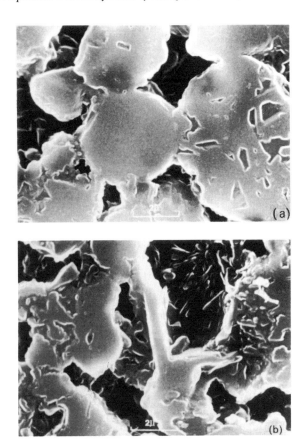

FIG. 8. SEM micrographs showing (a) neck formation due to liquid phase sintering (cf. Fig. 4), and (b) reprecipitation of Si_2N_2O from the liquid at 1400°C.

reduction of the oxide begin to form necks on hot-pressing at 1400°C (Fig. 8(a); cf. Fig. 4) and elongated, angular grains subsequently begin to grow (Fig. 8(b)). Figure 9 shows the morphology of fully dense material prepared at 1600°C. The elongated grains are shown by TEM to be silicon oxynitride and the darker regions (heavily etched in these photographs) are amorphous. Electron-beam microanalysis in TEM reveals that the glassy regions are rich in Ca and K which have been inherited from the original powder preparation. It may therefore be concluded that the glass has formed from an oxynitride liquid in which the impurities are concentrated as they are insoluble in silicon

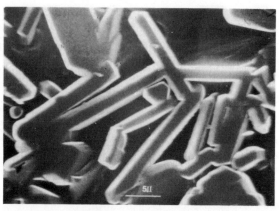

FIG. 9. SEM micrograph of fully dense silicon oxynitride prepared from rice-husk ash and hot-pressed at 1600°C.

oxynitride. The liquid is responsible for the formation of silicon oxynitride initially and also for densification during hot pressing.

Hampshire and Jack[10] have shown that densification is accompanied by phase transformation in liquid-phase sintering of high-purity silicon nitride and sialons but no phase change occurs during densification in the present case; the starting powder is impure silicon oxynitride and the final dense product is silicon oxynitride. There has been however a complete change in morphology of the oxynitride from spherical particles to elongated grains (Figs. 8 and 9) due to solution and reprecipitation in the liquid phase during hot-pressing and it is concluded that, in the absence of a phase change, reduction in total surface energy of the system must be the driving force for the process of densification.

It may also be possible to densify other pre-prepared sialons at relatively low temperature by liquid-phase sintering provided that a suitable quantity and composition of liquid phase is present. However such materials will only be suitable for low-stress applications and are unlikely to be useful high-temperature engineering ceramics due to the relatively high volume fraction of residual glass which will persist after densification. Post-densification heat treatment to remove the vitreous phase may present a possible means of improving high-temperature mechanical properties and provides scope for further development of these low-cost nitrogen ceramics.

5. CONCLUSIONS

Existing thermodynamic data predict the phase stabilities in the Si–C–O–N and related systems. However kinetic effects must also be considered in determining the phases formed under a given set of experimental conditions.

Sialons can be formed from impure alumino-silicates; for example, β'-sialon from kaolinite and silicon oxynitride from rice-husk ash. Silicon carbide is an important intermediate phase in the formation of β' and impurities in the raw materials play an important role in the reaction mechanisms.

Impurities also dictate the densification characteristics of oxynitride phases formed from oxides. Densification occurs via a solution–reprecipitation route where the liquid forming impurities essential for sintering are inherited from the starting materials. Densification occurs without the α–β phase change associated with reaction sintering.

ACKNOWLEDGEMENTS

The authors gratefully acknowledge the encouragement and stimulating advice given by Professor K. H. Jack and the financial support of the British Council and Morgan Materials Technology.

REFERENCES

1. VAN DIJEN, F. K., METSELAAR, R. and SISKENS C.A.M. *Journal of the American Ceramic Society*, **68** (1985) 16.
2. SIDDIQI, S. A. Ph.D. thesis, University of Newcastle upon Tyne, 1984. (See also SIDDIQI, S. A. and HENDRY, A. *Journal of Materials Science*, **20** (1985) 3230.
3. LHANG, S.-C. and CANNON, W. R. *Journal of the American Ceramic Society*, **68** (1985) 691.
4. UMEBAYASHI, S. In *Nitrogen Ceramics*, Riley, F. L. (Ed.), Noordhof, Leyden, p. 323, 1977.
5. SZWEDA, A. Ph.D. thesis, University of Newcastle upon Tyne, 1980. (See also Ref. 9.)
6. HIGGINS, I. *Internal Progress Report*, July (1985) Wolfson Laboratory, University of Newcastle upon Tyne.
7. SMITH, P. L. and WHITE, J. *Transactions and Journal of the British Ceramic Society*, **82** (1983) 23.

8. HOLDRIDGE, O. A. and VAUGHEN, F. *The Differential Thermal Investigation of Clays,* Mackenzie, T. C. (Ed.), Mineralogical Society, London, p. 102, 1957.
9. SZWEDA, A., HENDRY, A. and JACK, K. H. *Special Ceramics,* Vol. 7, Taylor, D. E. and Popper, P. (Eds), *Proceedings of the British Ceramic Society,* **31** (1981) 107.
10. HAMPSHIRE, S. and JACK, K. H. *Special Ceramics,* Vol. 7, Taylor, D. E. and Popper, P. (Eds), *Proceedings of the British Ceramic Society,* **31** (1981) 37.
11. KINGERY, W. D. *Journal of Applied Physics,* **30** (1959) 301.

COMMENTS AND DISCUSSION

Chairman: D. BROUSSAUD

B. C. Mutsuddy: What would be the effect of particle size on densification.

I. Higgins: We have not really investigated the effect of particle size on densification. In the case of the reaction for the production of silicon nitride from very pure silica and carbon, then clearly we have a solid/solid reaction and so the more interfaces between silica and carbon, the more efficient will be the reaction. In the reactions which involve a liquid phase we have not really seen any general trends.

Mutsuddy: Did you also investigate the nitrogen flow-rate problem?

Higgins: Yes, there is no problem for a very small scale process. However, if you start to scale up the process and you have a packed bed of clay and carbon or clay and coal then you will find that the rate of reaction is very much dependent on how low you can keep the partial pressure of CO. The higher the nitrogen flow rate, the lower will be the carbon monoxide partial pressure and the faster the reaction. In our system, using a very small pellet, the P_{CO} does not increase to a very high level and so it does not stop the reaction.

10

Preparation of Sialon Ceramics from Low Cost Raw Materials

C. J. SPACIE

Morgan Materials Technology Ltd, Bewdley Road, Stourport-on-Severn, Worcestershire DY13 8QR, UK

ABSTRACT

The production of sialons from relatively cheap and impure starting materials is thought to offer the potential for much wider use of these ceramics. It has been shown that the amounts that can be produced in a packed bed reactor are relatively large, and that a continuous production process is feasible. The powders can be sintered to high densities in either a protective powder bed or under a protective gas atmosphere without the use of additives. Some preliminary properties have been determined and suggest that the sialons derived from clay and coal have properties superior to those of many conventional refractories.

1. INTRODUCTION

Current commercial exploitation of the sialons is largely limited to the preparation of β'-sialon from high purity starting materials. However, the carbothermal reduction and nitriding of oxides is now a recognised process for the production of silicon nitride,[1,2] aluminium nitride[3] and β'-sialon.[4-7] The powders produced from naturally occurring minerals vary in quality according to the impurities present,[8,9] but the feasibility of producing relatively cheap sialon powders could lead to their far greater exploitation.

2. RAW MATERIALS

By using different minerals, such as pyrophyllite, sillimanite or kaolinite, a range of z value β'-sialons can be prepared.[10] For the current work kaolinite was used, and reacted according to the equation:

$$3(Al_2O_3 \cdot 2SiO_2 \cdot 2H_2O) + 15C + 5N_2 \rightarrow 2(Si_3Al_3O_3N_5) + 15CO + 6H_2O \quad (1)$$

The kaolin was a standard grade D (from English China Clays, St. Austell), with a low alkali oxide content since this is considered important if the high temperature properties are to be optimised. The carbon source was also a standard production grade (National Coal Board), originating from Coedbach in South Wales, and is reported as having a low impurity content. Chemical analysis of the raw material is given in Table 1.

TABLE 1
Chemical Analysis of Raw Materials

(a) Analysis of standard grade D clay (wt %)	
SiO_2	46·30
Al_2O_3	36·70
Fe_2O_3	1·10
TiO_2	0·08
CaO	0·08
MgO	0·30
K_2O	2·40
Na_2O	0·09
Loss on ignition	12·4

(b) Analysis of Coedbach coal (NCB) (wt %)	
Carbon source	
Fixed carbon	92·1–93·2
Volatiles	7·9–6·8
Total carbon	97·7
Ash	2·3
Ash composition	
SiO_2	0·71
Al_2O_3	0·81
Fe_2O_3	0·48
TiO_2	0·014
CaO	0·17
MgO	0·05
Na_2O	0·04
K_2O	0·014

3. PROCESSING

To ensure good reaction and mixing of the raw materials it is necessary to mill the coal to reduce its particle size prior to mixing and ball milling it with the clay and water for 24 h. Small pellets of the pre-sialon mix are prepared by extrusion, and dried before firing in flowing nitrogen at 1450°C. The resultant sialon pellets are milled to reduce them to fine powder, and leached to remove iron which is reported as being detrimental[7] to the high temperature properties of the material. An oxidation treatment is also used to remove any residual carbonaceous matter prior to adding binders and lubricants before grading the powder. Pressed bars or pellets are sintered (without the use of additives), in either a protective powder bed, or under a flowing nitrogen atmosphere, between 1650 and 1750°C.

For the chemical analysis the silicon, aluminium, calcium and magnesium contents are measured by atomic absorption, iron by spectrophotometry and sodium and potassium by flame photometry. The nitrogen content is determined by the formation of ammonium salts in a pressure vessel and their subsequent distillation. The loss on ignition figure is generally used as some measure of the carbon content; samples are burnt in air at 1000°C until constant weight is achieved, or if there is a weight gain then analysis is quoted as the weight gain at 800°C after 1 h. A more accurate carbon content can be determined by measuring the CO_2 gas evolved during the combustion in air of the sample.

4. POWDER PRODUCTION

The theoretical clay/coal ratio for the reaction to form β'-sialon is 81/19 wt %. However, this does not take into account the effect of the impurities in both the clay and the coal.

Much of the early work on 200 g batches of pre-sialon was to determine empirically the clay/coal ratio required; it was concluded that in general an extra 3-4% of coal is necessary. With too little coal a predominantly X-phase[11] powder is produced, with some β'-sialon. The quantity of X-phase was reduced as the amount of coal in the pre-sialon was increased. After optimisation of the clay/coal ratio the process was scaled up using gas-tight clay/graphite saggars (Fig. 1); the porous alumina foam is necessary to distribute the gas more evenly

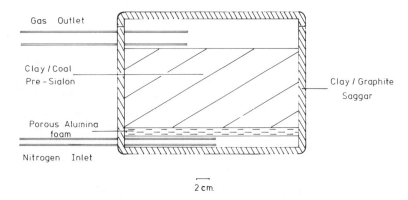

FIG. 1. Gas-tight clay/graphite saggars for powder production.

through the bed to prevent segregation. By using the saggars, up to 1·5 kg of pre-sialon could be produced; some typical results after the reaction are given in Table 2. Composition VIII contained 1% more coal than composition VII, and this clearly results in greater nitridation and less alumina, although traces of polytypoids are seen. It is interesting to note that by using a larger saggar and a gas-fired kiln, it was not difficult to produce 5 kg of sialon in a single batch.

Due to the short lifespan of the saggars and alumina tubes, a vertical tube furnace was developed (Fig. 2); again the porous alumina foam both supports the charge and helps to distribute the nitrogen more evenly. Using this apparatus 3 kg of pre-sialon could be fired in each batch, and no rapid deterioration of the equipment was seen. Most importantly, the reaction time was reduced from 42 to 20 h; the reaction was monitored by analysis of the carbon monoxide content of the exhaust gases. Batches 2 and 3 are typical of the material produced (Table 3). Clearly there is segregation of the bed, with the top being un-reacted whilst the bottom is predominantly β'-sialon. This segregation is due to the bed of pre-sialon extending beyond the hot-zone of the furnace; this is necessary both to minimise the heat loss by convection and to provide a reservoir of pre-sialon that can drop into the hot-zone as the pellets shrink, thereby increasing the volume of material reacted. From the chemical analysis it is evident that both sodium and potassium impurities tend to volatilise during the reaction, with the top of the bed being richer in these as they condense out.

By taking the 'tops' of 7 batches, and re-firing them it is evident

TABLE 2
Sialon Production Using Saggars

Sialon composition	Batch	Firing conditions		XRD phases (~%)			Chemical analysis							
		Soak time (h)	Gas input (litres min^{-1})	β'	Al_2O_3	Others	SiO_2	Al_2O_3	Fe_2O_3	CaO	MgO	Na_2O	K_2O	N
VIII	1	42	5·0	98	2	—	67·6	52·9	1·10	0·16	0·37	0·01	0·01	30·4
	2	42	5·0	97	2	tr.	66·6	51·0	1·30	0·21	0·46	0·14	0·01	25·5
	3	42	5·0	97	2	tr.	65·0	50·9	0·71	0·20	0·46	0·14	0·02	25·2
VII	1	42	2·0	85	15	—	63·5	50·6	1·1	0·19	0·42	0·01	0	26·0
	2	60	2·0	90	10	—	64·0	51·0	1·1	0·15	0·50	0·02	0	22·5
	3	60	2·0	85	15	—	65·0	49·0	1·1	0·17	0·06	0·02	0	22·8

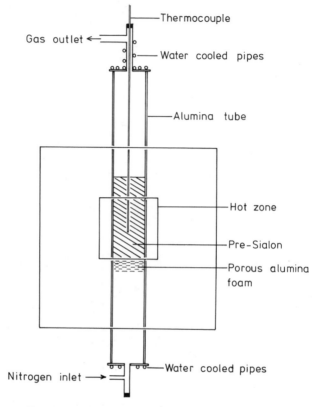

FIG. 2. Vertical tube furnace for powder production.

(Table 3) that the material can be largely converted to β'-sialon and much of the potassium removed; it is possible that a cycling process may be necessary to produce the optimum powder with a consistent chemical analysis.

By modifying the vertical tube furnace, such that it can be discharged whilst at temperature, it has been shown that 3 kg of sialon powder can readily be produced in 24 h; some difficulty is encountered due to the volatilisation and subsequent concentration of the potassium impurity, but modifications to the firing cycle can compensate for this.

As well as the standard grade D clay, three grades of experimental CHA clay were investigated. These are essentially the same clay, after

TABLE 3
Typical Results for VII Batches

Sialon	Batch	XRD phases (%)					Chemical analysis								
		β'	Al$_2$O$_3$	SiO$_2$	Mullite	Others	LOI	SiO$_2$	Al$_2$O$_3$	Fe$_2$O$_3$	CaO	MgO	Na$_2$O	K$_2$O	N
VI$_{STD}$	Pre-sialon 2	—	—	—	—	—	30·2	39	32	0·7	0·13	0·20	0·07	1·1	0
VI$_{STD}$	2 Top	—	—	35	50	15...Un.	20·4	41	33	0·9	0·13	0·20	0·20	2·8	0
	Bottom	100	—	—	—	—	+	64	50	1·3	0·17	0·36	Tr	Tr	24·2
VI$_{STD}$	Pre-sialon 3	—	—	—	—	—	30·8	36	31	0·7	0·12	0·19	0·09	1·7	0
VI$_{STD}$	3 Top	—	—	25	75	—	18	45	32	0·7	0·17	0·23	0·15	2·6	0·8
	Bottom	90	—	—	—	10...15R	+	63	48	1·3	0·19	0·33	Tr	Tr	23·8
VI$_{STD}$ 3	Mixed tops Batch 2–6 Pre-fire	55	—	—	30	15...Low X	14·5	46	36		0·12	0·35	0·29	5·0	5·6
	Fired	83	17	—	—	—	0·4	61	47·4	2·15	0·22	0·27	0·67	0·74	20·3

Un: unknown; 15R: polytypoid; +... slight wt. gain.

TABLE 4
Analysis of ECC Experimental Clays

Clay analysis	CHA screened	CHA refined	CHA refined + calcined (1050°C)
SiO_2	45·3	45·3	53·2
Al_2O_3	37·4	38·6	43·2
Fe_2O_3	1·7	0·72	0·77
TiO_2	0·82	0·30	0·38
CaO	0·16	0·09	0·09
MgO	0·19	0·14	0·18
K_2O	0·65	0·57	0·78
Na_2O	0·04	0·04	0·08
Loss on ignition	13·6	13·8	0·3

TABLE 5
Analysis of Lignite Clay

Chemical analysis	Before firing	After firing 24 h 5 litres min^{-1} gas i/p	
		Bottom	Top
LOI 800°C	50·8	9·8	22·1
SiO_2	27·8	63	45
Al_2O_3	17·1	40	28
Fe_2O_3	1·16	2·5	1·5
CaO	0·52	1·5	1·1
MgO	0·42	0·42	0·90
Na_2O	0·32	0·05	0·56
K_2O	1·37	0·37	6·6
N	0	28·1	11·1
XRD analysis (%)			
β'		—	Distorted
β-Si_3N_4		40	—
SiO_2		—	√
Mullite		—	√
Polytype		45 (27R)	—
Others		15% α'-Sialon	

different degrees of purification (Table 4). Each clay nitrided readily after processing, and formed typically 80–90% β'-sialon, indicating the flexibility of the process.

A clay with a high lignite content has also been processed. The lignite content (Table 5) corresponds to approximately 25% of carbon, which is higher than is currently used, and this is reflected in the degree of nitridation of the bed. Again there is segregation, with the top forming distorted β'-sialon and un-reacted clay, whilst the bottom formed nitrogen rich phases. It is interesting to note that α'-sialon was seen since its structure requires a third cation to stabilise it,[12] the general formula being:

$$M_x(Si, Al)_{12}(O, N)_{16}$$

where M can be Li, Y, Ca, Mg[13,14] and x cannot be greater than 2. Hence the impurities present in the clay must have stabilised the α'-sialon.

Although the clay inherently contained excess carbon, the possibility of controlled oxidation could allow this impure 'waste' material to be an important raw material for sialon production.

5. SINTERING

Prior to sintering the sialon powder, it is generally leached in 3 M hydrochloric acid to remove iron. There is some disagreement as to whether the iron should be removed before or after the nitridation; it is reported[4] as acting as a catalyst for the sialon reaction, and that leaching the sialon powder degrades it;[7] however, most workers agree that large quantities of iron result in ferrosilicon on sintering which degrades the high temperature properties of the material. The leaching of powder produced at M^2T clearly indicates that boiling acid removes more iron than does cold acid which can only remove 50% of the iron impurity (Fig. 3); however, the boiling acid adversely affects the silicon, aluminium and nitrogen in the sialon (Fig. 4). Elements such as sodium potassium, calcium and magnesium are unaffected by the leaching treatment; nor is there any evidence to support the observation that leaching promotes X-phase formation upon sintering, although the degradation caused by boiling acid is likely to result in oxide rich phases.

Early sintering experiments were performed using powder beds to

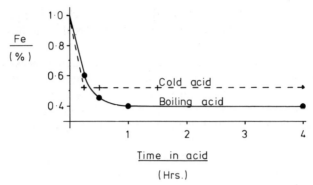

FIG. 3. Effect of time on iron removal by acid.

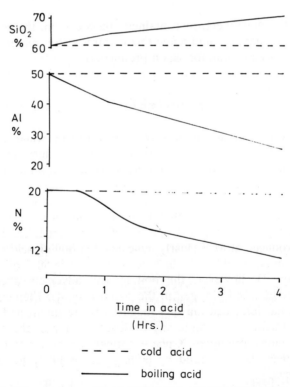

FIG. 4. Effect of time on Si, Al and N content of sialon after acid leaching.

TABLE 6
Typical Powder Bed Sintered Samples

Sialon powder	Sintering temp. (°C)	XRD phase analysis (~%)				% Theoretical density
		β'	X	Polytype	Other	
VIII$_1$	1 650	95	—	5	—	57
	1 700	80	—	20	—	53
VIII$_4$	1 750	90	10	—	—	96
VIII$_4{}^a$	1 750	80	20	—	—	92

^{a} Oxidised prior to pressing.

protect the samples from decomposition in the hostile environment of a gas-fired kiln; various beds were tried, including graphite, alumina, clay, sialon and silicon nitride. The latter bed proved the most satisfactory, and could be re-cycled for at least 4 firings. Using a silicon nitride/silica bed allowed densities of greater than 90% of the theoretical value to be achieved without the use of oxide additives. During the firing it was demonstrated how nitride rich powders such as VIII$_1$ (Table 6) could not be densified even at 1800°C or with the addition of 10 wt % oxide. Some typical properties of sialon sintered in a silicon nitride bed in a gas kiln are given in Table 7.

With a nitrogen atmosphere, samples have been sintered to full density (Table 8). In general, provided no polytypoid was present in the starting powder good densification is achieved, although X-phase or alumina are often seen. It is unclear as yet whether these phases are adversely affecting the physical properties of the sialon. The three

TABLE 7
Properties of Sialon ^{1}VIII$_3$ Sintered in a Powder Bed

Crystalline phases by XRD	90% β'
	10% X
Percentage of theoretical density	95%
Average hardness on R15 N scale	95
TBS (by three point bend) (MPa)	158 ± 20
K_{1c} (by indentation) (MN m$^{-3/2}$)	2·19
Thermal expansion coefficient (°C^{-1})	
20–400°C,	2·8 × 10^{-6}
20–600°C,	3·0 × 10^{-6}
20–1000°C,	3·4 × 10^{-6}

TABLE 8
Properties of Sialon Sintered in Nitrogen

Sialon		Sintering temp. (°C)	Soak time (hr.)	ASD (g cm^{-3})	BD (g cm^{-3})	XRD phases (%)			TBS (MPa)	Elastic modulus (GPa)	K_{IC} (MN m$^{-3/2}$)
						β'	Al$_2$O$_3$	Others			
2_{VIII_3}	OX	1650	2	3·16	3·00	—	—	—	—	—	—
2_{VIII_3}	NLE	1650	2	3·15	3·14	65	25	10...low X	288 399	50·9 49·5	2·97
1_{X_1}	NLE NOX	1650	4	3·23	3·22	63	25	2...Fe Si	207	85	3·00

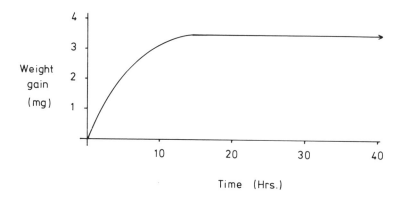

FIG. 5. Oxidation of sialon at 1380°C.

point bending strength on a 60 mm span is lower than expected,[15] but should be adequate for many applications.

A typical oxidation curve (by TGA) of the sialon (Fig. 5) shows the formation of a protective layer; whether this would be stable in a dynamic situation is unclear. XRD analysis of the surface reveals mullite as the only crystalline phase.

Some preliminary tests in EN16 steel with 'Scorialit S/Stg flux' at 1560°C for 30 min (Fig. 6) indicate that a standard refractory is

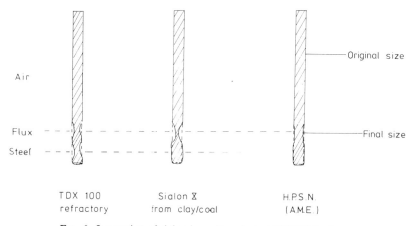

FIG. 6. Immersion of sialon in molten steel at 1560°C/30 min.

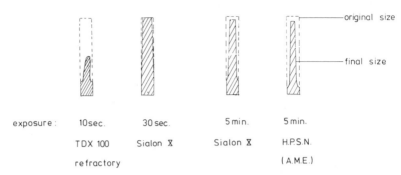

FIG. 7. Dip testing of sialon in NaOH at 900°C.

attacked at both the metal/slag and slag/air interfaces, whilst the cheap sialon appears to be only attacked at the slag/air interface. HPSN is wetted but appears otherwise unaffected.

Dip testing bars in sodium hydroxide at 900°C (Fig. 7) showed that the standard refractory lasts less than 10 s, and that after 5 min exposure there is significant attack of the HPSN and less so of the sialon.

6. CONCLUSIONS

The work has shown that the carbothermal reduction of aluminosilicates to form sialon phases is a relatively flexible process, and can accommodate a wide variety of impurities. Reproducible powder can be prepared and dense material sintered without the use of additives.

The mechanical properties are as yet below expectations, but the chemical stability and thermal shock resistance show the cheap sialon to be superior to the conventional refractories under the conditions used.

The relatively low cost of the production of sialon from clay/coal would indicate that even with the properties currently achieved, it should be a viable contender for the refractory market.

ACKNOWLEDGEMENTS

The work was funded partially by the EEC under contract number MSA-210-UK (H).

REFERENCES

1. HENDRY, A. and JACK, K. H. In *Special Ceramics*, Vol. 6, Popper, P. (Ed.), British Ceramic Research Association, Stoke-on-Trent, p. 199, 1975.
2. KOMEYA, K. and INOUE, H. *Journal of Materials Science*, **10** (1975) 1243.
3. ISH-SHALOM, M. *Journal of Materials Science Letters*, **1** (1982) 4.
4. LEE, J. G. and CUTLER, I. B. *American Ceramic Society Bulletin*, **58** (1979) 9.
5. GAVRISH, A. M., BOYARINA, I. L., DEGTYAREVA, E. V., PUCHKOV, A. B., ZKUKOVA, Z. D., GUL'KO, N. V. and TARASOVA, L. A. *Inorganic Materials*, **18** (1982) 46.
6. BALDO, J. P., PANDOLFELLI, V. C. and CASARINI, J. R. *Ceramica Sao Paulo*, **28** (1982) 83.
7. GORDON, R. S. and CUTLER, I. B. Sialon Refractories from Clay and Coal—Final Report, DOE Contract No. E-AS01-76ET10668, 1981.
8. CUTLER, I. B. Sialon Refractories from Clay and Coal, University of Utah Contract No. EX-76-S-01-2407, September 1978.
9. MOTO, S. and SASAKI, S. *Journal of the Ceramic Society of Japan*, **85** (1977) 537.
10. BALDO, J. B., PANDOLFELLI, V. C. and CASARINI, J. R. *Materials Science Monographs*, **16** (1983) 437.
11. THOMPSON, D. P. and KORGUL, P. In *Progress in Nitrogen Ceramics*, Riley, F. L. (Ed.), Martinus Nijhoff, The Hague, p. 375, 1983.
12. HAMPSHIRE, S., PARK, H. K., THOMPSON, D. P. and JACK, K. H. *Nature*, **274** (1978) 880.
13. MASAKI, H., OYAMA, Y. and KAMIGAITO, O. *Japanese Journal of Applied Physics*, **14** (1975) 301.
14. MITOMO, M. *Yogyo-Kyokai-Shi*, **85** (1977) 50.
15. HOGGARD, D. B., IKUMA, Y., CUTLER, I. B. and GORDON, R. S. Mechanical and Thermal Properties of Sialon Ceramics Fabricated from Inexpensive Clay and Carbon Raw Material Sources, Intrep 5000 183017, University of Utah.

COMMENTS AND DISCUSSION

Chairman: D. BROUSSAUD

K. H. Jack: Presumably the α'-sialon was a Ca α' since it is the only impurity that would go into the structure?

C. J. Spacie: It seems so, yes.

Jack: You didn't show this though.

Spacie: With 15% α', it is rather difficult to analyse it.

A. Hendry: What exactly is the substitution level of Al in the β-sialon.

Spacie: It is an expanded beta silicon nitride and therefore designated β' but the a and c values did not correspond to an individual z value.

11

Densification Behaviour of Sialon Powders Derived from Aluminosilicate Minerals

M. MOSTAGHACI, Q. FAN† AND F. L. RILEY

Department of Ceramics, University of Leeds, LS2 9JT, UK

AND

Y. BIGAY AND J. P. TORRE

Desmarquet Laboratoire de Recherches, Trappes, France

ABSTRACT

β'-sialon powders with a range of sialon z-values have been obtained by carbothermal reduction of kaolinite and illite minerals under nitrogen. The densification behaviour of these powders has been studied through the use of the hot-pressing technique. Information has been obtained regarding powder densification rates and product microstructure, and their relationships with the major materials variables of particle size and composition, and the type and level of liquid-forming additive.

1. INTRODUCTION

There are a number of routes for the production of dense β'-sialon materials, most of which involve sintering, or reaction sintering, powders of silicon nitride, silicon nitride oxide, silicon dioxide, aluminium oxide and/or aluminium nitride in appropriate proportions under controlled atmosphere. There is now considerable interest in the possibility of preparing sialon powders directly from cheap raw materials such as the aluminosilicate minerals. Such interest is due in part to the abundance of these minerals in nature, and in part to their inherently fine particle size. It might therefore be expected that β'-sialon powders with a similarly fine particle size (<1 μm) and thus

† Present address: Shandong Institute of Industrial Ceramics, Zi-Bo, Shandong, People's Republic of China.

high degree of sinterability might be prepared from these powders. Several methods[1-13] have been successfully employed to produce sialon powders from naturally occurring materials, a summary of which is given in Table 1.

Theoretically, it should be possible to obtain from kaolinite a β'-sialon of aluminium/silicon ratio 1·0, i.e. with a z-value of 3 ($Si_3Al_3O_3N_5$). Illite minerals enriched with silica yield sialon of considerably lower z-value. The degree of oxygen replacement is therefore critical from the point of view of the product composition. Over replacement of oxygen yields a mixture of β'-sialon with aluminium nitride polytypoid phases. Replacement of insufficient oxygen yields mixtures of β'-sialon, mullite, X-phase and aluminium oxide.[6] At temperatures above ~1450°C silicon nitride is unstable in the presence of carbon, and silicon carbide and aluminium nitride become the main products.[6]

The main objective of the present work is to assess the sintering behaviour of a range of β'-sialon powders prepared from high purity clay minerals with differing Al/Si ratios. As a means of rapidly assessing these powders, densification behaviour has been studied initially using hot-pressing. It is intended that studies will be extended to pressureless sintering.

TABLE 1
Methods for Production of Sialon Powders

Method of production	Reference
Nitridation of:	
naturally occurring silica and aluminium powders in nitrogen	1
silica sand and aluminium powder in the presence of carbon in nitrogen	2
silica, aluminium and silicon powders in nitrogen	3
volcanic ash and aluminium powder in nitrogen	4–5
kaolin and carbon black in nitrogen	6–8
metakaolin and silicon nitride in nitrogen	9
kaolin in ammonia	10
silicon and aluminium powders in nitrogen with controlled level of oxygen	11
Reduction–nitridation of aluminium hydroxide and silica gels under ammonia	12

Sialon powders derived from a kaolinite and an illite form the basis for this work. The influences of sialon z-value and particle size, and additive type and quantity, on densification behaviour, and on the development of microstructure in the dense sialons have been studied.

2. EXPERIMENTAL METHODS

2.1. Raw Materials

β'-sialon powders prepared by nitrogen substitution for oxygen in kaolinite and illite minerals have been supplied by Céramique Technique, Demarquest (France). The phase compositions of these powders are given qualitatively in Table 2.

The chemical compositions of the minerals used for the production of the sialon powders are given in Table 3.

The carbothermal reduction consists of heating a mineral–carbon mixture under flowing nitrogen at 1250–1500°C. The mixtures of mineral and carbon with an organic binder are normally compacted in the form of small cylinders or briquettes.[13]

2.2. Powder Milling

The sialon materials were generally supplied in the form of cylinders (75 mm length, 30 mm diameter). These were subsequently coarsely crushed (<10 mm) and then batch dry-milled (~25 g) in a tungsten carbide 'Shatter-box' (Glen Creston, Stanmore, UK) for varying lengths of time (from 20 to 600 s). A summary of specific surface area data (BET method), and of median Stokes' particle size values obtained by an automatic sedimentation technique (Sedigraph 5000ET, Micromeritics) is given in Table 4. Comparison of the data in Table 4 shows that sialon powders derived from illite are more easily milled than those powders from kaolinite.

TABLE 2
The Phase Composition of Sialons

Kaolinite-derived sialon	Illite-derived sialon
β'-sialon (major phase)	β'-sialon (major phase)
15R polytypoid (trace)	Aluminium nitride (trace)
Aluminium oxide (trace)	
Silicon nitride oxide (trace)	

TABLE 3
Chemical Compositions of the Minerals used for the Production of the Sialon Powders

	Kaolinite (wt %)	Illite (wt %)
SiO_2	46·30	66·90
Al_2O_3	34·90	21·45
Fe_2O_3	1·90	1·02
TiO_2	1·83	1·74
CaO	0·64	0·19
MgO	nd	0·13
K_2O	0·02	1·49
Na_2O	nd	0·12
P_2O_5	0·08	0·04
Ignition loss	13·80	6·90

As an alternative method to further reduce the powder particle size, milling was carried out under propan-2-ol with alumina elements in a polypropene container (McCrone Research Associates, London). The specific surface area and Stokes' diameter data together with the amount of alumina contamination arriving from milling are shown in Table 5 for the powders derived from kaolinite.

From the data shown in Table 5, it is evident that the extended milling with alumina elements produced a median Stoke's particle size

TABLE 4
Specific Surface Area and Stokes' Diameter Data for Milled Sialon Powders

Milling time (s)	Median Stokes' diameter (μm)	Specific surface area ($m^2 g^{-1}$)	Mean BET particle size (μm)
(i) 'Illite' sialon			
20	5·0	2·3	0·83
45	3·3	2·9	0·66
300	2·1	3·8	0·50
(ii) 'Kaolinite' sialon			
30	50	0·7	2·77
120	6·4	1·8	1·08
300	5·0	2·6	0·75
600	4·4	3·1	0·63

TABLE 5
Particle Analysis Data for Kaolinite-Derived Sialon

Milling time (min)	Median Stokes' diameter (μm)	Specific surface area ($m^2 g^{-1}$)	Al_2O_3 contamination (wt %)
0[a]	50	0·7	—
10	17	1·2	0·21
30	9	2·0	0·88
120	4	2·4	2·58

[a] Starting powder milled in a 'shatter-box' for 30 s.

of 4 μm, close to the median size of powder milled using tungsten carbide for 600 s, but at the expense of considerable contamination by alumina from milling media. Yttrium oxide (BDH Analar grade) used as sintering aid was normally wet-milled (for 600 s) into the sialon powders. Alumina-milled powders were not used in this work.

2.3. Hot-pressing

Sialon powders were sintered at 1700°C under 20 MPa pressure. The graphite dies and punches were protected from direct contact with the sialon powder by a thin coating of boron nitride. Shrinkage of the powder compact was monitored continuously by means of a linear displacement transducer. The theoretical density of the β'-sialon was assumed to be that corresponding to the observed z-value for the sialon obtained.

2.4. Microstructural Studies

Polished sections of hot-pressed sialon materials were prepared by rough grinding through 400, 600, 800 and 1000 grit silicon carbide powders on flat glass plates using water as lubricant. Rough polishing was carried out on paper (pan-W) laps with 6 μm and 1 μm diamond pastes, using an ethanol-based lubricant (Struers, Denmark). A 0·25 μm diamond micropolishing stage concluded the preparation. A mixture of 3:1 concentrated phosphoric acid/sulphuric acid at temperatures between 200 and 240°C for 20–120 s was used for etching the polished sections.

3. RESULTS

For the illite-derived powder, the effect of median Stokes' particle size and yttrium oxide addition on densification behaviour is shown in Figs. 1 and 2. As shown in Fig. 1, the powder milled for 300 s in tungsten carbide densifies to >98% theoretical density within 900 s. In Fig. 2, a considerable increase in densification rate due to the introduction of yttrium oxide into the sialon powder is evident. The effect of median Stoke's diameter on the densification behaviour of the kaolinite-derived sialon is shown in Fig. 3. As a means for comparison of the behaviour of illite and kaolinite-derived sialons, two powders having closely similar specific surface areas were chosen and hot-pressed under identical conditions. The result is illustrated in Fig. 4, and shows that the kaolinite-derived powder initially densifies faster than the illite, though it finally approaches a similar end-point density.

The microstructure development in dense (>98%) hot-pressed samples from both sialons without densification aid is shown in Figs. 5 and 6. There are two features of interest in these microstructures; firstly, the structure consists of equiaxed grains in both materials, and secondly, the volume of grain boundary phase in specimens made from the kaolinite-derived sialon appears to be larger than in those from the illite-derived sialons.

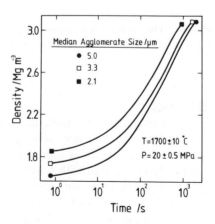

FIG. 1. The effect of apparent particle size on densification of the illite-derived sialons.

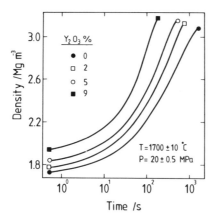

FIG. 2. The effect of yttrium oxide additions on densification of the illite-derived sialon powders.

4. DISCUSSION

It is evident from the data shown in Table 4 that tungsten carbide milling has a marked effect on median Stokes' diameter of both sialon powders, and also on the apparent primary particle dimensions as indicated by the specific surface area. It is also clear that the sets of

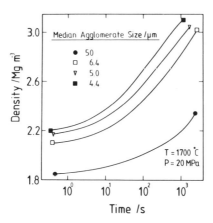

FIG. 3. The effect of apparent particle size on densification of the kaolinite-derived sialons.

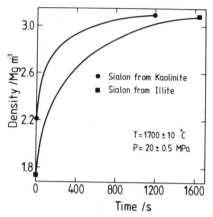

FIG. 4. Comparison of densification behaviour in the illite- and kaolinite-derived sialons.

particle size data obtained from sedimentation rate, and specific surface area measurements, are not in agreement with each other (see Table 4). This suggests that the sialon powders consist of particles which are themselves loosely agglomerated sub-micrometre particles. Powder particles of sub-micrometre dimensions might have been expected from the carbothermal nitridation of sub-micrometre aluminosilicate powders of the kaolinite type. That the actual particles appear to be agglomerated suggests that partial sintering of the clay minerals or their reduction products occurred during the carbothermal treatment. The degree of agglomeration in the powder derived from kaolinite appears to be higher than that derived from the illite. Assuming, for simplicity, 50% free space in the agglomerates, and using the data in Table 4, each agglomerate of sialon derived from illite (milled for 300 s in tungsten carbide) consists of an average ~40 primary particles, while each agglomerate of the kaolinite-derived powder (milled under the same conditions) consists of ~150 primary particles. This analysis is undoubtedly somewhat crude, but it serves the present purpose of a semi-quantitative assessment of the nature of the powders being handled.

In order to assess the strength of the agglomerates, powder derived from kaolinite was compacted in a steel die using an Instron Universal testing machine. The plot of % theoretical density as a function of applied pressure is shown in Fig. 7. In the construction of this curve, corrections were applied to take into account the separately measured

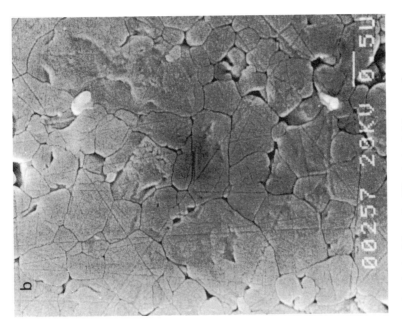
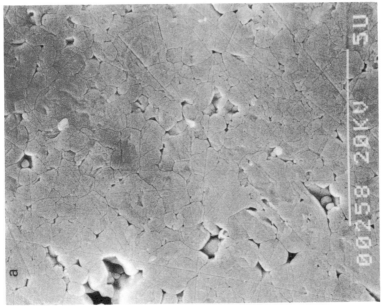

FIG. 5. SEM micrographs of polished and etched sialon ceramics obtained from the illite-derived sialon powders.

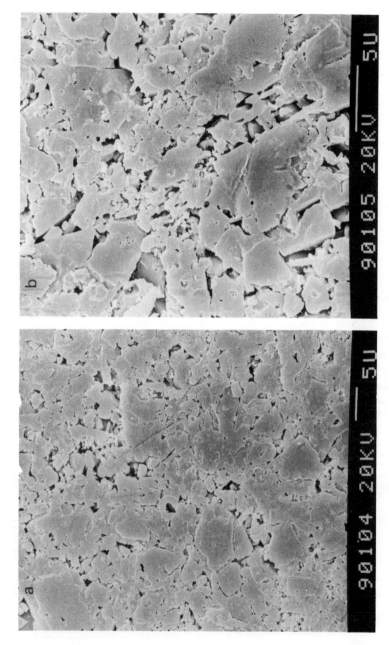

FIG. 6. SEM micrographs of polished and etched sialon ceramics obtained from the kaolinite-derived sialon powders.

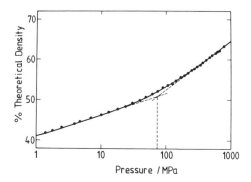

FIG. 7. Compaction behaviour of a sialon powder prepared from the kaolinite, milled 45 s in tungsten carbide.

deformation of the steel punches under load. The change in slope of the curve at ~75 MPa may be taken to be a rough measure of the crushing strength of an agglomerate. The particle size distributions in the powder, before and after this compaction, were determined and these data are shown in Fig. 8. The loss of agglomerates of the size range 15–25 μm following the application of 800 MPa pressure is clearly visible.

The influence of median agglomerate size on the instantaneous densification rate at a density of 2·6 Mg m^{-3} for both illite- and kaolinite-derived sialons is shown in Fig. 9. Under identical hot-pressing conditions, the dependence of the densification rates on agglomerate size is markedly different. For kaolinite-derived powders,

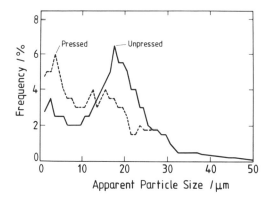

FIG. 8. Histograms of size distributions in uncompacted, and compacted powders.

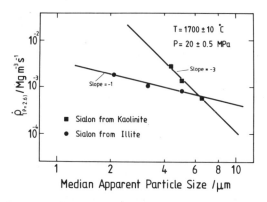

FIG. 9. The effect of median apparent particle size on densification rates of sialon powders obtained from the illite and kaolinite minerals.

the dependence of densification rate on apparent particle size shows the dependence expected from the Coble creep equation, i.e.

$$\frac{d\rho}{dt} = \frac{A\delta D_b \Omega}{kT\bar{G}^3} P_{appl}$$

where $\bar{G}$ is the mean grain size, ρ is the instantaneous relative density, D_b is the grain boundary diffusion coefficient, δ is the grain boundary width, Ω is the molar volume of rate controlling species, P_{appl} is the applied pressure, T is temperature, k is Boltzmann's constant, and A is a numerical constant. For the illite-derived powder, a slope of ~ -1; i.e. $(\dot{\rho}\alpha\bar{G}^{-1})$ is obtained. Such a grain size exponent is close to the value predicted for grain boundary sliding-controlled kinetics.[14]

The general pattern of behaviour is consistent with the conclusions of Rahaman, Riley and Brook,[15] who suggested that, in hot-pressing of sialon-forming systems, two essential densification processes (solution and reprecipitation and grain boundary sliding) are operating sequentially, the slower of which is rate-controlling. In such sialon systems containing relatively large quantities of transient liquid phase (as shown in high z-value sialons, or for O/N ratios corresponding to compositions above the theoretical β'-sialon line), the Coble diffusion step of the solution–diffusion–reprecipitation mechanism is rate-controlling. On the other hand, for systems with lesser quantities of transient liquid phase (as shown by lower z-value sialons, or for compositions with O/N slightly below the theoretical β'-sialon line), grain boundary sliding may become rate-controlling.

A further feature of interest is the dependence of densification rate on the equilibrium sialon z-value in sialon-forming systems. It has been shown that, under defined conditions, compositions corresponding to β'-sialons with higher equilibrium z-values densify faster than those with lower z-values.

In the present work, the β'-sialon powder derived from kaolinite $(z \sim 3)$ densified faster than that prepared from illite $(z \sim 2)$, consistent with the earlier studies. However, the densification of sialons from appropriate mixtures of nitrides and oxides appears to be faster than that of the pre-equilibrated powders for identical z-values, similar granulometry and pressing conditions. For example, a non-equilibrated sialon-forming powder mixture (with a final z-value of 3) densifies to full density in <300 s,[16] while a pre-equilibrated sialon powder with a similar z-value and under identical hot-pressing conditions, densifies to full density in ~1200 s (this study). It would be expected that, in a system consisting initially of silicon and aluminium nitrides, together with silicon oxide and aluminium oxide, the volume of transient liquid initially formed by rapid interaction between the oxides would be larger than the volume of liquid present in a system already partially pre-equilibrated during its formation.

The addition of 9 wt % yttrium oxide to the β'-sialon powder obtained from illite, accelerated the overall densification rate, and full density was attained in less than 200 s. This is now close to the densification behaviour of the non-equilibrated compositions (with a $z \sim 3$), hot-pressed under similar conditions.[16] A plot of instantaneous

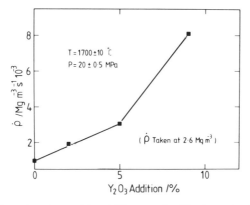

FIG. 10. The effect of yttrium oxide additions on densification rate of sialon powders prepared from the illite.

densification rate as a function of yttrium oxide content, at 1700°C for an instantaneous density of 2·6 Mg m^{-3}, is shown in Fig. 10. It is evident that the use of even 5 wt % yttrium oxide at 1700°C both enhances the instantaneous densification rate and shortens the time to attain full density by a factor of ~3. Yttrium oxide is well-established as a liquid-forming densification aid for silicon nitride and sialon systems.[17] These comparisons are made on powders considered to be comparable in mean particle size on the basis of Stoke's sedimentation data.

5. CONCLUSIONS

It has been shown that β'-sialon powders prepared from naturally occurring minerals can be hot-pressed to high density without the need for further liquid-forming additives. However, the rates of densification in these powders are overall lower than those in systems consisting initially of silicon and aluminium nitrides with silicon oxide and aluminium oxide (with similar z-values and under identical conditions).

Although β'-sialon powders are prepared from sub-micrometre raw materials, agglomeration occurs due to presintering, and therefore a careful control over the powder preparation conditions seems to be necessary for improving the densification behaviour of these powders.

Densification rates of the sialons derived either from illite or kaolinite were found to be dependent on the apparent particle size, and kaolinite-derived powder ($z \sim 3$) initially densifies faster than the powder obtained from the illite ($z \sim 2$), though they both finally attain densities >98% at 1700°C.

The addition of yttrium oxide (up to 9 wt %) to the sialon powders derived from illite was found to accelerate densification rate and shorten the time to attain full density by a factor of ~10.

ACKNOWLEDGEMENTS

This work has been supported in part by the Directorate-General for Science, Research and Development, of the Commission of the European Communities.

REFERENCES

1. UMEBAYASHI, S. and KOBAYASHI, K. *American Ceramic Society Bulletin,* **56** (1977) 578.
2. UMEBAYASHI, S. and KOBAYASHI, K. In *Factors in Densification of Oxide and Non-Oxide Ceramics,* Somiya, S. and Saito, S. (Eds), KTK Scientific Publishers, Tokyo, p. 480, 1979.
3. UMEBAYASHI, S., KOBAYASHI, K. and KATAOKA, R. *Yogyo-Kyokai-Shi,* **88** (1980) 469.
4. UMEBAYASHI, S. and KOBAYASHI, K. *Journal of the American Ceramic Society,* **58** (1975) 464.
5. UMEBAYASHI, S., KOBAYASHI, K., ISAYAMA, Y. and NAKAMURA, M. *Yogyo-Kyokai-Shi,* **88** (1980) 361.
6. LEE, J. G. and CUTLER, I. B. *American Ceramic Society Bulletin,* **58** (1979) 869.
7. GAVRISH, A. M., BOYARINA, I. L., DEGTYAREVA, E. V., PUCHKOV, A. B., ZKUKOVA, Z. D., GUL'KO, N. V. and TARASOVA, L. A. *Inorganic Materials,* **18** (1982) 46.
8. VAN DIJEN, F. K., SISKENS, C. A. M. and METSELAAR, R. *Science of Ceramics,* **12** (1984) 427.
9. WILD, S. In *Special Ceramics,* Vol. 6, Popper, P. (Ed.), British Ceramic Research Association, Stoke-on-Trent, p. 309, 1976.
10. WILD, S. *Journal of Materials Science,* **11** (1976) 1972.
11. TORRE, J. P. and MOCELLIN, A. *Journal of Materials Science,* **11** (1976) 1725.
12. HOCH, M. and NAIR, M. *American Ceramic Society Bulletin,* **58** (1979) 191.
13. GROLLIER-BARON, T. French Patent No. 2,462,405 (1981).
14. LANGDON, T. G. In *Deformation of Ceramic Materials,* Bradt, R. C. and Tressler, R. E. (Eds), Plenum Press, New York, p. 101, 1975.
15. RAHAMAN, M. N., RILEY, F. L. and BROOK, R. J. *Journal of the American Ceramic Society,* **63** (1980) 648.
16. RAIIAMAN, M. N. and RILEY, F. L. In *Special Ceramics,* Vol. 7, Taylor, D. E. and Popper, P. (Eds), *Proceedings of the British Ceramic Society,* **31** (1981) 63.
17. JACK, K. H. *Science of Ceramics,* **11** (1981) 125.

COMMENTS AND DISCUSSION

Chairman: M. J. POMEROY

S. Hampshire: Your densification kinetics show that the finer particle size densifies more rapidly or arrives at final density earlier. I was a little puzzled by your graphs which show that at the beginning the finer particle compacts appear to have a higher green density which is contrary to our experience.

M. Mostaghaci: This really depends on particle size distribution. Although the median particle size may be finer, the particle size range may be totally different.

A. Hendry: We had some discussion previously on the mechanism of densification where the importance of transformation in producing sialon was emphasised. The last three papers have discussed densification of three preformed oxynitrides. You have shown the effects of particle size and obviously the impurities present will have an effect. What are your views on the mechanism of densification of preformed sialon?

Mostaghaci: It will involve transient liquid phase but the volume is low.

12

Sintering and some Properties of Si_3N_4 Based Ceramics

S. BOSKOVIĆ AND E. KOSTIĆ
'Boris Kidrič' Institute of Nuclear Sciences, POB 522, Lab. 170, 11001 Belgrade, Yugoslavia

ABSTRACT

Sintering of Si_3N_4 was investigated in the temperature range 1550–1750°C. As additions which promote sintering of Si_3N_4 the following were used:
Y_2O_3, a mixture of $Y_2O_3 + Al_2O_3$, and the four component mixtures:
$Y_2O_3 + Al_2O_3 + AlN + Si_3N_4$,
$YAG + Al_2O_3 + AlN + Si_3N_4$,
$DyAG + Al_2O_3 + AlN + Si_3N_4$.
By using these additions, dense Si_3N_4 ceramics were obtained at 1750°C. Special attention was paid to phase composition of sintered samples which depends on the additive which allows sintering to take place. Furthermore, thermal treatment of some samples was performed at 1450°C with the intention of decreasing the amount of glassy phase formed during sintering. Changes of hardness, density and toughness were determined as a function of thermal treatment time.

1. INTRODUCTION

Silicon nitride is known to have excellent mechanical properties, especially high toughness. However, the hardness of this ceramic is no better than that of Al_2O_3. A number of investigations have been concerned with finding good additions which would increase the hardness of the material. On the other hand, these additives must fulfil another important condition, that is, to enable the densification of Si_3N_4 by liquid phase formation. Y_2O_3 and $Y_2O_3 + Al_2O_3$ are additions most commonly used for this purpose.[1,2] This paper also describes the use of the additions Y_2O_3 and $Y_2O_3 + Al_2O_3$ to silicon nitride and compares the four-component mixtures YAG + AlN + Al_2O_3 + Si_3N_4; DyAG + AlN + Al_2O_3 + Si_3N_4; and Y_2O_3 + Al_2O_3 + AlN + Si_3N_4 which will promote densification. Depending on additive composition, two kinds of ceramic, β-Si_3N_4 and β'-SiAlON were synthesized.

2. EXPERIMENTAL

α-Si_3N_4 powder was supplied by Starck, West Germany, with a specific surface area of 8 $m^2 g^{-1}$. The additives used were (wt %):

A_1: 10% Y_2O_3
A_2: 6% Y_2O_3 + 2% Al_2O_3
A_3: 5% Y_2O_3 + 5% Al_2O_3
A_4: 10% Y_2O_3 + 5% Al_2O_3
A_5: 10% Y_2O_3 + 10% Al_2O_3
B_1: 72% YAG + 6% AlN + 12·8% Al_2O_3 + 9·2% Si_3N_4
B_2: 72% DyAG + 6% AlN + 12·8% Al_2O_3 + 9·2% Si_3N_4
B_3: 41·2% Y_2O_3 + 43·6% Al_2O_3 + 6% AlN + 9·2% Si_3N_4

Liquid formation occurs below 1550°C.

Homogenization of α-Si_3N_4 with previously homogenized mixtures was performed in isopropyl alcohol. Green pellets were isostatically pressed under 147 MPa. Sintering was carried out in the temperature range 1450–1750°C for 1 hour with the pellets packed in a powder bed of the same composition.

3. RESULTS AND DISCUSSION

Additions which promote sintering of Si_3N_4 in the presence of the liquid phase bring about chemical reaction development. By choosing appropriate compositions of additions, either β-Si_3N_4 or β'-SiAlON can be produced as a major phase within the material.

For consolidation of Si_3N_4 the volume of liquid as well as its

TABLE 1
Densities of Si_3N_4 (g cm^{-3}) With Different Additions

Temperature (°C)	A_1 10% Y_2O_3	A_2 6% Y_2O_3 + 2% Al_2O_3	A_3 5% Y_2O_3 + 5% Al_2O_3
1450	—	2·30	—
1550	—	2·47	2·82
1650	—	2·59	2·83
1750	2·97	2·71	2·65

composition is of great importance. This will be illustrated by the data given in Tables 1, 2 and 3.

The degree of densification in Si_3N_4 samples sintered with 8–10% of additions is low, with large amounts of open porosity remaining. Moreover, decomposition of Si_3N_4 takes place at 1650°C which is evident from X-ray results (formation of free silicon), given in Table 2.

This is the consequence of a low volume of liquid phase [3,4] and very likely of its high viscosity. Results show that 10 wt % of additions form a low volume of liquid for bringing about the densification of the Si_3N_4. The next step was to increase the amount of liquid phase (Table 3) by increasing the level of additions to the range 15–20%.

Comparing with data in Table 1, the degrees of densification obtained were higher and at the same time Si_3N_4 decomposition was suppressed. This suggests that the amount of liquid was enough to cover the Si_3N_4 particles before decomposition started.

In all cases mentioned, the first liquid is formed at Y_2O_3–SiO_2 particle contacts (surface silica). Thereafter the other constituents (Si_3N_4 and Al_2O_3) are dissolved in it and reprecipitate. During sintering of these samples the starting α-Si_3N_4 transforms to β-Si_3N_4 via a solution–reprecipitation process through the liquid phase, and

TABLE 2
Phases Detected in Samples Sintered With Different Additions

Temperature (°C)	A_1	A_2	A_3
1450	—	α, β	—
1550	—	β, Y_2SiAlO_5N	α, β
1650	—	β, Si, Y_2SiAlO_5N	β, Y_2SiAlO_5N
1750	β, $YSiO_2N$	β, Si, Y_2SiAlO_5N	β, Si, Y_2SiAlO_5N

TABLE 3
Densities and Phase Composition of Samples With Additions A_4 and A_5

Temperature (°C)	A_4 10% Y_2O_3 + 5% Al_2O_3		A_5 10% Y_2O_3 + 10% Al_2O_3	
	Density (g cm^{-3})	phases present	Density (g cm^{-3})	phases present
1 550	2·98	$\alpha, \beta, Y_2SiAlO_5N$	2·94	$\alpha, \beta, Y_2SiAlO_5N$
1 650	3·05	β, Y_2SiAlO_5N	3·07	β, Y_2SiAlO_5N
1 750	3·08	β, Y_2SiAlO_5N	3·09	β, Y_2SiAlO_5N

Y_2SiAlO_5N phase appears (except in the case when 10% Y_2O_3 was added) and $YSiO_2N$ is observed. This is in accordance with published data,[5] bearing in mind that the α-Si_3N_4 powder contained 3·7 wt % surface SiO_2.

On the basis of earlier results[3] dense SiAlON ceramics can be obtained by using additions from the YAG–Si_3N_4–SiAlON system.[6] For these investigations, additions B_1, B_2 and B_3 were chosen. The amount of these additions was in the range 20–25%. Densities obtained by using 25% B_1, 20% B_2 and 25% B_3, as well as the phase composition of sintered samples are shown in Tables 4 and 5.

Due to different additive compositions, i.e. different liquid phase compositions, from those of samples with additions of group A, different reaction sequences can be observed. Namely, by using B_1 and B_2 additions as sintering aids, the first liquid appears at YAG–SiO_2 particle contacts, into which other liquid constituents are dissolved. β'-SiAlON is formed by crystallization from the melt, whereas with additions of group A, β-Si_3N_4 is the major phase.

Lattice parameters of SiAlON as a function of sintering temperature

TABLE 4
Densities of Samples Sintered With Additions B_1, B_2, B_3 (g cm^{-3})

Temperature (°C)	B_1	B_2	B_3
1 450	2·87	2·46	2·29
1 550	2·93	2·96	3·00
1 650	3·00	3·09	3·11
1 750	3·02	3·22	3·10

TABLE 5
Phases Detected After Sintering With Additions B_1, B_2, B_3

Temperature (°C)	B_1	B_2	B_3
1 450	α, SiAlON, YAG	α, SiAlON, DyAG	YAG,
1 550	SiAlON, α, YAG Y_2SiAlO_5N	SiAlON, α, DyAG, same lines as Y_2SiAlO_5N	SiAlON, α, YAG
1 650	SiAlON, Y_2SiAlO_5N	SiAlON, same lines as Y_2SiAlO_5N	SiAlON, Y_2SiAlO_5N
1 750	SiAlON, Y_2SiAlO_5N	SiAlON, same lines as Y_2SiAlO_5N	SiAlON, Y_2SiAlO_5N

are given in Table 6. Lattice parameter values indicate the particular SiAlON composition. In the case of addition B_1, $z = 0.6$, and in the case of addition B_2, $z = 0.5$.

With samples with additions B_1 and B_2 respectively, DyAG ($3Dy_2O_3.5Al_2O_3$) X-ray diffraction line intensities decrease with increasing sintering temperature, because garnet forms a liquid phase at these temperatures. With addition B_3, garnet forms at lower temperatures from Y_2O_3 and Al_2O_3 which are present in the additive mixture and with increasing temperature its line intensities decrease (Fig. 1).

At higher temperatures besides SiAlON, Y_2SiAlO_5N appears in small amounts within the samples with additions B_1 and B_3. In samples with DyAG (B_2 additions) a phase with the same diffraction lines appears (Fig. 2), and it is very likely the same wollastonite phase[7] containing Dy instead of Y.

TABLE 6
Lattice Parameters of Samples with Additions B_1 and B_2

Lattice parameters (10^{-10} m)		Sintering temperature (°C)			
		1 450	1 550	1 650	1 750
B_1	a	7·66	7·64	7·63	7·62
	c	—	2·94	2·93	2·92
B_2	a	—	7·63	7·62	7·61$_5$
	c	—	2·92	2·92	2·91$_6$

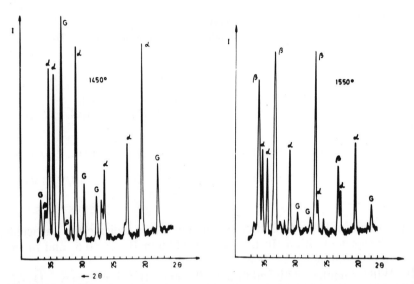

FIG. 1. X-ray diffraction pattern from samples sintered with addition B_3 at 1450 and 1550°C.

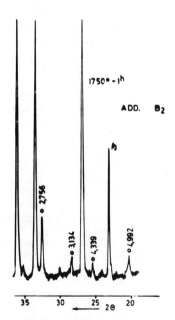

FIG. 2. X-ray diffraction pattern from sample sintered with addition B_2 at 1750°C, 1 h.

TABLE 7
Hardness and Toughness of Samples Sintered at 1750°C With Different Additions

Addition	B_1	B_2
H_V, Vickers hardness (MPa)	11 400	19 200
K_{IC} (toughness) (MN m$^{-3/2}$)	4·15	6·39

Samples obtained during this work by pressureless sintering were characterized by measuring hardness and toughness and the results are given in Table 7. By applying thermal treatment procedures, an increase in the hardness of these materials can be achieved, because during annealing of these compositions at subsolidus temperatures, YAG crystallizes from the glass phase[4,6,7] and the overall amount of crystalline phase increases. Annealing was performed at 1400°C for 5 and 10 h on samples with additions B_1 and B_2. After annealing, hardness and toughness of the samples were measured by the

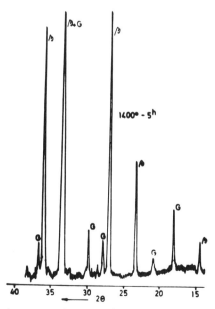

FIG. 3. X-ray diffraction pattern from sample with B_1 addition sintered at 1750°C for 1 h, and annealed at 1400°C for 6 h.

TABLE 8
Hardness and Toughness of Annealed Samples at 1400°C

Sample	H_V (MPa)		K_{IC} (MN m$^{-3/2}$)	
	5 h	10 h	5 h	10 h
B_1	14 000	14 500	4·40	4·50
B_2	19 250	20 650	6·22	6·19

indentation method as before. X-ray analysis was carried out and the data show that even after 6 h of annealing time (Fig. 3) YAG lines appear with samples having additions B_1 and B_2.

On the basis of the data given in Table 8 and Fig. 3, it can be observed that with SiAlON samples hardness increases due to YAG crystallization from the melt. These results indicate the possibility of synthesizing Si_3N_4–YAG composites having improved hardness.

Synthesis of a ceramic composite which was thought to have better hardness compared with samples with addition B_1 was carried out starting with a 52% Si_3N_4 + 48% YAG mixture, to which 15% of addition B_1 was added to enable densification. This material densifies very easily at 1550°C and the density achieved was 3·45 g cm^{-3} without open porosity. Above this temperature the sample melted completely. Hardness and toughness were measured and the results are shown in Table 9.

Results in Table 9 for the composite sintered at 1550°C, as well as for samples annealed at 1400°C for 5 and 10 h indicate the possibility of obtaining ceramic composites of improved properties.

TABLE 9
Hardness and Toughness of Si_3N_4–YAG Composite

52% Si_3N_4–48% YAG +15 wt % B_1	As sintered 1 550°C	After annealing	
		1 400°C—5 h	1 400°C—10 h
H_V (MPa)	17 500	19 200	20 000
K_{IC} (MN m$^{-3/2}$)	3·42	4·15	4·53

4. CONCLUSIONS

1. Using additions from group A, β-Si_3N_4 ceramics were synthesized.
2. Using additions from group B, β'-SiAlON ceramic materials were synthesized.
3. Applying a thermal treatment procedure at 1400°C results in improvement of hardness.
4. A Si_3N_4–YAG ceramic composite was synthesized with improved hardness.

REFERENCES

1. TSUGE, A. and NISHIDA, K. *American Ceramic Society Bulletin*, **57**, No. 4 (1978) 424.
2. BULJAN, S. T. and SARIN, V. K. In *Sintered Metal–Ceramic Composites*, Materials Science Monograph, 25, Upadhyaya, G. S. (Ed.), Elsevier, Amsterdam, p. 455, 1984.
3. BOSKOVIĆ, S. In *Sintering Theory and Practice*, Kolar, D., Pejovnik, S. and Ristić, M. (Eds), Elsevier, Amsterdam, p. 245, 1982.
4. BOSKOVIĆ, S. and KOSTIĆ, E. *Science of Ceramics*, **12** (1983) 391.
5. JACK, K. H. *Science of Ceramics*, **11** (1981) 125.
6. HONKE, H. and TIEN, T. Y. *American Ceramic Society Bulletin*, **58** (1978) 349.
7. JACK, K. H. In *Phase Diagrams: Materials Science & Technology*, Vol. V, Alper, A. M. (Ed.), Academic Press, New York, p. 241, 1978.

COMMENTS AND DISCUSSION

Chairman: M. FARMER

S. Hampshire: I notice that the density you achieved with YAG additions as yttria and alumina were slightly higher than with preformed YAG. In what form did you introduce the YAG into the samples?
S. Bosković: It was previously synthesised and then ground so the particle size may be larger than the yttria/alumina which must influence the densification.
Hampshire: Have you looked at the microstructure?
Bosković: Not yet.

M. Farmer: What load is used for the Vickers hardness test?
Bosković: 30 kg/20 s.
D. P. Thompson: Do you know what effect the DyAG has on the high temperature strength?
Bosković: No. We have not examined that?

13

Syalon Ceramic for Application at High Temperature and Stress

M. H. Lewis and S. Mason

Centre for Advanced Materials Technology, University of Warwick, Coventry CV4 7AL, UK

AND

A. Szweda

Lucas Cookson Syalon Ltd, Cranmore Boulevard, Shirley, Solihull, West Midlands, UK

ABSTRACT

In the quest for a Si_3N_4-based ceramic to meet the requirements of hot-gas flow components in future turbines, the microstructure of a syalon ceramic has been refined by post-sintering heat treatments.

The achievement of oxidation resistance and non-cavitational creep at stress/temperature combinations of 100 MPa/1300°C has previously been demonstrated for a syalon ceramic pressureless sintered with a Y–Si–Al–O–N liquid phase in which the matrix phase has subsequently been crystallised as YAG ($3Y_2O_3.5Al_2O_3$).

The temperature ceiling for application in oxidising environments is determined by the reversion of YAG to eutectic liquid by reaction with a SiO_2-rich oxidation layer.

In this paper a description is given of enhanced high-temperature performance via improved processing, optimisation of crystallisation and the formation of a YAG-free surface microstructure composed of β'-Si_3N_4 and Si_2N_2O. In particular a stress/temperature combination, unmatched by other comparable Si_3N_4 ceramics at 1375°C and 400 MPa has been achieved in bend and compressive creep at low strain rates via a diffusional mechanism, without cavitation.

1. INTRODUCTION

The gain in thermal efficiency or output power of advanced gas turbines at temperatures above that limited by superalloy components is well established. Turbine designers are currently requesting that in substituting ceramics for metallic alloys, the hottest components (blades, stator vanes) should be stable at temperatures up to 1375°C for peak stresses over 100 MPa. This performance requirement has also to be achieved with oxidising gas-flow conditions in the presence of corrosive impurities from fuel and environment.

Oxynitride ceramics, based on the β'-Si_3N_4 structure, have long been recognised as having the necessary intrinsic properties for such applications, but have suffered from various problems associated with the fabrication process. Porosity and inhomogeneity are typical problems associated with liquid-phase sintering in multicomponent systems and have a sensitive influence on brittle-fracture stress (MOR). However, the major problems at elevated temperatures (>1000°C) are creep-rupture and oxidation resistance; properties which are most sensitively linked to sintering liquid residues at grain boundaries.

An important advance in the development of β'-Si_3N_4 based ceramics was the demonstration of a diffusional creep mechanism in the absence of intergranular cavitation via reduction in triple-function glassy residues to a 'sub-critical' size for cavity nucleation.[1,2] This may be achieved by control of the final liquid content during sintering via the O/N ratio within the sintering additives and may be further refined via post-sintering heat treatment in an oxidising environment.[3,4] Both processes are facilitated by the ability to partition oxygen and nitrogen atoms between the major phase and liquid residue. This is achievable in β'-Si–Al–O–N ceramics in which β-Si_3N_4 adopts the substituted composition $Si_{3-x}Al_xO_xN_{4-x}$.[1] These principles were initially applied to β' ceramics which were hot-pressed with minimal liquid volumes, resulting in an essentially monophase ceramic with exceptional creep, creep-rupture and oxidation resistance to temperatures in excess of 1400°C. However a recognition of the economic importance of 'near-net-shaping' of engine components has resulted in the eclipse of this high cost ceramic in favour of those produced by pressureless sintering. The latter are fabricated to near theoretical density with comparatively large sintering-liquid volumes (5–15%) and offer an additional advantage of higher fracture-toughness (K_{Ic}), and hence MOR due to smaller β' grain sizes which are morphologically anisotropic. The high-temperature problems have been partially resolved, in the case of β'-Syalon ceramics, by nearly-complete crystallisation of the residual liquid matrix during post-sintering heat treatment.[5]

2. MICROSTRUCTURAL DEVELOPMENT

2.1. The Sintered Microstructure and Matrix Crystallisation

Sintering additions based on the Y_2O_3–Al_2O_3–SiO_2 eutectic and now commonly used in the pressureless-sintering of β and β'-Si_3N_4 ceramics. This eutectic offers a combination of moderately-low liquidus (the ternary oxide eutectic is at $\sim 1320°C$) with a comparatively-high residual liquid viscosity and glass-transition temperature, beneficial to high-temperature properties. A major feature of the Syalon range of ceramics is the use of nitride additions which serve to control the amount of liquid residue and, in particular, the liquid composition such that it may be converted via a 'glass–ceramic' route to a desirable crystalline matrix phase. Thus, small additions of AlN (added as a polytypoid phase, near to the AlN in composition) are used to manufacture glass-matrix ceramics with limited temperature capability ($\sim 1000°C$) whereas large additions (8–12 wt % in relation to $\sim 6\%$ each of Y_2O_3 and Al_2O_3) are used as a basis for matrix crystallisation. The crystallisation process has been described in previous publications, the main phase in high-temperature ceramics being yttro-garnet ($3Y_2O_3.5Al_2O_3$–YAG).[5] Following a separate programme of research on crystallisation of matrix glass compositions in bulk-synthesised form we have a more complete understanding of the importance of the N/O ratio in the glass in determining the crystallisation products.[6] Figure 1 is a Jänecke prism representation of compositional relations between the various stoichiometric source materials and product phases and will be used in illustrating the crystallisation sequence.

Bulk-synthesised glasses with Si/Al/Y ratios similar to Syalon ceramics (Al-rich with respect to the oxide eutectic) when crystallised with differing N/O ratios show a gradation in product phase which is consistent with a shift from the pure oxide prism face to a tie-line between oxynitride and oxide phases. For example, yttrium disilicate ($Y_2Si_2O_7$) + alumina (Al_2O_3) + mullite ($Al_6Si_2O_{13}$) crystallise at zero nitrogen content whereas YAG + silicon oxynitride (Si_2N_2O) crystallise from a glass near the nitrogen solubility limit (~ 30 equivalent per

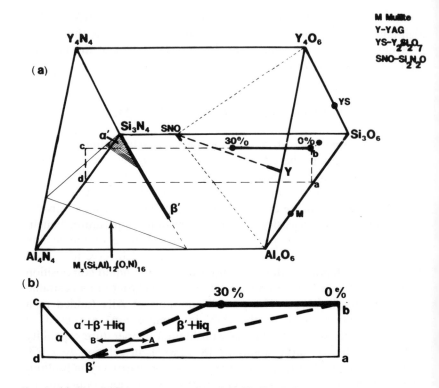

FIG. 1. (a) Jänecke Prism representation of the Y–Si–Al–O–N system. Glass compositions from 0 equiv. % up to the nitrogen saturation level—30 equiv. % indicated, with the possible crystallisation products for low and high nitrogen glasses. (b) A shift of the mean composition due to increased nitrogen levels, A to B, during sintering may result in the stabilisation of α'-Si_3N_4, resulting in an $\alpha' + \beta'$ + liquid ceramic with a high nitrogen ⩾30 equiv. % residual glass.

cent (Fig. 1(a)). In the 'composite' β' + glass sintered ceramic there is previous evidence[7] from electron energy-loss spectroscopy (ELS) for a change in glass composition from ~zero nitrogen to the saturation limit with increasing polytypoid content (0–6%). There is a corresponding change from $Y_2Si_2O_7$ to YAG as the primary crystallisation product. Si_2N_2O is not detected in this case and there is evidence, from the reduced β' substitution level, of incorporation of excess Si + N in β'.

In the current range of Syalon ceramics two compositional modifications have been made with the object of increasing the completeness of YAG crystallisation and hence avoiding glassy cavity nuclei to extend the high temperature operational stress. Firstly, an increase in polytypoid content to maximise total nitrogen content and to increase the Al + O substitution in β', which has an indirect effect of increasing the N/O ratio in the glass phase. Secondly, to avoid impurities in initial components which segregate to the residual glass phase and inhibit complete crystallisation.

Evidence for the effectiveness of these modifications is contained in the 'as-sintered' and crystallised microstructures (Fig. 2). The first appearance of $\alpha' Si_3N_4$ (yttrium stabilised) may be conveniently used as an indication of maximum glass nitrogen content; this may be viewed as a shift in mean composition from A to B in the schematic compositional section (Fig. 1(b)). There is also direct evidence from ELS and X-ray energy dispersive analysis (EDS) (using a light-element detector—Fig. 2 inset) of enhanced N/O ratio over that previously measured in Syalon matrix glasses. This is also higher than the maximum bulk solubility for glasses synthesised under 1 atmosphere of N_2 gas and hence is closer to, or may extend beyond, the Si_2N_2O–YAG tie line (Fig. 1(a)). Crystallised microstructures are essentially bi-phase β'-YAG mixtures with rarely detectable triple function glass residues.

In addition to minor constitutional changes it is preferable to modify the crystallisation kinetics via a 'nucleation' heat treatment prior to a higher-temperature 'growth' stage, i.e. a typical glass–ceramic sequence. This two-step treatment is appropriate in reducing the tendency to heterogeneous nucleation on sintering micro-cavity surfaces and rapid growth of large dendritic YAG crystals with associated inter-dendritic impurity segregation. In extreme cases this may be observed in reflected-light with the naked eye. The preferred β'-YAG microstructures result in a homogeneous black, highly reflective, surface when diamond polished.

2.2. Surface Microstructure

There is a particular interest in differences in near-surface microstructure from the 'bulk' state. This stems from the increasing requirement for component sintering to near-final shape and the susceptibility to oxidation or corrosion of the normal bulk microstructure. The latter problem is of interest here since we have previously identified an upper limiting temperature at which oxidising environments cause a reversion of the YAG matrix to a ternary Y_2O_3–Al_2O_3–

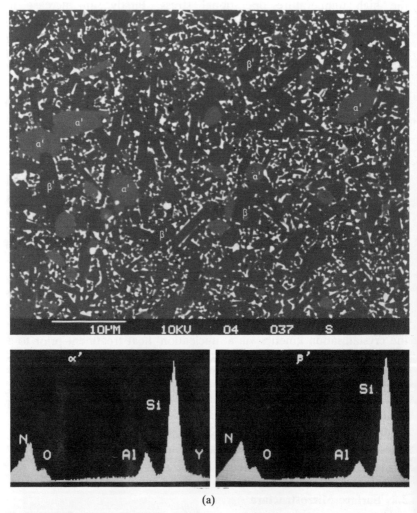

FIG. 2. (a) Backscattered scanning electron micrograph of a 'high nitrogen' Syalon. Increasing grey levels indicate higher mean atomic number: black, β'-Si_3N_4; grey, yttrium stabilised α'-Si_3N_4; white, yttro-garnet (YAG). (b) Transmission electron micrograph of 'as-sintered' β'/glass ceramic with EDS analysis of the glass matrix phase. (c) Transmission electron micrograph of crystallised β'/YAG ceramic with minimal residual glass.

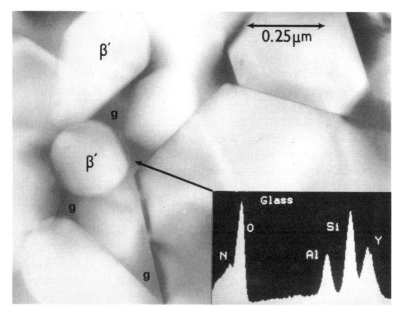

(b)

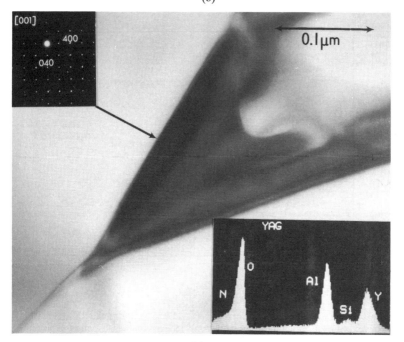

(c)

FIG. 2—*contd*.

SiO$_2$ eutectic liquid by reaction with a surface SiO$_2$-rich layer.[8,9] Although the kinetics of this surface reaction are reduced in the current range of Syalon ceramics, presumably due to increased N content and purity level, the problem is still significant above 1350°C. This is exemplified by a comparison of oxidation kinetics at this temperature for a series of different surface microstructures (Fig. 3). β' + YAG surfaces exhibit comparatively rapid and discontinuous kinetics due to oxide-spalling. The liquid-reversion reaction causes a rapid out-diffusion of metallic ions (mainly Y and Al) into the SiO$_2$ film, reducing its viscosity. Other ceramic combinations and the monophase β' microstructure are stable in oxidising conditions and maintain a high-viscosity SiO$_2$ surface film.

In a previous publication[9] we have described how the out-diffusion of Y and Al at temperatures for controlled oxidation (~1300°C) can lead to a reaction of this type; β'-Si$_3$N$_4$ + SiO$_2$ → O'-Si$_2$N$_2$O within a subsurface zone. β' and O' are the substituted forms of nitride and

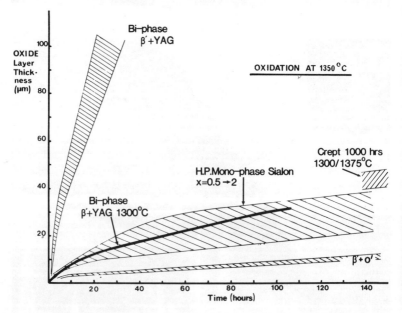

FIG. 3. Oxidation kinetics at 1300°C and 1350°C for β'/YAG ceramic in comparison with H.P. monophase levels, and β'/O' surface transformed layer. The oxide layer thickness of a crept β'/YAG specimen at 1300–1375°C indicates the formation of an O' protective layer during long-term oxidation, increasing the oxidation resistance at temperatures above 1300°C.

oxynitride and SiO_2 derives partly from the intergranular glass and partly from the oxidation layer. We have subsequently studied various ways of accelerating this sluggish, diffusion-controlled reaction and an extreme example of a surface-transformed microstructure of $\beta' + O'$ is illustrated in Fig. 4. This appears to be totally effective in suppressing the liquid-reversion reaction and exhibits oxidation kinetics comparable with the monophase ceramics. In the next section the results of creep deformation in oxidising conditions up to 1375°C confirm the effectiveness of this surface treatment.

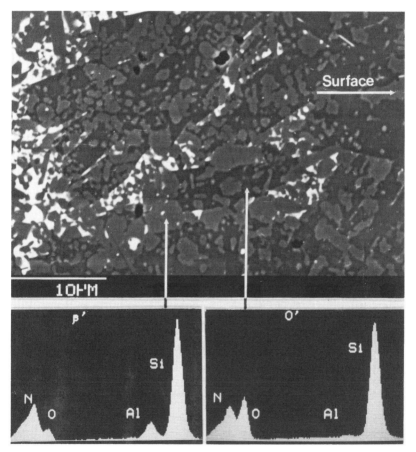

FIG. 4. Backscattered scanning electron micrograph of the β'/O' protective layer (O'-dark contrast) at the surface of the β'/YAG ceramic.

3. HIGH-TEMPERATURE DEFORMATION AND FRACTURE

A characteristic property of ceramics containing glassy intercrystalline residues is the transient rise in fracture-toughness (K_{lc}) near the glass-softening point. This is associated with the onset of sub-critical crack growth (SCG) within a creep cavitation zone at the primary crack tip and is a convenient indicator of the susceptibility of a ceramic to creep-rupture. The K_{lc} rise is due to energy absorption by plastic deformation, mainly in the form of grain-boundary sliding and viscous deformation of residual glass bridging the crack surfaces.

The importance of matrix glass crystallisation is illustrated in the K_{lc}-temperature relation for a β'-syalon ceramic in the 'as-sintered' state and after crystallising heat treatment (Fig. 5). In the latter condition there is no evidence for a zone of SCG preceding rapid fracture at temperatures in excess of 1450°C, in a non-oxidising

FIG. 5. Fracture toughness (K_{lc}-T) plot for β'/YAG and β'/glass ceramics. Inset: optical micrograph of the fracture surface of a β'/glass ceramic indicating an extensive region of sub-critical crack growth (SCG) which occurs at temperatures in excess of 1300°C. Crystallised β'/YAG ceramics show no evidence of SCG up to 1450°C.

environment, whereas the zone of creep-cavitation is clearly visible on fracture surfaces of glass-containing ceramics (Fig. 5 inset). The change in high-temperature creep and creep-rupture behaviour associated with the two microstructural states cannot be defined so precisely because of matrix crystallisation during the longer-term test. However, prior heat treatment is required to suppress creep cavitation and minimise creep rate at the highest stress levels. A comparison of creep data for bend and compressive loading with that for other β' ceramics is exemplified in Fig. 6. At 1300°C, following a prolonged period of non-steady-state creep, the deformation rate is linear and has a stress-sensitivity appropriate for a diffusional creep mechanism over a wide range of stress. Hence this terminal steady-state creep is comparable with a hot-pressed monophase β' ceramic (Fig. 6) in which Coble, grain-boundary diffusional, creep may be described by

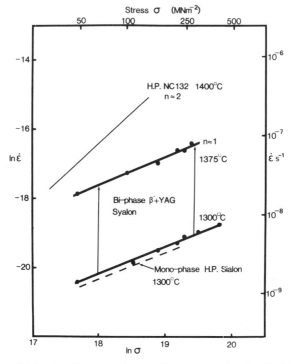

FIG. 6. Four point bend, and compressive steady state creep data for the 'high nitrogen' Syalon at 1300°C and 1375°C up to 400 MN m^{-2} compared with H.P. monophase Syalon and NC132.

the general creep equation;

$$\text{Creep rate } (\dot{\varepsilon}) = \text{Constant } \sigma^n d^m \exp(-Q/kT)$$

in which the stress exponent $n = 1$ and Q is an activation energy for grain boundary diffusion. The grain-size exponent m is not normally measurable in β' ceramics because of the limited range of grain size (d) attainable without varying other microstructural characteristics.

The transient creep behaviour of β'-YAG ceramics is due to a superposition of the primary visco-elastic creep and a time-dependent microstructural change. This may be circumvented by long prior heat treatment at 1300–1400°C following the normal crystallisation treatment (Fig. 7). The loss of a partially recoverable, visco-elastic component of creep is due to continued crystallisation of residual glass. Direct microstructural evidence for this is obtained from high resolution electron microscopy of crept or heat-treated specimens. There is an accompanying change in β'-YAG morphology following equilibration of interfacial tensions at β'/YAG/β' triple junctions (Fig. 8). This contrasts with the locally faceted β'/glass/β' junctions

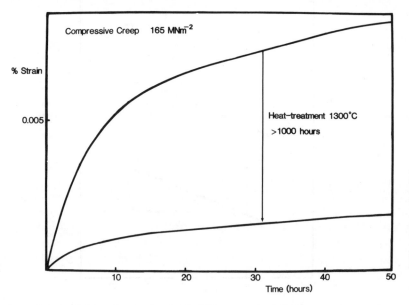

FIG. 7. Compressive creep curves for β'/YAG ceramic showing the reduction in the visco-elastic, primary stage of creep due to the elimination of residual glass by long-term heat treatment.

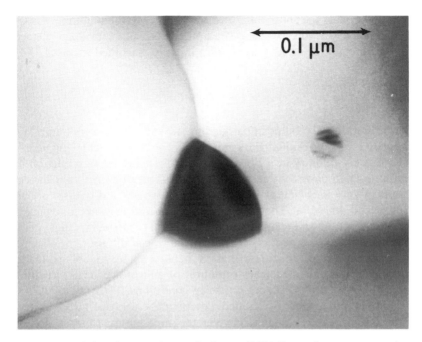

FIG. 8. Transmission electron micrograph of crept β'/YAG ceramic at temperatures up to 1375°C and stresses of 400 MN m^{-2} showing the absence of creep cavitation.

initially present due to anisotropy in β'/liquid interfacial energy. The activation energy for steady-state diffusion creep is $\sim$1000 kJ mol^{-1} and is believed to be that for β'/YAG interfacial diffusion. This is the rate-controlling step with a higher activation energy than for β'/β' grain-boundary diffusion. The latter process must operate in series but has previously been studied in the monophase ceramics with a limiting value of $Q \sim 850$ kJ mol^{-1} (Refs. 2, 3).

The importance of the creep data presented here is the unambiguous identity of a diffusional mechanism without creep-cavitation in both compressive and bend specimens. Creep deformation with a cavitational component normally results in non-integral stress exponents greater than unity and has different rates in compressive and bend creep. In this work the rates are similar, there is no microscopically detectable cavitation or creep-rupture to the normal limits of strain ($\sim$2%) in bend specimens up to 400 MPa. Further, this behaviour has been demonstrated to 1375°C following the $\beta' + O'$ surface-transformation treatment described in Section 2. 400 MPa is a large fraction of the short-time fracture stress (MOR) at about 1300°C, and is proof that the oxidising treatment does not introduce surface flaws of much greater severity than in initially machined surfaces.

4. CONCLUSIONS

Control of initial composition leading to high N/O ratio and low impurity levels in residual glass matrices is essential in achieving nearly complete crystallinity of a β' + YAG microstructure. The reduction in phase-function glass residues to a sub-critical size for cavity nucleation results in very long times to failure at temperatures and stresses above 1300°C and 400 MPa. The creep deformation mechanism, following a transient period with minor microstructural change, is that of a steady-state Coble creep, controlled by interfacial diffusion.

An extension of this deformation behaviour to 1375°C is made possible by a controlled transformation of subsurface microstructure from β'-YAG to β'-O' under oxidising conditions.

Ceramic processing of shaped components for operation at the highest temperatures is possible via the pressureless-sintered route. Post-sintering treatments involve a two-step nucleation and growth sequence for glass-matrix crystallisation followed by a surface transformation to provide protection in oxidising environments. Hence oxynitride ceramic alloys have reached a degree of microstructural refinement, achieved by complex processing and transformation in parallel with that of many of the metallic alloys they are destined to replace in severe thermal conditions.

ACKNOWLEDGEMENTS

M. H. L. and S. M. wish to thank the directors of Lucas Cookson Syalon for financial support for this research.

REFERENCES

1. LEWIS, M. H., POWELL, B. D., DREW, P., LUMBY, R. J., NORTH, B. and TAYLOR, A. J. *Journal of Materials Science,* **12** (1977) 61.
2. KARUNARATNE, B. S. B. and LEWIS, M. H. *Journal of Materials Science,* **15** (1980) 449.

3. KARUNARATNE, B. S. B. and LEWIS, M. H. *Journal of Materials Science*, **15** (1980) 1781.
4. LEWIS, M. H. and KARUNARATNE, B. S. B. In *Fracture Mechanics for Ceramics, Rocks and Concrete*, Freiman, S. W. and Fuller, E. R. (Eds), American Society for Testing and Materials, Philadelphia, Pa, STP 745, p. 13, 1981.
5. LEWIS, M. H., BHATTI, A. R., LUMBY, R. J. and NORTH, B. *Journal of Materials Science*, **15** (1980) 103.
6. LENG-WARD, G. and LEWIS, M. H. *Materials Science and Engineering*, **71** (1985) 101.
7. WINDER, S. M. and LEWIS, M. H. *Journal of Materials Science Letters*, **4** (1985) 241.
8. LEWIS, M. H., KARUNARATNE, B. S. B., MEREDITH, J. and PICKERING, C. In *Creep and Fracture of Engineering Materials*, Wilsher, B. and Evans, R. C. (Eds), Pineridge Press, Swansea, p. 365, 1985.
9. LEWIS, M. H., HEATH, G. R., WINDER, S. M. and LUMBY, R. J. In *Deformation of Ceramics*, Vol. II, Plenum Press, New York, p. 605, 1984.

COMMENTS AND DISCUSSION

Chairman: M. J. POMEROY

K. H. Jack: The fact that you get very good properties by this controlled oxidation and heat treatment to produce β' and O'. How much O' is there in that?

A. Szweda: Probably not quite 50–50 but there is an appreciable amount there.

Jack: So this supports our evidence that a composite of O'/β' is a very good ceramic because on heat treatment of the different kinds of glass we do get good oxidation resistance right up above 1350°C.

Szweda: The importance of that particular technique is to remove the YAG from the surface layer so that you cannot get liquid reacting with the silica layer on the surface. Once the liquid is formed, there is rapid diffusion of Y and Al outwards.

Jack: And the same thing presumably would occur if you had yttrium silicate, but not to the same extent?

Szweda: I am not certain. If you only had yttrium disilicate there, this would tend to react with the silica on the surface to produce the eutectic liquid at a lower temperature. It is unfortunate that the YAG and silica are on opposite sides of the eutectic liquid in the ternary system. If you have YAG and silica then the liquid will form.

Jack: Once the liquid is formed there will be rapid diffusion.

M. J. Pomeroy: I would like to comment. We have carried out oxidation studies[1] with yttria-densified sialon (β', $z = 3$) which is different from the one discussed here. At the end of an 80 h oxidation trial we examined it and found that the glass had devitrified to produce YAG. Comparing with your β' + YAG material, our oxidation kinetics were far lower than those observed with a lower z-value sialon. But the important thing was the stage of reaction sequences on the surface which went from β' to silicon oxynitride to X-phase and then through to mullite + silica. So we were essentially getting the same thing. An interesting point is that in your work, heat treatment for 1000 h in air gave you very good results and I wonder if you are crystallising out the YAG totally after such a long time?

Szweda: This is probably the case. During that oxidising treatment it is inevitable that more YAG will be produced in the bulk material.

Pomeroy: Does this mean that you would have to commission a component for 1000 h?

Szweda: With this system, if you put a crystallised bar into a creep rig and run the creep test in air, that oxidising treatment will automatically occur so the creep rate should diminish with increasing time.

R. W. Davidge: What is the YAG grain size initially and does it grow during the oxidation surface treatment?

Szweda: You would not be able to see it under a microscope. Virtually each triple junction in the material would have its own orientation. So we try not to get that kind of microstructure.

D. Broussaud: Without the heat treatment have you only partially crystallised the grain boundaries? Are they glassy pockets or YAG?

Szweda: You do tend to get some crystallisation on cooling after sintering the material. You cannot cool instantaneously so obviously you do tend to get some crystallisation of YAG on cooling.

Broussaud: But you will have glass remaining?

Szweda: Yes, after sintering, the material consists of β', O', glass and a little bit of YAG but then when you re-heat the material in two stages you recrystallise the glass completely and the creep rate diminishes.

Reference

1. POMEROY, M. J. and HAMPSHIRE, S. *Materials Chemistry and Physics,* **13** (1985) 437.

14

Structural Evolution under High Pressure and at High Temperature of a β'-Sialon Phase

G. ROULT

Centre d'Etudes Nucléaires de Grenoble, DRF-G/SPh/S-85X, 38041 Grenoble, France

AND

M. BROSSARD, J. C. LABBE AND P. GOURSAT

University of Limoges, Laboratoire de Céramiques Nouvelles, 87060 Limoges, France

ABSTRACT

The structure of a β'-sialon phase of general formula $Si_{6-z}Al_zO_zN_{8-z}$ (with $z = 2 \cdot 13$) was studied by time-of-flight neutron diffraction.

The sample contained more than one phase and at the beginning this was thought to be Si_3N_4. In fact, due to the accuracy of the time of flight technique it was found that it was another β'-sialon phase richer in nitrogen.

After a short description of the time-of-flight neutron diffraction and of the jointed profile analysis technique, the structure is presented. Its evolution under high pressure and at high temperature is given in terms of motion of the $[Si(Al), N(O)]_4$ tetrahedra. This motion is compared with the motion in other nitrides or oxynitrides such as Si_3N_4, AlN and Si_2N_2O of the SiN_4 and $SiON_3$ tetrahedra.

1. INTRODUCTION

The β'-sialon phases with the general formula $Si_{6-z}Al_zO_zN_{8-z}$ are well known as being structurally isomorphous with β-Si_3N_4. Two space groups are possible. The aim of the present work is to determine which space group is the correct one and what is the evolution of the structure under high pressure and at high temperature.

The experiments were performed on the time-of-flight powder neutron diffraction spectrometer at the reactor Melusine (Centre d'Etudes Nucleaires de Grenoble). A particular advantage of neutron diffraction for the structural study of sialon is that the difference in the scattering amplitude of nitrogen and oxygen atoms is greatly enhanced compared with X-ray diffraction.

Time of flight gives the advantages:

— of having a good resolution on a large range of d spacings,
— that the diffraction peak shape is approximately gaussian,
— of having no limitation for the number of diffraction peaks (in fact we can use up to 300 peaks),
— that it is a fixed-angle technique which is very convenient for special sample environments such as pressure cells.

The data collection is made on a time multichannel analyser so that patterns are plotted against channel number.

2. EXPERIMENTAL METHOD

2.1. Data Analysis

The neutron diffraction intensities data were analysed by the T.O.F. Profile analysis technique described by Worlton et al.[1] A function F is fitted on the observed pattern point by point, the least square refinement being monitored by the reliability factor.

$$R = \sum_i \frac{|Y_{o_i} - F_i|}{\sum_i Y_{o_i}} \quad (1)$$

where Y_{o_i} is the observed value in the channel i and F_i is the value of the fitting function for this channel.

The function F is defined by:

$$F(i) = \alpha + Y \frac{\beta}{d_i^\gamma} \exp\left(\frac{-\delta}{d_i^2}\right) + \text{Sc}_e \sum_j m \, |F_j hkl|^2 \exp\left(-4 \log 2 \frac{(d_i - d_j)^2}{\Delta d_j^2}\right)$$

+ some expression as above for each impurity (2)

d_i being the d spacing for the channel i. In this expression the first line defines the background. The second line is the contribution in channel i of each reflection peak of the sample with Sc_e being a scale factor, m the multiplicity factor, $F_j hkl$ the structure factor and d_j the d spacing of

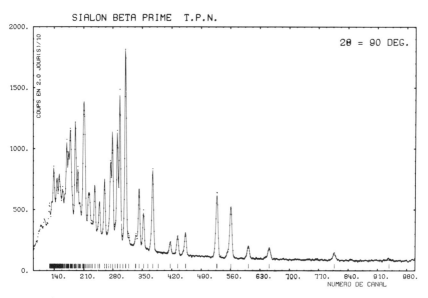

FIG. 1. Neutron diffraction pattern at R.T.P., counts per day plotted versus channel number.

the jth peak. α, β, γ, δ, Sc, the cell parameters, the atomic positions, the occupation of sites, the temperature factor B are variable parameters. An example of the pattern obtained and the fit is shown in Fig. 1.

At room temperature the sample was contained in a thin-walled vanadium cylinder. For high pressure the powder sample is contained in a Teflon tube and set in the pressure cell described by Buevoz.[2] Pressure was varied from 1 kbar to 29 kbars. For high temperature work the sample is contained in an alumina tube set in the furnace described by Alderbert et al.[3] Temperature was varied between room temperature and 1125°C.

3. RESULTS

3.1. The Structure

Being an isomorph of β-Si_3N_4, two space groups are possible $P6_{3/m}$
with the metal M(Al, Si) in

$6(h) x_1, y_1, 1/4$
and the anions X(O, N) in
$6(h) x_2, y_2, 1/4$
$2(c) 1/3, 2/3, 1/4$

$P6_3$

with the metal in
$6(c) x_1, y_1, z_1$
and the anion in
$6(c) x_2, y_2, z_2$
and
$2(b) 1/3, 2/3, z_3$

the only difference between those two groups being that z_1, z_2 and z_3 are 1/4 with $P6_{3/m}$.

The structure is based on tetrahedral atomic grouping MX_4, each tetrahedron having an horizontal edge and a vertical edge. Three of those tetrahedra are positioned by the three-fold axis around the atom in the special position 1/3, 2/3, 1/4. Three others are positioned at one half cell distance around the atom at 2/3, 1/3, 3/4 (Fig. 2). The

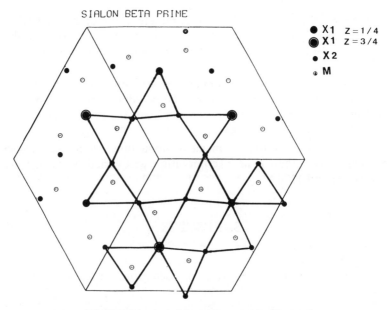

FIG. 2. Projection of the structure along the c-axis.

vertical sequence of these groups of three tetrahedra gives some pillars along the c axis, the vertical edges of the tetrahedra making a straight line parallel to the c axis. The gathering of three cells by the three-fold symmetry gives a long channel along the x axis around the origin. This structure is a very strong one. It explains its evolution with high temperature and under pressure.

3.2. The Material

The β'-SiAlON was prepared by solid state reaction between silicon nitride and α-alumina. These products, provided respectively by Alfa Ventron and Merck, have a purity level of 99·5% and a grain size lower than 20 μm. Starting powders were dry mixed and cold pressed (200 MPa) without an organic binder. Specimens of 17 mm in diameter and 5 mm thick were heated for 6 h at 1650°C in a nitrogen atmosphere. After the heat treatment, compacts were crushed and crystalline phases identified by X-ray diffraction. If alumina remained, the samples were heated again until complete reaction occurred.

The composition was determined by atomic absorption for silicon and aluminium and by neutron activation for oxygen and nitrogen. The results obtained are summarised in Table 1.

Figure 1 is the diffraction pattern at R.T.P. A shoulder appears at the left of the peaks which is due to an impurity isomorph of the sample. This impurity cannot be β-Si_3N_4 because its cell parameters do not correspond (Table 2). It must be considered as being another β'-sialon phase, the composition of the two phases being in agreement with the general composition given previously. The scale factor gives a ratio of 13·3% of the second phase to the principal phase so, if z_1 and z_2 are the quantities of aluminium in the formulas we have

$$0·117z_2 + 0·883z_1 = 2·15 \qquad (1)$$

According to Gillott et al.[4] the cell parameters vary linearly with z.

TABLE 1
Sialon Starting Composition and Analysis

Starting materials (wt %)		Composition (wt %)				Formula
Si_3N_4	Al_2O_3	Si	Al	O	N	
70	30	38·6	20·2	12·2	29·0	$Si_{3·87}Al_{2·13}O_{2·15}N_{5·85}$

TABLE 2
Lattice Parameters of Phases Observed

	$a(\text{Å})$	$c(\text{Å})$
Sample	7·669	2·963
Impurity	7·611	2·918
$\beta\text{-}Si_3N_4$	7·303	2·906

That gives some new equations with which we could determine z_1 and z_2 with

$$z_1 = 2\cdot 34$$
$$z_2 = 0\cdot 62$$

Concerning the space group the first refinements were made with group $P6_3$ leaving the z coordinates of the atoms unassigned. All of them come close to a value of 0·25 which fits the space group $P6_{3/m}$. The refinement of the diffusion factor (occupation of sites) left these unchanged meaning that we have occupation of sites at random. These results must be compared with those of Gillott et al.[4] who worked with three compositions of sialon (the quantities of aluminium, z, being 2·0, 2·9 and 4·0). They found a preferential occupation of sites $6(h)$ by nitrogen atoms and some vacancies on metallic sites. This difference can be explained by a difference in the preparation of the samples giving a composition different to the formula $Si_{6-z}Al_zO_zN_{8-z}$.

3.3. High Pressure and High Temperature

The consequence of high pressure is that the WC pistons give their own neutron diffraction pattern (Fig. 3) while at high temperature the Al_2O_3 containers, again, give their own pattern (Fig. 4).

Figures 5 and 6 give the relative variation of the cell parameter against pressure and temperature. We can note that the sialon exhibits anisotropic behaviour under pressure while it is isotropic at high temperature.

Concerning the structure itself, of interest is the shape change of the MX_4 tetrahedra and the change of their relative positions. Table 3 gives the values of the six X_1MX_2 angles which have the value 109·47° for a regular tetrahedron.

Evolution at High Pressure and High Temperature of a β'-Sialon Phase 197

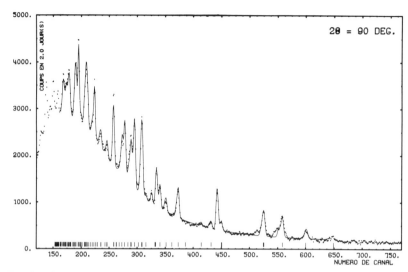

FIG. 3. High-pressure neutron diffraction pattern, counts per day plotted versus channel number (β' sialon pressure = 29 kbars).

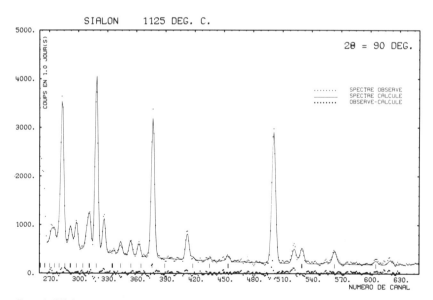

FIG. 4. High-temperature neutron diffraction pattern, counts per day plotted versus channel number (+ + + +, observed spectrum; ———, calculated spectrum; ■ ■ ■ ■, observed − calculated).

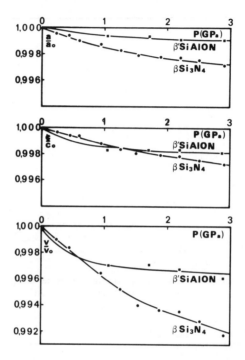

FIG. 5. Relative variation of cell parameters and volume at high pressure.

We can see that at R.T.P. the tetrahedron has 4 angles very close to the ideal for a regular one. At high temperature those angles are spread over 10°, each of which are now farther from the ideal value. At high pressure the values are spread over 10° but two angles are close to the ideal one. High temperature contributes to making the tetrahedra less symmetrical.

Table 4 gives the relative variations of the tetrahedron volume and of the cell volume for β'-sialon, α and β-Si_3N_4 and for Si_2N_2O. Space groups of each of them are also given.

We can note that as long as we have a high symmetry (a very strong structure such as with the space group $P6_{3/m}$) the relative variations are the same for the cell and for the tetrahedra (even if we have some oxygen bonds). The tetrahedra are perfectly aligned by their vertical edges and gather around the atoms in 2(c) positions. The alignment cannot be changed.

But as soon as the structure is not so symmetrical the angle between

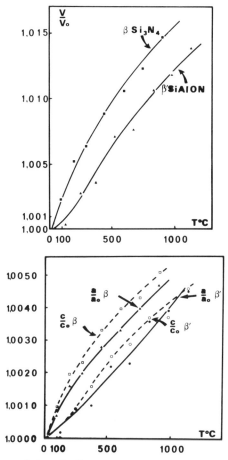

FIG. 6. Relative variation of cell parameters and volume at high temperature.

TABLE 3
Tetrahedral Angles (°) in Sialon

	25 kbars	R.T.P.	1125°C
X_1MX_2	100·5	104·7	105·5
X_1MX_3	109·2	109·5	115·7
X_1MX_4	109·2	109·5	115·7
X_2MX_3	107·7	108·6	102·4
X_2MX_4	107·7	108·6	102·4
X_3MX_4	120·5	115·4	112·9

TABLE 4
Relative Variations of Cell and Tetrahedral Volumes

	Cell $\Delta v/v_0$	Tetrahedron $\Delta v/v_0$		Tetrahedron	Space group
			H.P.		
Sialon	0·997	0·996		MX_4	$P6_{3/m}$
			H.T.		
Sialon	1·014	1·013		MX_4	$P6_{3/m}$
β-Si_3N_4	1·015	1·015		SiN_4	$P6_{3/m}$ (5)
α-Si_3N_4	1·015	0·996	1·05	SiN_4	$P3_1C$ (5)
Si_2N_2O	1·012	1·007		$SiON_3$	$CmC2_1$ (6)

M = (Si, Al), X = (N, O)

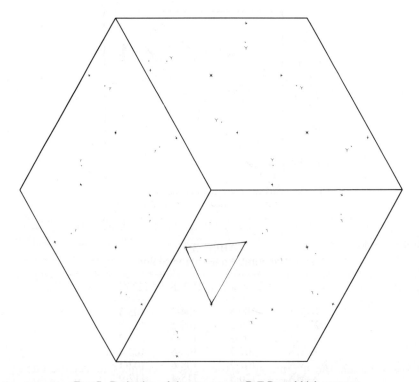

FIG. 7. Projection of the structure at R.T.P. and high pressure.

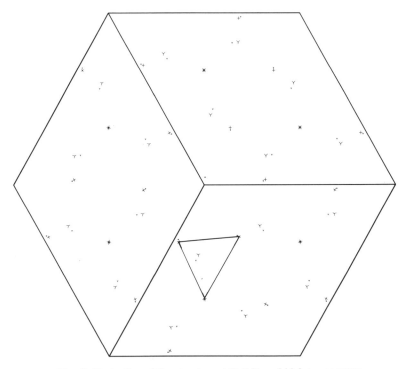

FIG. 8. Projection of the structure at R.T.P. and high temperature.

tetrahedra can change. This is the case with α-Si_3N_4 where the tetrahedra are not aligned and also with Si_2N_2O where we have a mirror plane through the oxygen atom. $SiON_3$ tetrahedra are joined by the oxygen atom and the angle between them can change.

Figures 7 and 8 are respectively the projection of the R.T.P. structure (large symbols) and the projection of the high-pressure and high-temperature structures (small symbols). We can see that in both cases anions in $6(h)$ do not move (or very slightly) while the metal atoms move considerably, due to the fact that the tetrahedra become more or less regular.

6. CONCLUSIONS

Depending on the symmetry of the crystal, changes in cell volume under pressure or at high temperature are due essentially to the modification of bond lengths as to some evolution of the structure itself. It would be interesting to study this further with some other space groups or with some sialons richer in oxygen to see if the modification is connected with the type of bond.

REFERENCES

1. WORLTON, T. G., JORGENSEN, J. D., BEYERLEIN, R. A. and DECKER, D. L. *Nuclear Instruments and Methods*, **137** (1976) 331.
2. BUEVOZ, J. L. Thesis, Grenoble, 1975.
3. ALDERBERT, P., BADIE, J. M., TRAVERSE, J. P., BUEVOS, J. L. and ROULT, G. *Revue Internationale des Hautes Temperature et Refractaires*, **12** (1975) 307.
4. GILLOTT, L., COKLAM, N. and BACON, G. E. *Journal of Materials Science*, **16** (1981) 2263.
5. BILLY, M., LABBE, J. C., SELVARAJ, A. and ROULT, G. *Materials Research Bulletin*, **18** (1983) 921.
6. BILLY, M., LABBE, J. C., SELVARAJ, A. and ROULT, G. *Materials Research Bulletin*, **15** (1980) 1207.

15

Multianion Glasses

W. K. TREDWAY† AND S. H. RISBUD‡

Department of Ceramic Engineering and Materials Research Laboratory, University of Illinois at Urbana-Champaign, Urbana, Illinois 61801, USA

ABSTRACT

The preparation, characterization, and properties of a variety of metal–SiON and metal–SiAlON glasses has been the subject of worldwide current interest in the field of non-oxide ceramic materials. The nitrogen based glass systems are, perhaps, only one set of a broader class of anion substituted amorphous solids. The present paper summarizes results of research on the synthesis and characterization of several multianion glasses being investigated in our laboratory. Examples of the preparation of melt solidified oxynitride, oxyfluoronitride, and oxycarbide glasses are presented. Data on the thermal stability, crystallization behavior, and microscopic characterization of selected glass systems and the effects of anion substitution on glass properties are discussed.

1. INTRODUCTION

Non-oxide technical ceramics for emerging engineering applications have been the subject of widespread research in the last two decades. Silicon nitride, silicon carbide, and the sialons are among the most

---

† Present address: United Technologies Research Center, MS 24, Silver Lane, East Hartford, Connecticut 06108, USA.
‡ Present address: Department of Materials Science and Engineering, University of Arizona, Tucson, Arizona 85721, USA.

popular of these new materials. Parallel research on thin amorphous phases in densified silicon nitride and formation of bulk oxynitride glasses has also led to the development of novel glasses, glass–ceramics, and composites based on non-oxide systems.[1-8]

The oxynitride glasses are, perhaps, only one example of new materials obtained via single or multiple anion exchanges. Thus, multianion inorganic glasses can be viewed as structures lacking long range order and derived by the replacement of oxygen anions in the glass network by nitrogen, fluorine or carbon. The additional linkages possible due to the new anion introduced leads to cross-linking that is thought to result in improved glass properties. In the present work we summarize the preparation, characterization, and properties of multianion glasses in several non-oxide systems including Ba–Si–Al–O–N, metal–Al–O–N, Mg–Si–Al–O–C, and Mg–Si–Al–O–N–C.

2. EXPERIMENTAL PROCEDURE

2.1. Glass Preparation

Multianion glasses in the systems Ba–Si–Al–O–N, Mg–Si–Al–O–N, Mg–Si–Al–O–C, metal–Al–O–N, and Mg–Si–Al–O–N–C were prepared by melting mechanically blended mixtures of oxides (e.g. SiO_2, Al_2O_3, BaO, MgO) and Si_3N_4, AlN, or SiC in molybdenum crucibles under an atmosphere of either nitrogen or argon. The furnace used was fitted with a graphite heating element. Melting temperatures ranged from 1650°C to 1900°C, depending on the particular composition, while the soak times varied from 6 to 8 h. The glasses were allowed to cool at the natural cooling rate of the furnace (~40–50°C per min through the transformation range).

2.2. Glass Characterization

All glasses were subjected to X-ray diffraction to confirm their amorphous nature. Chemical analyses of the constituent elements in the glasses were performed by a commercial laboratory. Thermal analysis of the glasses was carried out using a DuPont 1090 system. Microstructural characterization of the glasses both before and after secondary heat treatment was performed using electron microscopy techniques on polished sections that were etched in dilute HF solution.

2.3. Heat Treatments

Glass samples were given heat treatments in a dry nitrogen atmosphere at temperatures ranging from 1000°C to 1150°C. The samples were placed in Al_2O_3 boats and positioned to within 5 mm of the thermocouple. A dry atmosphere was used since we have shown[1] that water vapor in the atmosphere can result in decomposition of oxynitride glasses when they are heated to temperatures above T_g. This decomposition is presumably due to the evolution of some gaseous species, possibly NH_3, that forms as the result of a reaction between nitrogen in the glass and H_2O vapor in the atmosphere.

3. RESULTS AND DISCUSSION

3.1. Si-Oxynitride Glasses

The oxynitride glasses formed in both systems were gray to brown in color and translucent to opaque in transparency.[1,2] All glasses were optically heterogeneous to some degree, with the Mg-Si-Al-O-N glasses being of better quality than those in the Ba-Si-Al-O-N system. Metallic spheres ~1 mm in diameter were normally found on the melt surface. EDAX analysis in the SEM showed these spheres to consist mainly of Fe and Si.[2] The high temperatures and long melting times required to produce homogeneous, amorphous materials generally resulted in a 20–60% loss of nitrogen during melting.

In an attempt to produce oxynitride glasses of better quality, we have recently[3] used chemically pre-mixed oxide batches and high purity Si_3N_4 to produce oxynitride glasses in the Ba-Si-Al-O-N system. The lower melting temperatures and shorter soak times afforded by the highly reactive gel-derived powders resulted in glasses which were clear and homogeneous, free of metallic inclusions, and which retained considerably more nitrogen (<25% loss) than conventional Ba-Si-Al-O-N glasses.

Glass transition temperatures (T_g) obtained from differential thermal analysis (DTA) traces indicated that addition of nitrogen increased the thermal stability of glasses in both systems[1,2] (Fig. 1). Similar effects have been reported[5,6] in several oxynitride glass systems for other properties such as elastic modulus, indentation hardness, and density, which all increase with increasing nitrogen content. This behavior is consistent with the present thinking about oxynitride glass structure in which trivalent nitrogen substitutes for divalent oxygen in the glass network to produce a more highly polymerized glass structure.

Incorporation of nitrogen in the network of an oxide glass produced different effects in the two systems with regard to their crystallization

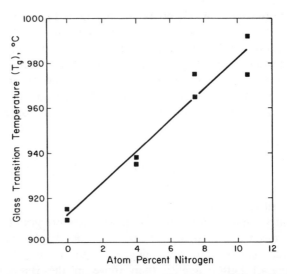

FIG. 1. T_g versus % N data showing the effect of N on glass stability in Ba–Si–Al–O–N glasses.

behavior. In the Mg–Si–Al–O–N system, nitriding resulted in glasses which were more resistant to devitrification, with TiO_2 additions being required in order for nucleation and crystal growth to occur.[1] Glasses in the $BaO-Al_2O_3-SiO_2$ system, however, did not exhibit any appreciable crystallization until Si_3N_4 was added. These Ba–Si–Al–O–N glasses were then self-nucleating and crystallized quite readily when subjected to heat treatment.

In the case of the Mg–Si–Al–O–N glasses, nitrogen is thought to increase the viscosity to a point where the amorphous structure is more stable and rearrangement to a crystalline structure is resisted. The ability of nitrogen in improving the glass-forming tendency in other Mg-containing systems has been reported.[5,6] In the Ba–Si–Al–O–N glasses, it is thought that nitrogen probably induces a phase separation process which in turn enables nucleation and crystal growth to take place.[4] The substantially different behavior exhibited by nitrogen in the two systems is not yet completely understood.

A representative microstructure of a Ba–Si–Al–O–N glass after heat treating at 1020°C for 2 h is shown in Fig. 2. The crystalline phase was identified by X-ray diffraction as a solid solution of cymrite ($BaAlSi_3O_8(OH)$) and hexagonal celsian ($BaAl_2Si_2O_8$). The lath-like oxide crystals are embedded in a glassy matrix which was found by

FIG. 2. SEM micrograph of cymrite–hexacelsian solid solution crystals obtained upon devitrification of Ba–Si–Al–O–N glass.

scanning Auger spectroscopy to be enriched in nitrogen.[4] This microstructure is desirable since the nitrogen-rich glassy matrix will improve the overall refractoriness of the glass–ceramic material. Prospects for the use of this glass–ceramic as a matrix for novel ceramic–ceramic composites based on an oxynitride–SiC fiber system have been recently explored.[7,8]

3.2. Aluminate Oxynitride Glasses

The difficulties with Si inclusions in Si-oxynitride glass systems lead us to experiment with the preparation of calcium–aluminate oxynitride glasses. Glass formation, albeit reluctant, was achieved[9] by quenching 20 g melts batched from BaO, CaO, MgO, Al_2O_3, and AlN compositions. Figure 3 shows some of the bulk samples obtained. SEM analysis revealed that sample 3 (see Fig. 3), although X-ray amorphous, contained some dendritic crystals indicating that melt devitrification had occurred to a slight extent. The effect of N incorporation on the properties of these glasses is summarized in Table 1.

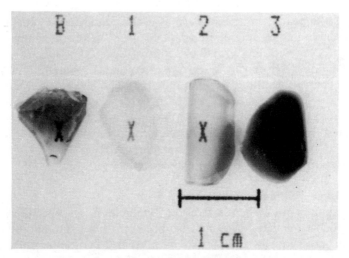

FIG. 3. Photograph of bulk samples in aluminate oxynitride systems.

Low silica aluminate oxynitride glasses have also been synthesized in our laboratory[10] from batches similar to the above calcium aluminate melts but modified with low amounts (~3 wt %) of silica. Systematic increases in glass properties were observed in these systems as well. Nitrogen dissolution is apparently related to the presence of some amount of SiO_2 in the batch.

3.3. Oxycarbide Glasses

Multianion glasses in the Mg–Si–Al–O–C system derived from C substitution of O were obtained in our recent work.[11] 15 g batches of MgO, Al_2O_3, SiO_2, and SiC were melted at 1800°C (2 h) and solidified into X-ray amorphous materials. Dissolution of C into the glass melts

TABLE 1
Effect of Nitrogen on Glass Transition Temperature and Thermal Expansion Coefficient of Aluminate Glasses

Glass	Atom % N	T_g (°C)	Thermal expansion coefficient, 500–650°C ($\mu m\, m^{-1}\, °C^{-1}$)
Base	0	723	8·37
1	0·3	752	8·06
2	0·9	780	7·79
3[a]	1·8	777	6·70

[a] Partially devitrified.

does not appear to be as extensive as in the oxynitride systems but, as shown in Table 2, microhardness and glass transition temperatures improve even with these small amounts of C in the glasses.

In analogy with oxynitride glass structures, it seems reasonable to expect that C substitutes for O in the glass. The higher valence of C(4) over N(3), however, may permit more extensive crosslinking and thus lend greater coherency to the network structure. The structural role of C needs to be explored by spectroscopic techniques and we are currently pursuing[12] magic angle spinning NMR methods to probe the structure of several multianion glass systems.

3.4. Other Multianion Glasses

A number of other interesting multianion glasses have been prepared in our laboratory and we present a brief summary of these efforts here. Vaughn and Risbud[13] synthesized fluoronitride glasses in Zr–metal–F–N systems by nitriding a fluoride glass made from Zr, Y, Al, and Ba fluorides. Chemical analyses showed that ~0·5–1 wt % N could be retained in these glasses but processing refinements were needed to reduce the oxygen levels in what is mainly a F glass. The improved thermal stability of F glasses via N incorporation may be worthy of pursuit in the context of fiber optic applications of these halide glasses; N does not appear to significantly alter the desirable IR optical transmission of fluoronitride glasses.

The effect of N on oxyfluoronitride glasses based on Al–Ba–P–F–O–N compositions was recently explored by Fletcher and Risbud.[14] The fluorophosphate glass systems are of interest in glass-to-metal seals, IR material technology, and as laser hosts. We are addressing the role of N in improving both the viscous and chemical durability behavior of these melts and glasses.

Finally, a family of multianion glasses designated as oxycarbonitride glasses have been prepared in our laboratory recently.[15] The prelimi-

TABLE 2
Effect of Carbon Incorporation on Vickers Hardness and Glass Transition Temperatures of Mg–Si–Al–O–C Glasses

Wt % C (batched)	Vickers hardness (kg mm^{-2})	T_g (°C)
0	623	790
0·5	688	840
1·0	723	880

nary indication is that both C and N can substitute for O to form Mg–Si–Al–O–C–N glasses which can be crystallized to yield glass–ceramic materials containing crystalline phases which have been identified by X-ray diffraction as enstatite and forsterite. The potential of these glass–ceramics may be worthwhile especially since they may be of utility as matrices in high performance ceramic–ceramic composites.

ACKNOWLEDGEMENTS

We would like to acknowledge the support of the Ceramics and Electronic Materials Program, Division of Materials Research at NSF under grant DMR-84-04013. W. K. Tredway is grateful to the IBM Corporation for a Pre-Doctoral Fellowship for his graduate studies.

REFERENCES

1. TREDWAY, W. K. and RISBUD, S. H. Influence of atmospheres and TiO_2 nucleant on the crystallization of Mg–SiAlON glasses, *Journal of Materials Science Letters*, **4** (1985) 31–3.
2. TREDWAY, W. K. and RISBUD, S. H. Melt processing and properties of barium–sialon glasses, *Journal of the American Ceramic Society*, **66** (1983) 324–7.
3. TREDWAY, W. K. and RISBUD, S. H. Use of chemically pre-mixed oxide batches for oxynitride glass synthesis, *Materials Letters*, **3** (1985) 435.
4. TREDWAY, W. K. and RISBUD, S. H. Preparation and controlled crystallization of Si–Ba–Al–O–N oxynitride glasses, *Journal of Non-Crystalline Solids*, **56**(1–3) (1983) 135–40.
5. TREDWAY, W. K. and LOEHMAN, R. E. Scandium-containing oxynitride glasses, *Journal of the American Ceramic Society*, **68**(5) (1985) C-131–3.
6. LOEHMAN, R. E. Oxynitride glasses, *Journal of Non-Crystalline Solids*, **42** (1980) 433–6.
7. HERRON, M. A., RISBUD, S. H. and BRENNAN, J. J. *Ceramic Engineering & Science Proceedings*, **6** (1985) 622.
8. HERRON, M. A. and RISBUD, S. H. *American Ceramic Society Bulletin*, **65** (1986) 342.
9. BAGAASEN, L. M. and RISBUD, S. H. Silicon-free oxynitride glasses via nitridation of aluminate glassmelts, *Journal of the American Ceramic Society*, **66**(4) (1983) C-69–71.
10. DURHAM, J. A., BAGAASEN, L. M. and RISBUD, S. H. Calcium Aluminate Oxynitride Glasses with Low Silica Contents, unpublished work.
11. HOMENY, J. and RISBUD, S. H. Novel multianion Mg–Si–Al–O–C oxycarbide glasses, *Materials Letters*, **3** (1985) 432.

12. RISBUD, S. H., KIRKPATRICK, R. J. and OLDFIELD, E. University of Illinois at Urbana, unpublished continuing research.
13. VAUGHN, W. L. and RISBUD, S. H. New fluoronitride glasses in Zr–M–F–N systems, *Journal of Materials Science Letters*, **3**(2) (1984) 162.
14. FLETCHER, J. P. and RISBUD, S. H. Formation and characterization of nitrogen-containing fluorophosphate glasses, *Materials Science Forum*, **5** (1985) 167.
15. IMON, M. M. and RISBUD, S. H. *Journal of Materials Science Letters*, **5** (1986) 397.

COMMENTS AND DISCUSSION

Chairman: E. GUGEL

D. P. Thompson: What is the co-ordination of the carbon in these glasses? In the carbon-containing glasses, if you have carbon and nitrogen together, would that reduce the amount of nitrogen that you can include?
W. K. Tredway: This is very recent work which has just been completed. You are right about the reduction of the amount of nitrogen in the glass if you have the carbon present. What the carbon has done so far is enable us to crystallize the magnesium sialon glasses which we were not able to do before. We can heat treat these glasses now and this allows a very fine crystal size to precipitate out. As for the carbon co-ordination in the glass network, I feel that it is probably down to 4 in the glass structure, but that is a supposition; it has not been proved yet. We are carrying out some NMR work to determine the local structure of these glasses and we have been trying to make good quality glass which can be used with NMR to determine the co-ordination of the different atoms in the glass. So far, there are a number of peaks in the spectrum which we cannot identify so it may prove to be difficult.
S. Hampshire: Is there a possibility that in carbon-containing glasses the carbon is actually acting as a cation rather than an anion or is that not really possible?
Tredway: Yes, I guess it could be acting as a cation rather than an anion. We assumed that it would show the same behavior as oxygen or nitrogen but this is not conclusive.
A. Hendry: The observation of carbo-oxynitride glasses is extremely good news to those of us who are interested in forming sialons from oxides by carbo-thermal reduction because I have never seen any evidence that you could dissolve silicon carbide, for example, into

oxynitride glasses. But is the 2% C that you quoted the maximum solubility or as far as you have investigated?
Tredway: That is as far as we have been able to go. The cooling rate is only about $40°C\,min^{-1}$, which is very slow, and at that cooling rate that is all the carbon we have been able to retain in the glass. If we were able to quench the glass more rapidly we would probably be able to dissolve more carbon into the glass and for it to remain amorphous.
Hendry: What temperature do you need to dissolve the carbon?
Tredway: The melting temperature is 1600°C or may be lower.
K. H. Jack: We have never been able to prove that you could put carbon, nitrogen and oxygen together in either any crystalline or any amorphous phase. You can put two of these in but we have no evidence for all of them combined and I am not sure that you have any evidence either. Maybe the NMR will give you that evidence but as we have discovered, it is quite difficult to interpret NMR spectra. One thing you have not mentioned is that the devitrification of the nitrogen glasses, as in oxide glasses, will give entirely new crystalline phases. Dr Thompson found, many years ago, a nitrogen petallite[1] and then also if you have a magnesium sialon glass close to the forsterite composition, on devitrification you get an expanded β' sialon.[2,3] For the structures of phenacite and forsterite, the free energies of formation are so similar that you only have to replace a little bit of Si by Al and a little bit of oxygen by nitrogen and the stability changes.
Tredway: With reference to the NMR results[4] the chemical shifts in the spectra provide strong evidence that most, if not all, the N is linked to Si atoms in the glasses. Also, nitride and oxynitride environments can be readily distinguished in the MASNMR-spectra.
Jack: There may be a large number of applications for these glasses but improvement in properties may be small compared with the disadvantages of working with them.

References
1. THOMPSON, D. P. University of Newcastle upon Tyne, unpublished work.
2. DREW, R. A. L., HAMPSHIRE, S. and JACK, K. H. In *Special Ceramics,* Vol. 7, Taylor, D. E. and Popper, P. (Eds), *Proceedings of the British Ceramic Society,* **31** (1981) 119.
3. WILD, S., LENG-WARD, G. and LEWIS, M. H. *Journal of Materials Science Letters,* **3** (1984) 83.
4. TURNER, G. L., KIRKPATRICK, R. J., RISBUD, S. H. and OLDFIELD, E. *Journal of the American Ceramic Society* (in press).

16

Non-destructive Evaluation of Ceramic Surfaces and Sub-surfaces

L. McDonnell

Tekscan Ltd, Cork RTC Campus, Bishopstown, Cork, Ireland

AND

E. M. Cashell

Department of Physics, Regional Technical College, Bishopstown, Cork, Ireland

ABSTRACT

A novel instrument—the PEAT (Photo Electro Acoustic Thermal wave) microscope has been designed and constructed to image and analyse the surfaces and sub-surfaces of solid materials. The PEAT microscope combines five analytical techniques: Scanning Electron Microscopy, Auger Electron Spectroscopy, Scanning Auger Microscopy, Thermal Wave Microscopy and Photo Acoustic Spectroscopy within an ultra-high vacuum environment.

The basic concept of the microscope is discussed and its application is described in general and with specific reference to engineering ceramics. Experiments performed using a prototype microscope on a silicon nitride sample, that contained an open crack in the surface, resolved both this feature and sub-surface lateral cracks that were not evident by surface inspection.

1. INTRODUCTION

In recent years, surface sensitive analytical techniques have found increasing application in materials science and technology.[1] Corrosion, catalysis and wear are just a few of the many surface/interface-specific materials phenomena for which bulk-sensitive techniques, for example the X-ray microprobe analyser, are either ineffective or inappropriate.

An array of surface sensitive techniques are available to chemically analyse the top atomic layers of a solid surface and provide information concerning the presence, identity, quantity and spatial distribution of elements. The most important requirement for a surface technique is the provision of an ultra-high vacuum environment in which a given surface condition can be established and maintained.

Many materials investigations need in addition to this chemical information the physical analysis of the surface and sub-surface to determine the presence, location, size and spatial distribution of, for example, cracks, voids and grain boundaries within the material. While there are a range of non-destructive evaluation (NDE) techniques available for the physical analysis of the bulk of the material, for example microfocus X-ray and a range of acoustic techniques, they have not been operated in conjunction with surface techniques.

We have designed and constructed a novel instrument—the PEAT microscope—that allows integrated studies of this nature and we believe that this instrument will prove to be extremely valuable in advanced materials development. Within the field of engineering ceramics there is considerable scope for its application both for materials characterisation and at the product stage where performance and reliability have to be optimised. In addition to the clear need to chemically characterise surfaces there is the need to locate physical defects of the order of 10 μm in size within the sub-surfaces of the material.

In this paper the basic concept of the microscope will be discussed and its application will be described with particular reference to engineering ceramics. Preliminary results obtained from ceramic surfaces will be described.

2. THE PEAT MICROSCOPE

The PEAT microscope is shown schematically in Fig. 1 and will be described in detail elsewhere.[2] Briefly, the microscope combines five analytical techniques:

Scanning Electron Microscopy (SEM)
Auger Electron Spectroscopy (AES)
Scanning Auger Microscopy (SAM)
Thermal Wave Microscopy (TWM)
Photo Acoustic Spectroscopy (PAS)

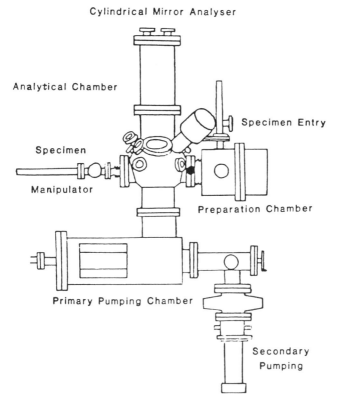

FIG. 1. Schematic of PEAT microscope.

within an ultra-high vacuum environment. While the last technique (PAS) requires the provision of a laser or some other optical source the other four techniques can all be operated with the same electron beam probe. This results in a very convenient single probe experimental arrangement for those experimental studies where the PAS facility is not required.

As shown in Fig. 1 the microscope is equipped with specimen entry locks to facilitate sample entry and removal without disturbing the ultra-high vacuum of the analytical chamber. A number of samples can be entered in a single loading operation and the maximum sample size is 100 mm. The sample preparation chamber provides facilities for heat treatment, ion beam erosion, fracture, etc., and is equipped with versatile manipulators to assist in such operations.

The general arrangement and disposition of the analytical techniques is shown in Fig. 2. This figure indicates for electron beam probing the region of the surface and sub-surface penetrated by the electron beam. This tear-drop shaped region develops as the primary beam generates secondary electrons which then generate further secondary electrons in the solid. This cascade process is characterised by increasing electron density with depth but decreasing electron energies with depth. Eventually, at about 1–2 μm depth, the electron energies are insufficient to continue this process. The relative sampling depths for AES, SEM and X-rays are indicated in this figure and clearly illustrate that the X-ray microprobe is inappropriate for surface studies.

The generation of thermal waves is shown schematically in Fig. 2. For this particular technique the electron beam is chopped with a square wave and the resultant periodic heating of the sample generates critically damped thermal waves whose range is of the order of the thermal diffusion length μ.[3] By scanning the probe, or the sample, thermal wave images of the sub-surface can be obtained. Image contrast results from reflection and scattering interactions of the thermal wave with for example mechanical and crystallographic artefacts within the imaging field. As local heat flow is drastically altered by the absence of material TWM is particularly sensitive to physical discontinuities such as empty cracks, porosity and voids.

In circumstances where neither the probe diameter nor beam spreading in the sample produce limitations, the spatial resolution of

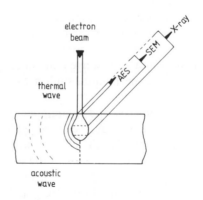

FIG. 2. Disposition of technique.

TWM is thermally limited by the thermal diffusion length:

$$\mu = \frac{\{K\}^{1/2}}{\{\pi\rho cf\}^{1/2}}$$

where μ is the thermal diffusion length, K is the thermal conductivity, ρ is the density, c is the specific heat capacity and f is the chopping frequency.

One important feature of this equation is the dependence of the thermal wavelength on the chopping frequency, f. A selection of thermal wavelengths at three different chopping frequencies are shown in Table 1 for thermal conductors and for thermal insulators. Table 1 shows the potential for the technique at high frequencies and in particular for thermal insulators.

There are three TWM detection schemes namely:

Infra-red sensor;
piezoelectric transducer (PZT); and
capacitive sensor.

The infra-red sensor is a non-contact method requiring direct viewing, through a suitable lens, of the sample region being probed by the beam. The PZT requires direct contact with the sample while the capacitive sensor has to be positioned close to the sample surface. For the latter two detection schemes the short-range thermal wave data are carried by long-range acoustic waves to the sample surfaces. It is possible to detect the acoustic carriers at either the front or rear surfaces depending on the experimental geometry required. For chopping frequencies below about 50 MHz the acoustic wavelength is much larger than the thermal wavelength.

TABLE 1
The Variation of Thermal Wavelength With Chopping Frequency for Thermal Conductors and Thermal Insulators

Modulation frequency	Thermal conductor (μm)	Thermal insulator (μm)
100 Hz	200–300	20–30
10 kHz	20–30	2–3
1 MHz	2–3	0·2–0·3

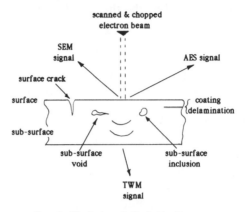

FIG. 3. Typical analytical situation.

A typical analytical situation that the PEAT microscope has been designed for is shown in Fig. 3. This figure illustrates the physical features that may be associated with a sample surface and sub-surface in addition to the chemical variations that may exist. By using the PEAT microscope in a single experimental study we can obtain:

via SEM—an image of surface topography;
via TWM—the location of sub-surface defects and surface cracks;
via AES—the chemistry of the surface; and
via AES *and argon ion etching that is targeted by* TWM—the chemistry of defects and their environment and of control regions on the same sample.

3. APPLICATIONS

The prototype PEAT microscope is currently operating with approximately 25 μm probe diameter for both the electron beam and laser beam probes. A number of tests on ceramic samples have been made in the laser mode and these will be reported here to demonstrate the viability and potential of the microscope. 1 μm diameter probes are now available.

The ability of the microscope to sense sub-surface defects is shown by the TWM line scans of Fig. 4 which were taken from an aluminium sample in which two horizontal holes were drilled under the front surface.

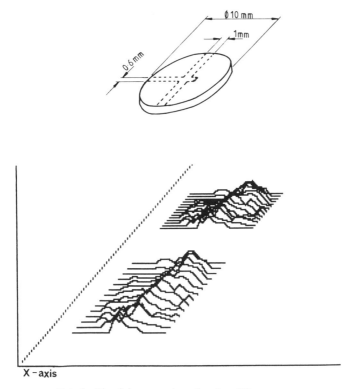

FIG. 4. Aluminium sample and series of line scans.

Further experiments were performed on a silicon nitride sample that contained an open crack in the surface. The TWM line scan of Fig. 5 clearly resolve the open crack which was estimated optically to be approximately 100 μm in width. Optical inspection of the surface of this sample did not reveal any other visible defects. The TWM line scans however do contain, in addition to the open surface crack, considerable structure which we believe to be due to variations in thermal properties in the sub-surface region.

One of the most probable causes of these variations are sub-surface cracks or voids but confirmation of this cause will require sectioning of the sample along the line scan directions and subsequent optical inspection. We have however been able to show the existence of lateral sub-surface cracks by observing the sides of the disc-shaped sample when the sample surface is illuminated by the He–Ne laser.

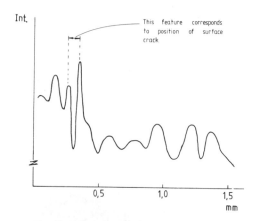

FIG. 5. Line scan of silicon nitride sample.

Considerable laser light was emitted from the sides of the sample in a way that visually highlighted crack locations and orientations.

4. SUMMARY

A novel instrument—the PEAT microscope—has been developed for the investigation of the physical and chemical characteristics of material surfaces and sub-surfaces. A critical feature of the microscope is its ability to probe in a non-destructive manner the microstructure and other physical properties of the sub-surfaces of materials. The basic concept and potential of the microscope has been demonstrated for ceramic samples containing surface and sub-surface defects. A particular advantage of the microscope is that real-world engineering samples may be handled since prior chemical or mechanical preparation of sample surfaces is not required. Higher resolution studies using all the techniques available within the microscope are now being undertaken.

ACKNOWLEDGEMENTS

The authors wish to thank Dr S. Hampshire, NIHE, Limerick for providing the silicon nitride ceramic used in this work. This work is based upon contract research supported by the National Enterprise

Agency Limited under NEA Project No. P4:057 of the NEA/NBST Joint Strategic Research Programme.

REFERENCES

1. BRIGGS, D. and SEAH, M. P. (Eds) *Practical Surface Analysis,* John Wiley, New York, 1983.
2. MCDONNELL, L. and CASHELL, E. M. *Proceedings of ECASIA International Conference,* Holland, October, 1985, to be published.
3. ROSENCRAIG, A. and WHITE, R. M. *Applied Physics Letters,* **38** (1981) 165–7.

COMMENTS AND DISCUSSION

Chairman: M. FARMER

R. W. Davidge: What area of specimen can you feasibly examine in your apparatus?

L. McDonnell: We can examine samples up to 100 mm.

A. Kennedy: In your preparation chamber, is it possible to subject your specimens to flowing gas treatment?

McDonnell: It can be done. The gate valves as designed are made for a particular differential so they could take pressures greater than one atmosphere. A special treatment chamber could be added which would be off-line from the preparation or sample chamber. Otherwise, on a treatment you find you cannot use the microscope. We designed the larger one in such a way with interchangeable modules so that you can bolt on the kind of thing you are talking about but still stay in business, so if you have to 'cook' something for three days or three weeks, you can.

17
Preparation and Characterization of Yttrium α'-Sialons

S. SLASOR AND D. P. THOMPSON

Wolfson Laboratory, Department of Metallurgy and Engineering Materials, University of Newcastle upon Tyne, NE1 7RU, UK

ABSTRACT

α'-Sialons are attractive materials for high-temperature engineering applications because they offer the potential of incorporating the densifying additive present in the starting mix into the crystal structure after densification. In this way, essentially single-phase ceramics can be produced which contain a minimum of grain-boundary glass; such materials would be expected to exhibit good mechanical properties at both high and low temperatures.

Detailed phase relationships have been determined in the α'-plane of the yttrium sialon system and the results show a very limited region over which α' occurs in equilibrium with a liquid phase. The fabrication of single-phase α'-sialons is further complicated by the small amount and the high viscosity of the liquid phase which occurs in equilibrium with α' and this makes the dual requirement of complete densification and complete reaction difficult.

These results are compared with similar studies in other α'-sialon systems and the feasibility of preparing dense single-phase α'-sialons is discussed.

1. INTRODUCTION

β'-Sialon materials are used commercially in a wide range of applications of which cutting tools for machining cast iron and nickel-based alloys are the most important.[1,2] They are formed from β-Si_3N_4 by substituting equal amounts of aluminium and oxygen for silicon and nitrogen and are represented by the formula $Si_{6-z}Al_zO_zN_{8-z}$ where $0 < z < 4 \cdot 2$. When β'-sialons are formed by reaction of Si_3N_4, AlN, Al_2O_3 or other Si–Al–O–N containing powders, it is impossible to make fully dense materials by pressureless-sintering because of the high viscosity of the sialon liquid phase. Oxide additives are therefore employed to reduce the viscosity and yttrium oxide is currently the best choice. However, yttrium cannot be accommodated into the structure of β'-sialon and therefore remains in the grain-boundary as a yttrium sialon glass. This gives the material very good mechanical properties at low temperatures but above 1000°C the glass starts to soften and the mechanical properties deteriorate rapidly. A possible way of improving the high-temperature properties is by devitrifying the glass to give refractory crystalline phases, for example, YAG,[3] but this reduces the room temperature strength and also requires very careful control of the starting composition and firing schedule.[4] It would obviously be desirable to incorporate the yttrium into the structure of the final ceramic and a way of doing this is provided by the range of α'-sialons.

The structure of α-silicon nitride is similar to that of the β, but with a c-glide plane relating the upper and lower halves of the unit cell. The β-structure contains long vacant channels parallel to the z-axis but the effect of the c-glide plane in the α-structure is to break up these channels into large holes repeated at unit cell intervals. The unit cell contains two such holes and the general formula for α'-sialons is therefore:

$$M_xSi_{12-(m+n)}Al_{m+n}O_nN_{16-n} \quad (x<2)$$

where x large metal atoms (M = Li, Ca, Y and Ln with $z > 60$) occupy the holes and valency balance is achieved by substituting m(Si–N) bonds by Al–N and n(Si–N) bonds by Al–O in α-silicon nitride. A detailed discussion of the crystal chemistry of α'-sialons has been given by Hampshire et al.[5] These materials offer possibilities for development as single phase ceramics because the metal oxide present in the starting mix promotes the formation of a liquid phase which densifies the material and then at firing temperatures can be incorporated into the α'-structure. The difficulty of achieving this in practice is twofold. Firstly, the overall composition is nitrogen-rich and the amount of oxynitride liquid available for densification is therefore small. Secondly, as α' forms during firing, the amount of liquid is reduced still further until in principle the final drop of liquid reacts with the last grain of unreacted α-Si_3N_4 to complete the transformation to α'. This picture is obviously idealistic and in practice it is to be expected that some residual porosity and some unreacted material will remain in the sample.

The purpose of the present paper is to evaluate whether essentially dense single phase α'-sialons can be produced and whether the properties of these materials offer advantages over equivalent β'-sialons.

2. RESULTS AND DISCUSSION

2.1. Phase Relationships in the α'-Yttrium Sialon Plane

α'-Sialon compositions are represented on the plane of the yttrium sialon system which has a $(Si + Al):(O + N)$ atomic ratio of $3:4$. This plane extends from the β'-sialon line on the basal Si–Al–O–N plane to intersect the YN–Si_3N_4–AlN ternary nitride plane as shown for the general case in Fig. 1. For ease of preparation it is desirable to use yttria rather than yttrium nitride as the starting source of yttrium, but this is quite restrictive on the range of α' compositions which can be prepared, especially when the oxide content of the starting nitride powders is considered. Figure 2 shows the α'-region and the Si_3N_4–Y_2O_3.9AlN join above which compositions can be prepared using yttria as a starting ingredient. This line becomes the dashed line when 3 wt% silica on silicon nitride and 4 wt% alumina on aluminium

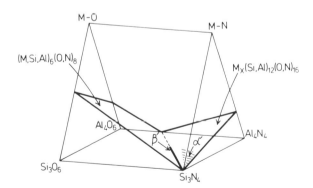

FIG. 1. Phase regions for α' and β' sialons in metal sialon systems.

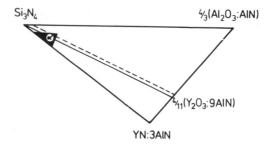

FIG. 2. The yttrium α'-sialon plane showing the limit of compositions which can be prepared using yttria as a starting material (solid line). The dashed line shows the position of this line when surface oxides of the nitride powders are taken into account.

nitride are taken into account and this only just intersects the α'-region.

Figure 3 shows phase relationships in the yttrium α'-sialon plane at 1750°C. The size of the single phase α' region is smaller than in the calcium sialon system[6] with a maximum of one yttrium atom and a minimum of 0·33 yttrium atoms being accommodated in the two interstices of the unit cell.[7] The maximum oxygen content is approximately 1·5 atoms per cell as compared with 2·5 in the calcium sialon system. The two phase $\alpha' + \beta'$ region is quite small, restricted by a limiting β' z-value of ~1·2. However, this region is in equilibrium with a range of liquid compositions which readily facilitates the preparation of dense $\alpha'-\beta'$ sialon composites. Most of the α'-sialon plane is dominated by sialon polytypoid phases which occur when the maximum solubility of aluminium in α'-sialon is exceeded. It should

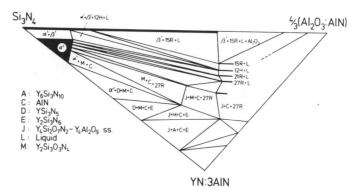

FIG. 3. Phase relationships in the yttrium α'-sialon plane at 1750°C.

be noted that whereas β' is in equilibrium with 15R and 12H on this plane and with the other sialon polytypoid phases on the sialon base plane, α' sialon cannot occur in equilibrium with 15R. At nitrogen rich compositions it occurs in equilibrium with N-melilite and the appearance of this phase in the preparation of α' is discussed later.

From Fig. 3 it is clear that single phase α'-sialon materials can be prepared and that from the point of view of using yttrium oxide as a starting ingredient, the oxygen-rich corner of the α' region is the most convenient. Densification studies carried out on this region showed that samples with densities exceeding 98% of theoretical could be achieved by sintering mixtures of Si_3N_4 (Grade LC10, Starck-Berlin, W. Germany), AlN (Grade E, Starck-Berlin, W. Germany), Al_2O_3 (Grade A17, Alcoa, USA) and Y_2O_3 (99·9%, Rare Earth Products Ltd, UK) powders for 5 h at 1800°C with an initial rise time of 2 h. Since the range of α' compositions facing β' is in equilibrium with an yttrium sialon liquid phase it is possible to choose a starting composition slightly above the plane and along the α'-liquid tie line to provide a slight excess of liquid which ensures against residual porosity. The presence of 1–2% of residual glass in the product can be tolerated and is in any case a significant improvement over the large (>10%) amounts of glass present in commercial β'-glass ceramics.

Because of the limited size of the α'-liquid phase field in the yttrium sialon system, it is impossible to prepare α'-glass materials which can then be devitrified to give α' plus a single crystalline phase. Attempts to prepare α'-YAG ceramics in this way always result in the formation of additional sialon polytypoid phases.

2.2. Reaction Sequences in the Formation of Yttrium and Neodymium α'-Sialons

Previous work[8] has shown that the reaction sequence to produce α' generally passes through an intermediate stage where a nitrogen melilite phase occurs. This is especially marked when neodymium is used as the densifying additive.

In the preparation of an yttrium α'-sialon of composition $Y_{0·5}Si_9Al_3N_{14·5}O_{1·5}$, using mixed Y_2O_3, Al_2O_3, AlN and Si_3N_4 powders, a liquid phase starts to form at 1500°C. This allows α-silicon nitride to dissolve and α'-sialon to precipitate but also facilitates the formation of N-melilite ($Y_2Si_3O_3N_4$) (see Fig. 4). The effect of this is to reduce the quantity of liquid (and particularly the yttrium content of the liquid) so that its effectiveness in promoting densification is

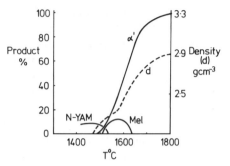

FIG. 4. Product phases and densification behaviour for yttrium α'-sialon compositions sintered for 30 min at increasing temperatures.

reduced. As the temperature increases, the proportion of N-melilite passes through a maximum at 1600°C and then starts to redissolve in the liquid with subsequent precipitation of α'. The reduced viscosity of the liquid results in a rapid increase in densification and by 1750°C α'-sialon is the only product and if left for 5 h can reach theoretical density (3·30 g cm^{-3}).

This behaviour is even more marked in the preparation of neodymium α'-sialons. Here, liquid starts to form at 1400°C and some initial densification is accompanied by the production of Nd–N–melilite. However, by 1650°C less than half the total densification has occurred and only a trace of α' is present in the product (see Fig. 5). Final conversion to α' takes place between 1650 and 1800°C by which time the density has also increased; long firing times are again needed to achieve full density. These conclusions are in good agreement with

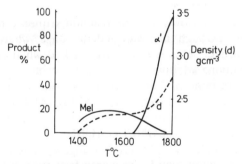

FIG. 5. Product phases and densification behaviour for neodymium α'-sialon compositions sintered for 30 min at increasing temperatures.

TABLE 1
Mechanical Properties of Some α'-Sialons

	Ca	Y	SYALON 101[a]
Density (g cm^{-3})	3·14	3·27	3·23–3·26
Modulus of rupture (MPa)	450	750	945
Hardness (Rockwell A)	92·0	93·5	91·2
K_{1C} (MN m$^{-3/2}$)	5·9	5·3	7·7

[a] SYALON is a registered trademark of Lucas Cookson Syalon Ltd.

Hampshire[8] who showed that the melilite problem could be substantially avoided by rapid heating to the firing temperature.

2.3. Mechanical Properties of α'-Sialons

Some mechanical properties have been determined for calcium and yttrium α'-sialons of composition $Ca_{0.5}Si_9Al_3N_{14.5}O_{1.5}$ and $Y_{0.33}Si_9Al_3N_{14.5}O_{1.5}$ respectively and prepared as described above (see Table 1). These samples contained traces of the 21R polytypoid phase and some intergranular glass and were at least 98% of theoretical density. Room temperature modulus of rupture, determined by four-point bend on 16 test bars of each specimen, gave results which are low compared with commercial SYALON 101[9] but are still quite good for the early stages of development. The improved result for calcium is due to the lower viscosity of the calcium sialon intergranular liquid which promotes easier densification and fewer pores in the final microstructure. Improved powder processing should improve both results. The hardness results (Rockwell A, using a 3 kg load) reflect the higher hardness of α'-sialons as compared with β' as also observed in cutting tool materials. K_{1C} values (determined by indentation using a 30 kg load) are similar for both calcium and yttrium samples and are larger than for hot-pressed silicon nitride (3–5 MN m$^{-3/2}$) but are smaller than commercial β'-sialon materials. Again these results are promising and improved powder processing techniques should yield significant improvements.

3. CONCLUSIONS

Dense yttrium α'-sialons can be prepared free from other crystalline phases and residual intergranular glass but very careful powder processing is required to avoid some residual porosity in the final product. The range of α' compositions in equilibrium with β' sialon is also in equilibrium with a small range of yttrium sialon liquid but α'-glass compositions prepared in this region cannot be devitrified to give α' plus a single refractory crystalline phase as in β'-YAG ceramics. In both the yttrium and neodymium systems, N-melilite occurs as an intermediate phase in α' formation but heating for long enough times and at high enough temperatures causes it to redissolve in the grain-boundary liquid and reprecipitate as α'.

The hardness of α'-sialons is greater than commercial β'-sialons and this alone merits further exploration of these materials. Other mechanical properties (MOR and K_{1C}) offer no particular improvements over commercial β' materials but with further development will probably yield comparable values.

ACKNOWLEDGEMENTS

One of us (S.S.) acknowledges a SERC CASE award with Lucas Cookson Syalon for the period during which the work was carried out.

REFERENCES

1. JACK, K. H. In *The Chemical Industry,* Chapter 22, Sharp, D. H. and West, T. I. (Eds), published for Society of Chemical Industry, London by Ellis Horwood Ltd., Chichester, p. 271, 1982.
2. COTHER, N. E. and HODGSON, P. *Transactions and Journal of the British Ceramic Society,* **81** (1982) 141.
3. HOHNKE, H. and TIEN, T. Y. In *Progress in Nitrogen Ceramics,* Riley, F. L. (Ed.), Martinus Nijhoff, The Hague, p. 101, 1983.
4. SPACIE, C. J., JAMEEL, N. S. and THOMPSON, D. P. In *Proceedings of the First International Symposium on Ceramic Components for Engine,* Hakone, 1983, Somiya, S., Kanai, E. and Ando, K. (Eds), KTK Scientific Publishers, Tokyo, p. 343, 1984.
5. HAMPSHIRE, S., PARK, H. K., THOMPSON, D. P. and JACK, K. H. *Nature,* **274** (1978) 880.
6. PARK, H. K. Unpublished work, Wolfson Research Group for High-Strength Materials, University of Newcastle upon Tyne, 1978.
7. JAMEEL, N. S. Ph.D. thesis, University of Newcastle upon Tyne, 1984.
8. HAMPSHIRE, S. Unpublished work, Wolfson Research Group for High-Strength Materials, University of Newcastle upon Tyne, 1979.
9. LUCAS COOKSON SYALON, *Technical Information Sheet No. 1,* 1984.

18

Characterisation and Properties of Sialon Materials

T. EKSTRÖM AND N. INGELSTRÖM

AB Sandvik Hard Materials, Box 42056, S-126 12 Stockholm, Sweden

ABSTRACT

Different types of sialon materials have been characterised by X-ray diffraction and scanning electron microscopy. It is shown that by changing the overall composition and the process parameters several microstructural features can be affected. The phase composition, the grain size and the grain boundary phase can be altered. The observed changes in the microstructure are discussed, and related to physical properties, such as hardness and indentation fracture toughness. Finally, it is illustrated how some selected sialon materials perform as metal cutting tools.

Process parameters, like heating temperature, mainly affect the grain growth. It is shown how coarser or finer grained microstructures can be obtained. The grain shape can be changed by lowering the amounts of sintering aids to reduce the amount of the high-viscous liquid at the annealing temperature. In this way crystals will not be able to grow in the typical elongated form. However, full density will not be obtained for such sintered bodies if HIP techniques are not used to help densification.

Hardness and fracture toughness change significantly as the phase composition or the microstructure alters. However, the possibility to use these easily measured parameters to predict the cutting tool behaviour in different materials is difficult. The tool wear mechanisms are very complex involving several simultaneously acting processes causing deterioration of the cutting edge.

1. INTRODUCTION

Because of the many excellent inherent properties, silicon nitride (Si_3N_4) has obtained great attention as an engineering ceramic material. Difficulties in producing dense sintered Si_3N_4-materials with properties close to the parent material itself have so far limited the use in engineering constructions. However, one commercially successful application has been the use of sialon as a cutting tool material for machining metals, in particular, when cutting cast iron or nickel based alloys. The high thermal shock resistance of silicon nitride-based materials makes them especially suitable for intermittent cutting at high speeds, depth of cuts or feed rates, all important factors in achieving a better machining economy in rough metal cutting.

The tool life of a nitride ceramic cutting tool is not only dependent upon the operating conditions, but also upon the machined material. The reason for this is that the tool wear mechanisms are very complex, involving several simultaneously acting processes. One of these, the one that proceeds most quickly, determines the life of the tool. A careful study when machining the nickel based alloy Incoloy 901 and an attempt to relate the microstructure of the used sialon materials to the observed wear and metal cutting performance has been given by Aucote and Foster.[1] In addition, another study of the wear mechanisms of sialon cutting tools has been made by Bhattacharyya et al.[2] In this paper, however, such a detailed analysis of the wear mechanisms will be postponed, although a few examples of cutting tool performances will be given.

The emphasis of this paper is to illustrate how, by changing the composition of sialon materials, several microstructural and physical properties can be affected. As will be shown, important factors like crystalline phase composition, grain size, hardness and indentation fracture toughness can be varied.

2. EXPERIMENTAL

In this study the different overall compositions have been produced by careful milling of appropriate amounts of the raw materials Si_3N_4, SiO_2, Al_2O_3, AlN and Y_2O_3. The sample compositions are given as equivalents of aluminium and oxygen in Fig. 1. The equivalents are

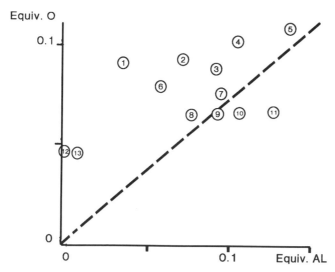

FIG. 1. Overall compositions of the samples referred to in this paper. The dashed line represents the β-sialon compositions $Si_{6-z}Al_zO_zN_{8-z}$.

defined as

$$\text{equiv. Al} = \frac{3 \times \text{at \% Al}}{3 \times \text{at \% Al} + 4 \times \text{at \% Si}}$$

$$\text{equiv. O} = \frac{2 \times \text{at \% O}}{2 \times \text{at \% O} + 3 \times \text{at \% N}}$$

which includes the oxygen from the added Y_2O_3. All samples contain 6 wt % Y_2O_3, except for samples No. 12 and 13 which contain 1 wt % and 7 wt %, respectively.

The powders were compacted to blanks and sintered at temperatures between 1650 and 1850°C in a nitrogen atmosphere. Selected samples were sintered at 1750°C and a pressure of 200 MPa (HIP) using a glass encapsulation technique.

Specimens were prepared for microstructural characterisation by grinding and polishing using standard techniques. The specimens for scanning electron microscopy (SEM) were either studied after application of a coating of carbon or after chemically etching in a mixture of molten sodium and potassium carbonate to remove the grain boundary phase. X-ray diffraction (XRD) spectra were recorded using copper radiation and a small amount of a standard substance dusted on the polished specimen surface. Lattice parameters were calculated using a least squares programme on the measured data. The z-value defining the β-sialon phase, $Si_{6-z}Al_zO_zN_{8-z}$, is obtained from the lattice parameter shift. Hardness (HV10) and indentation fracture toughness (K_{IC}) were obtained using a 98 N (10 kg) load and by using the formula by Anstis et al.[3] The thermal diffusivity was determined using a laser flash diffusivity method.[4] Finally, the metal cutting tests were performed on a high-power lathe using flood coolant. The test inserts were of the shape ISO SNGN 120416.

3. PHASE COMPOSITION

The crystalline phases present, as obtained from the XRD recordings, are summarised in Table 1. A comparison with Fig. 1 shows that in compositions with 6% Y_2O_3 as a sintering aid, the phase field of α- + β-sialon is slightly shifted compared with the β-sialon line, indicated on the basal plane with no Y_2O_3 additions. As expected, the general trend of the β-sialon composition $Si_{6-z}Al_zO_zN_{8-z}$ is that the z-value increases as the equivalents of aluminium increases. Finally, it can be noted from the very small lattice parameter shifts of the

TABLE 1
The Crystalline Phases Observed by X-ray Diffraction in Sialon Samples Sintered at 1750°C

Sample no.	Relative amounts by XRD(%)				β-sialon z-value	α-sialon unit cell	
	α-sialon	β-sialon	b-phase[a]	others		a (Å)	c (Å)
1	—	100	—	—	0·08	—	—
2	—	100	—	—	0·35	—	—
3	—	100	—	—	0·62	—	—
4	—	100	—	—	0·60	—	—
5	—	97	—	3 12H[a]	0·69	—	—
6	—	100	—	—	0·24	—	—
7	9	90	1	—	0·54	7·794	5·677
8	13	87	—	—	0·49	7·796	5·675
9	28	70	2	—	0·58	7·799	5·680
10	35	62	3	—	0·62	7·801	5·683
11	45	46	7	2 12H[a]	0·69	7·803	5·684
12	—	100	—	—	0	—	—
13	—	97	—	3 Si_2N_2O	0	—	—

[a] b-phase denotes the compound Y_2SiAlO_5N and 12H denotes the 12H sialon polytype phase.

α-sialon phase, Y_x (Si, Al)$_{12}$(N, O)$_{16}$, that the composition of this phase varies very little.

4. MICROSTRUCTURAL

4.1. General

The microstructure of each sample was characterised by two types of SEM pictures as illustrated in Fig. 2 showing a composition with an α–β sialon phase of about ratio 1:1 (sample No. 11). Figure 2a micrograph shows a polished, carbon coated section registered by the backscatter detector, which is sensitive to atom number contrasts. Thereby the yttrium distribution becomes visible and the β-sialon grains, that contain no yttrium, appear dark. The α-sialon grains are medium grey and the intergranular phase, containing most yttrium, is bright. The distribution of the α- and β-sialon can be estimated from this micrograph, and the relative amount of intergranular phase can be obtained by a point counting technique. The latter is especially important, as the intergranular phase in most sialons is a glass, not detectable by XRD. However, to determine overall grain size and shape (aspect ratio or length to diameter ratio), this is more easily carried out on a chemically etched sample as shown in Fig. 2b.

4.2. Intergranular Phase

The formation of a liquid phase that helps the densification is of utmost importance for the sialon materials. In the β-sialon compositions, samples 1–6 and 13, this liquid forms a glass upon normal cooling from the sintering temperature. However, it is possible by a post heat treatment to crystallise this glass, but this has not been attempted here. The amount of glass in samples 1–6 is considerable, falling in the range 18–25 vol. %.

The samples 12 and 13 have a fairly similar composition, but the major difference is that the added amount of Y_2O_3 is 1 and 7 wt % Y_2O_3, respectively. Both compositions need HIP to densify. The obtained microstructures are shown in Fig. 3. First of all, it is seen that additions of 7 wt % Y_2O_3 allows roughly 20 vol. % of a silicate glass to form. In this glass the silicon nitride grains are dissolved and the sialon grains are precipitated. The amount of glass is more than enough to allow a considerable grain growth with the typical elongated shape of grains that is beneficial for the fracture behaviour. However,

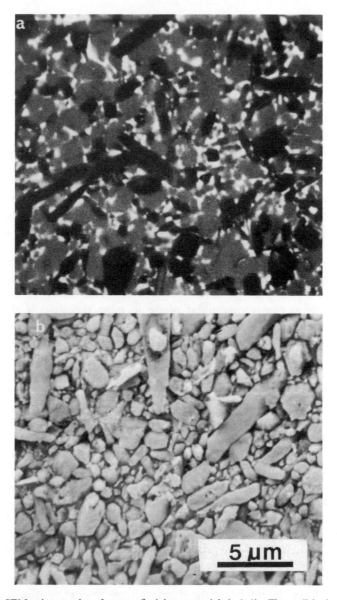

FIG. 2. SEM micrographs of an $\alpha-\beta$ sialon material (~1:1). The polished section recorded by backscattered electrons shows mainly the yttrium distribution (a). The chemically etched surface with the intergranular phase removed, reveals the grain morphology (b).

FIG. 3. Examples of the importance of enough liquid phase at the sintering of two samples fairly close in composition. In (a) and (b) a sample with 7 wt % Y_2O_3 is illustrated and in (c) a sample with only 1 wt % Y_2O_3. Considerable amounts of glass phase are seen in the structure of the former material and the grains have grown to a rodlike form.

in the sample containing only 1 wt % Y_2O_3 very little liquid phase (or glass) has been formed, clearly too little to allow a significant reconstructional grain growth to occur. The grains are small and more or less equiaxed.

Samples 7–11 containing a mixture of the α- + β-sialon phases have overall compositions away from the liquid phase region in the phase diagram. Consequently, the estimated amount of the intergranular phase using a point counting technique is smaller, being 12–19 vol. %. In this case the designation 'intergranular phase' is more correct, as a part of the glass crystallises into b-phase, Y_2SiAlO_5N.

In fact, in sample 11 the estimated amount of intergranular phase from SEM micrographs is about 12 vol. % and XRD indicates that most of this has crystallised to b-phase. Complementary studies by TEM supports the assumption that only very little glass is to be found in this material.[5]

4.3. α- and β-Sialon Phases

From inspection of SEM micrographs it appears that both sialon phases have a binomial type of grain size distribution. There are two classes of sialon grains in the structure. The smaller grains, about 1 μm large and below, and the larger grains up to 10 μm. Both small and large grains are seen for the α- and β-sialon phases and no clear difference in grain size distribution between the two phases can be detected.

The aspect ratios of the sialon grains can be estimated in a tentative way by visual inspection of a large number of SEM micrographs. In the β-sialon materials, containing only one sialon phase, this was preferably done on chemically etched samples. For this group of materials the aspect ratio of most β-sialon grains fall in the range 5–9 and no clear systematic trend was observed between different compositions (Nos. 1–6). However, a slight tendency was observed that the finer grains had a somewhat higher aspect ratio. A similar tendency was noted for the α-sialon grains in the mixed α–β sialon materials, but here all α-sialon grains had a more platelike shape compared with β-sialon. Especially the coarser α-sialon grains having aspect ratios of about 3–4.

4.4. Effect of Temperature and Time

As expected both temperature and extended heating times promote grain growth. In particular, the temperature is more effective in this respect, as well as for obtaining a high density of the sintered materials. For temperatures below about 1700°C a required density higher than 99% of theoretical density was very difficult to obtain regardless of sintering time. On the other hand, very high temperatures, such as 1850°C, promote decomposition of Si_3N_4 and loss of nitrogen and SiO from the samples. In Fig. 4, the effect of heating time and temperature on the microstructure for a β- and an $\alpha-\beta$ sialon material is illustrated. It can be seen that despite a shorter heating time, the effect of grain coarsening at high temperature proceeds quicker in the $\alpha-\beta$ sialon material. From XRD investigations of the mixed $\alpha-\beta$ sialon materials it could be seen that the z-value of the β-sialon phase, as well as the lattice parameters of the α-sialon phase, was not altered by the heating temperature. However, in the pure β-sialon material a slight shift in z-value was noted, cf. Table 2.

4.5. Raw Material

Figure 5 shows the microstructure of a material of the same composition (sample No. 11) using exactly the same process parameters, but starting with two different Si_3N_4 raw materials. These are two commercially available materials of high quality, having low impurity levels, about the same BET-values (around 10) and the same α/β Si_3N_4 ratio (95/5). Still a clear effect is noted on the observed microstructure. The phase composition by XRD is about the same.

5. PHYSICAL PROPERTIES

5.1. Hardness

The measured data are summarised in Table 3 and Fig. 6. It is seen that the presence of an increasing amount of α-sialon phase increases the hardness. Materials comprising β-sialon only fall in hardness into an extended interval. However, plotting the hardness against β-sialon composition, i.e. the z-value in $Si_{6-z}Al_zO_zN_{8-z}$ as in Fig. 7, reveals a trend with increasing hardness by higher z-value. Materials with $z = 0$ or close to 0, prepared by HIP technique, fall outside this trend. Finally, it can be said that the hardness of the very same composition varies somewhat by grain size (or sintering temperature and time). As a typical example, sample No. 9 heated for 2 h at 1750°, 1800° and 1850°C, shows a decrease of the hardness of about 80–100 units for each increase in temperature level.

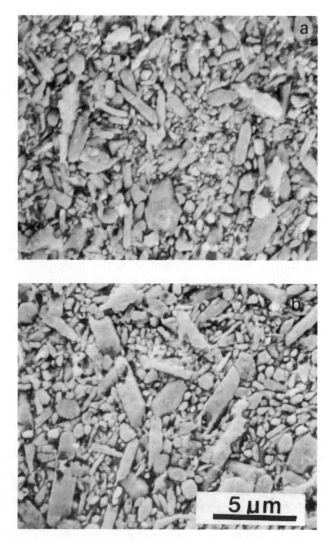

FIG. 4. The effect of heating temperature is shown for a β-sialon material (No. 5)—(a) and (b) and a mixed $\alpha-\beta$ sialon material (No. 9).

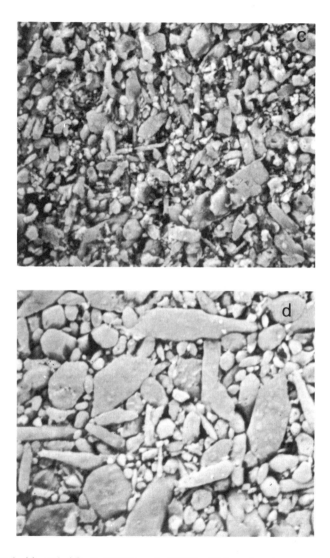

FIG. 4—contd. (c) and (d) at 1750°C and 1850°C. Higher sintering temperature promotes grain growth.

TABLE 2
z-Value of Sialons. The Obtained Shifts in the Calculated z-Values From XRD Data of Three β-Sialon Materials. These are Heated at Three Different Temperatures for 5 h and it can be Seen That the Grain Growth is Associated With a Slight Composition Shift of the β-Sialon Phase $Si_{6-z}Al_zO_zN_{8-z}$

Sample no.	z-value at heating temp.		
	1650°C	1750°C	1850°C
2	0·34	0·35	0·39
4	0·60	0·60	0·62
5	0·66	0·69	0·70

5.2. Indentation Fracture Toughness

A summary of the obtained data is found in Table 3 and Fig. 8. An increased content of the α-sialon phase in the mixed α–β sialon material decreases the fracture toughness K_{IC}. However, plotting the K_{IC}-value against hardness, as in Fig. 9, still shows that the α–β sialon materials have a more positive combination of these properties than the pure β-sialon materials.

5.3. Thermal Diffusivity

Figure 10 illustrates how the thermal diffusivity of a β-sialon with $z = 0·2$ and an α–β sialon material change with temperature. The presence of the α-sialon phase decreases the thermal diffusivity. In the same figure are also plotted the findings by others for $z = 0$ and $z = 3$, which will be discussed below.

6. METAL CUTTING PERFORMANCE

6.1. General

The deterioration of the cutting tool edge is a very complex phenomenon with several different wear mechanisms operating simultaneously. Besides high wear resistance, other features, such as high hot hardness and good toughness behaviour are very important. When machining heat resistant alloys many different wear patterns have been observed:

— flank wear,

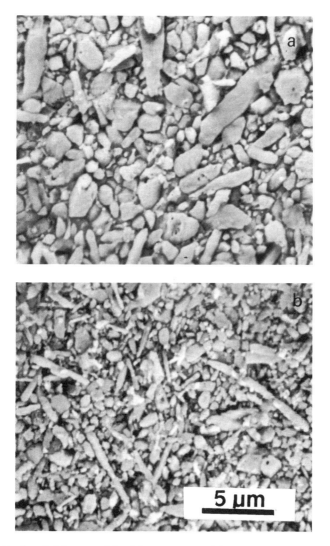

FIG. 5. Mixed α–β sialon materials of the same composition and processing, but starting from two different commercial silicon nitride raw materials.

TABLE 3
Properties of Sialons. Indentation Fracture Toughness K_{IC} and Hardness HV10, Measured on Sialon Samples Sintered at 1750°C. The Ratio $(\alpha/\alpha + \beta)$ of the Sialon Phases is Obtained from the XRD Findings

Sample no.	Phase ratio $\alpha/\alpha + \beta$	K_{IC}	HV10
1	—	5·2	1 550
2	—	5·1	1 460
3	—	4·8	1 530
4	—	4·9	1 520
5	—	4·8	1 530
6	—	5·2	1 440
7	0·09	4·8	1 560
8	0·13	5·0	1 570
9	0·29	4·8	1 630
10	0·36	4·6	1 670
11	0·50	4·5	1 730
12	—	5·7	1 550
13	—	3·6	1 790

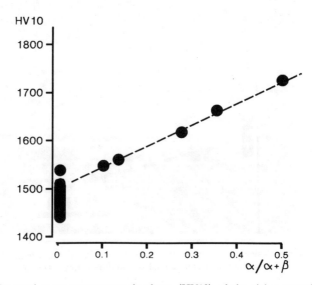

FIG. 6. Measured room temperature hardness (HV10) of the sialon materials, as a function of the relative amount of α-sialon phase.

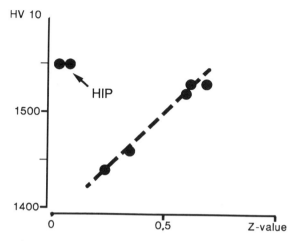

FIG. 7. Measured room temperature hardness (HV10) of pure β-sialon materials as a function of β-phase composition as represented by the z-value.

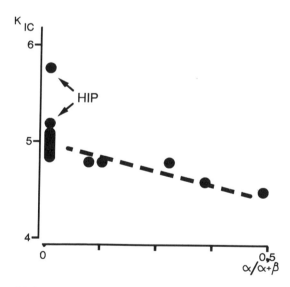

FIG. 8. Observed indentation fracture toughness K_{IC} as a function of relative amount of α-sialon phase.

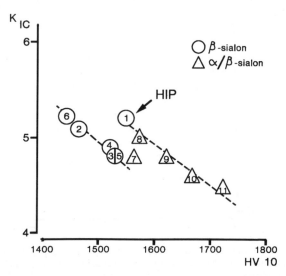

FIG. 9. Indentation fracture toughness K_{IC} against hardness HV10. The β-sialon materials and the mixed $\alpha-\beta$ sialon materials seem to fall upon two different curves.

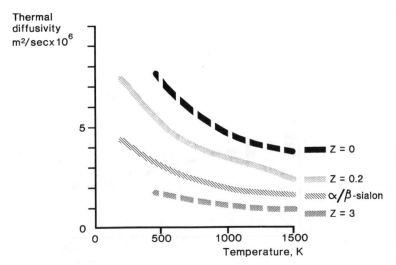

FIG. 10. The measured thermal diffusivity of a β-sialon material with $z = 0.2$ and a mixed $\alpha-\beta$ sialon material (~1:1). The results by Howlett et al.[10] for a β-phase material with $z = 0$ and $z = 3$ are also indicated and is discussed in the text.

— depth of cut notch wear,
— trailing edge notch wear,
— rake face flaking,
— overall nose fracture,
— crater wear.

The tool life is dependent on one or several of the above-mentioned wear patterns and these may be changed when the machining conditions are altered. In addition, other work-piece materials may have totally different failure mechanisms. It is therefore very important to assess which wear mechanism is determining the ultimate tool life.

The wear pictures for the cast irons are not as complex as for the heat resistant alloys. As long as the toughness of the tool material is good enough to withstand edge fracture, the tool life to a large extent is determined by flank wear.

The underlying deterioration mechanisms for all these wear patterns are not fully understood. For instance, silicon nitride and sialon materials are prone to dissolution type of wear (silicon and nitrogen are readily dissolved in iron). It seems, however, that in some cases intermediate, wear resistant layers are formed on the cutting tool delaying such wear.

6.2. Nickel Based Alloys

In rough metal cutting of nickel-based alloys sialon materials have outperformed other cutting tool materials, such as cemented carbide or alumina based ceramics. By using sialon ceramics the cutting speed can be increased several hundred per cent compared to cemented carbides. The edge security is also very good compared to alumina ceramics.

The mixed $\alpha-\beta$ sialon materials are shown to be especially good for this purpose. For instance, the flank wear decreases with increasing α-sialon content, see Fig. 11.

Different sialon compositions perform slightly different, depending upon which material is machined. However, it has been found that in order to get an extended overall tool life, regardless of failure mechanism, compositions with a certain amount of α-sialon added should be used. The amount is determined by a compromise between toughness behaviour and wear resistance.

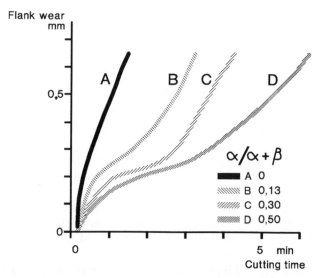

FIG. 11. Measured flank wear of different sialon materials when machining the nickel-based alloy INCOLOY 901. In this example from Aucote and Foster,[1] the feed is 2 mm per rev, the depth of cut 0·15 mm per rev and the speed 150 m min^{-1}.

6.3. Cast Iron

Ceramic cutting tool materials have now become of interest in cast iron machining as modern, stable CNC lathes allow these materials to be better exploited because of higher spindle speeds and greater flexibility. Due to their ability to withstand higher temperatures ceramics can be used at much higher cutting speeds than cemented carbides. When a good surface finish is required, Al_2O_3 based ceramics are often used. However, in roughing especially when intermittent cutting is involved, or in semi-finishing operations the sialon ceramics are clearly a good choice.

A typical flank wear behaviour when turning grey cast iron is shown in Fig. 12. Despite the higher wear rate, the sialon materials are more reliable against flaking and do not suffer from sudden edge fractures, the overall useful tool life becoming considerably prolonged.

6.4. Predicting Cutting Performance

A great number of metal cutting tests on sialon materials has shown that there is not a simple relationship between properties, such as

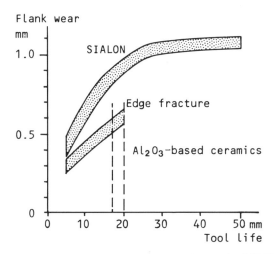

Continuous turning of cast iron (SS0125)
Feed: 0.3 mm/rev.
Depth of cut: 2.0 mm
Cutting speed: 800 m/min

FIG. 12. A typical flank wear behaviour of an α–β sialon material (~1:1) when used in cast iron machining at high cutting speeds compared with an Al_2O_3-based ceramics.

hardness, fracture toughness, grain size, etc., and the cutting performance. The obvious reason for this is the complex tool wear involving several simultaneously acting deteriorating processes.

During metal cutting the tool region close to the edge gets very hot. Important properties are therefore to retain the hardness and strength of a material to high temperatures. As shown above, the presence of the α-sialon phase in a sialon material improves the metal cutting performance when machining nickel based alloys. One reason for this might be the improvement in relative hardness at high temperature, see Fig. 13. However, such simple arguments do not hold in all instances as shown by the two examples illustrated in Figs. 14(a) and 14(b). Here a β-sialon material, with $z = 0.69$ (sample No. 5), is compared with an α–β sialon material (sample No. 11), also having the z-value 0.69 for the β-sialon phase. The β-sialon material has been sintered at two temperatures, 1750°C and 1850°C, giving respectively a fine grained and a more coarse grained structure. The hardness of the low temperature (fine grained) variant was about 100 units higher than the other.

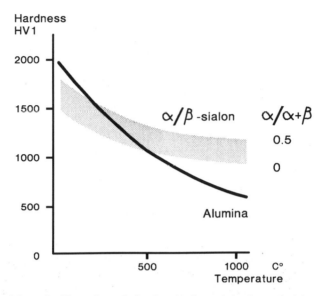

FIG. 13. Schematic illustration of the hot hardness behaviour of sialon materials compared with an alumina material containing 4% zirconia.

When machining Incoloy 901 the best flank wear behaviour was obtained by the α–β sialon material, which is the hardest of the three materials tested, see Fig. 14(a). However, the order between the fine and coarse grained β-sialon material is somewhat unexpected. Normally one expects a finer grain size and a higher hardness to improve abrasive wear behaviour, which is not seen here. In addition, when machining Inconel 718 the observed flank wear behaviour is even more complex, Fig. 14(b). A deeper understanding of the underlying processes for the wear is therefore needed in order to predict the cutting tool performance. As shown by Aucote and Foster[1] a very important factor for the flank wear is the chemical and dissolution wear resistance, which partly explains the observed wear behaviour.

7. DISCUSSION

By a number of examples it has been shown how phase composition, microstructure or physical properties can be affected. Some of these changes are certainly determined already in the very first sintering

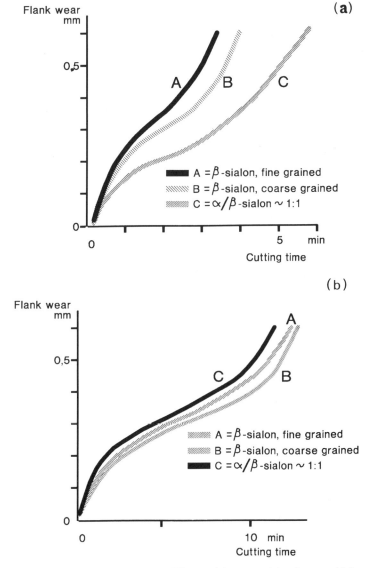

FIG. 14. The measured flank wear of different sialon materials when machining two nickel-based alloys. The feed was 2 mm per rev, the depth of cut 0·15 mm per rev and the speed 150 m min^{-1} (Ref. 11).

step. This involves the formation of a liquid phase and the precipitation of the α- or β-sialon grains that subsequently will form the major part of the microstructure. We believe that the microstructural differences that can be seen by using different raw materials mainly falls back to this first step, cf. Fig. 5. Depending on the history of preparation, different raw materials will have different amounts of surface silica or other trace elements which may affect the formation of the liquid phase and thus the microstructure. It has also previously been shown that the α- and β-sialon crystals in most cases seem to start to grow on a nuclei of α- or β-Si_3N_4 present in the raw materials.[5] Therefore the particle size distribution of the silicon nitride phases is believed to affect the final sialon grain size distribution in the sintered material.

In the subsequent sintering or densifying of the sialon material, a longer heating time and especially a higher temperature gives rise to coarsening of the sialon grains. In the β-sialon materials this is followed by a slight change of the β-phase composition or the z-value. Thus the composition of the liquid phase must also shift somewhat. A similar behaviour has not been noted for the α-β sialon materials.

α-Si_3N_4 is known to have a higher hardness than β-Si_3N_4.[6] Therefore it was not unexpected to note that the hardness of the α-β sialon materials appears to increase with the amount of α-sialon. Factors not taken into account here that might affect the hardness are the reduction in the amount of glass phase and changes in its composition, as well as the presence of some crystalline b-phase. However, we do believe that the major part of the increase in hardness is caused by the increased α-sialon phase content.

The variation of the hardness of β-sialon materials as a function of the β-sialon composition (z-value) and amount of glass phase has previously been reported by Lumby.[7] The hardness increases with z-value or reduced amount of glass. In this paper we have also noted a similar tendency of increased hardness due to higher z-values.

The change of thermal diffusivity of the sialon ceramics with the α-phase content is in agreement with similar findings by others[8,9] and the results by Howlett et al.[10] The latter authors measured the thermal diffusivity of β-phase materials with different z-values from 0 up to 3. It was shown that the diffusivity decreased by increasing z-value or the more aluminium and oxygen that was substituted into the β-phase structure. In Fig. 10 two selected data for $z = 0$ and $z = 3$ from this investigation are compared with the findings for the two materials reported here, the β-sialon material with $z = 0.2$ and the α–β sialon material. The general trend of decreasing diffusivity with increasing substitution of aluminium and oxygen is accomplished. For the α–β sialon material the z-value of the β-sialon phase is about 0.7 and it should be pointed out that the α-sialon phase of approximate composition $Y_{0.3}Si_{10.2}Al_{1.8}O_{1.2}N_{14.8}$ contains quite a lot of aluminium.

The fracture toughness is dependent on different material parameters, one of which certainly is the aspect ratio of the rodlike sialon grains. This is clearly seen by comparing the two samples Nos. 12 and 13, the microstructures being illustrated in Fig. 3. Here the sample with rodlike grains had a significantly higher toughness. Another factor that cannot be excluded is that the glass phase content in sample 12 also might be too small for a good 'bonding' of the silicon nitride grains.

The decrease in fracture toughness with increasing α-phase content can only be speculated upon, as other parameters also are changed simultaneously. One reason might be that the α-phase grains are more plate-like and thus decrease the overall aspect ratio.

ACKNOWLEDGEMENTS

AB Sandvik Hard Materials is thanked for permission to publish this article and J. Aucote and S. R. Foster for the use of their cutting tool performance data for sialon materials.

REFERENCES

1. AUCOTE, J. and FOSTER, S. R. *Materials Science and Technology* **2** (1986) 700.
2. BHATTACHARYYA, S. K., JAWAID, A., LEWIS, M. H. and WALLBANK, J. *Journal of Metallurgical Technology*, **10** (1983) 482.
3. ANSTIS, G. R., CHANTIKUL, P., LAWN, B. R. and MARSHALL, D. B. *Journal of the American Ceramic Society*, **64** (1981) 533.
4. TAYLOR, R. E. *High Temperature and High Pressure*, **11** (1979) 43.
5. CHATFIELD, C., EKSTRÖM, T. and MIKUS, M. *Journal of Materials Science*, **21** (1986) 2297.
6. GRESKOVICH, C. and GAZZA, G. E. *Journal of Materials Science Letters*, **4** (1985) 195.
7. LUMBY, R. J. *Journal of Materials Science Letters*, **2** (1983) 345.

8. XI, T. G., CHEN, O. T., NI, H. L., WU, F. Y. and YEN, T. S. *Transactions and Journal of the British Ceramic Society*, **82** (1983) 175.
9. MITOMO, M., HIROSAKI, N. and MITSUHASHI, T. *Journal of Materials Science Letters*, **3** (1984) 915.
10. HOWLETT, S. P., TAYLOR, R. and MORRELL, R. *Journal of Materials Science Letters*, **4** (1985) 227.
11. AUCOTE, J. and FOSTER, S. R. Personal communication.

19

Comparison of CBN, Sialons and Tungsten Carbides in the Machining of Surgical Implant Alloys and Cast Irons

M. Austin, J. Tooher and J. Monaghan

Department of Mechanical Engineering, Trinity College, Dublin 2, Ireland

AND

M. El-Baradie

King Saud University, Riyadh, Saudi Arabia

ABSTRACT

This paper investigates the machining performance of CBN and sialons compared to conventional carbide cutting tools, in particular, with reference to tool life, surface finish and metal removal rate. Two materials were used in the cutting tests; Vitallium, a cobalt-based alloy used in the manufacture of orthopaedic implants, and grey cast iron, a common engineering material. The machining operations used were turning of grey cast iron and Vitallium and face milling of Vitallium.

Grey cast iron can be machined relatively easily using carbide cutting tools. However, the greater tool life and superior surface finish claimed for CBN and sialons make the use of these materials attractive in mass production systems involving long batch runs. In this way tool changes can be reduced to a minimum with a consequent increase in productivity. The results of a series of tests on grey cast iron confirm the claims of increased tool life and an improved surface finish.

In the case of turning of Vitallium it has been found that CBN tools allow greater metal removal rates and give improved surface finish and longer tool life compared to all other cutting tools including sialons. Tests on the milling of Vitallium are at present underway but as yet no significant trend has been established.

In general while CBN and sialons do give improved cutting performance in comparison to standard carbides, the economics of using these tools must also be taken into consideration for a given application.

1. INTRODUCTION

Over the past twenty years major changes have taken place in the metal cutting industry due to the development of existing and new cutting tool material. This is particularly so in the case of tungsten carbides, coupled with the introduction of newer cutting materials such as ceramics and boron nitride. The developments of the carbides has resulted in the introduction of such features as multilayered coatings, various substrate matrices and a wide range of geometrically shaped inserts. These developments have led to the situation in which carbides can now be considered as 'Universal Cutting Tools' in so far as they can be used very efficiently to machine a wide range of materials under various cutting conditions. In addition, parallel improvements in machine tool design have taken place such that high cutting speeds and high metal removal rates are now possible. These design changes have led to considerable research interest into the provision of new cutting materials capable of operating at the high temperatures associated with high cutting speeds and metal removal rates. It is these factors which have led to the increased activity in the development of ceramic and cubic boron nitride (CBN) metal cutting tooling.

1.1. Ceramics

The introduction of ceramics as cutting tools dates back approximately 20 years. Over this period these materials have been simultaneously praised for the ability to operate at high machining speeds and criticised for their fragility. It is true, that due to their physical and chemical make-up ceramic cutting tools have a low tolerance for improper use.[1] The older ceramic materials are usually classified as either hot-pressed or cold-pressed.

The cold-pressed material is usually white in colour and is based on pure aluminium oxide (99% Al_2O_3). This material has a high hot hardness and resistance to edge and crater wear. Cold-pressed ceramics are recommended for cutting cast irons having a hardness less than 235 BHN and steel of hardness less than 35 RC.

The hot-pressed material is usually black in colour due to the addition of approximately 30% tungsten carbide. These materials have a greater fracture toughness compared to the cold pressed variety. Overall they are tougher and have excellent wear resistance. They are recommended for use on cast irons having a hardness above 235 BHN and for steel having a hardness within the range 35–66 RC.[1] Hot-pressed ceramics have also been used successfully to machine nickel-based alloys at cutting speeds up to five times greater than those possible with carbides.[2] In addition, because of their good shock resistance and transverse rupture strength hot-pressed ceramics are recommended for milling and rough-turning operations.

The use of silicon nitride (Si_3N_4) as a cutting tool originated from research work carried out by both Lucas Industries in Britain and the Ford Motor Co. in the USA into the use of Silicon Nitride for structural components in internal combustion engines, particularly gas turbines. Two companies, Sandvik of Sweden, with Sylon, and Kennametal of the USA, with Kyon 2000, have been licensed by Lucas Industries to supply the Lucas Material to the metal cutting tool industry. The Lucas sialon involves the use of a high α-phase silicon nitride powder, milled with aluminium nitride, aluminium and yttria, this latter material acting as a liquid phase sintering aid for full densification. The material produced by the Ford Motor Co. of the USA, code named S-8, is handled under licence by Iscar Metals as Iscanite. This material is a hot-pressed silicon nitride with additions of alumina, yttria and tungsten carbide. The high resistance of silicon nitride to both thermal and mechanical shock compared to alumina is put forward as the reason for selecting these materials for cutting cast irons and nickel-based alloys. In general they are not recommended for machining low and medium carbon steels.[3-5]

1.2. Cubic Boron Nitride, CBN

CBN is produced by subjecting hexagonal boron nitride to extremely high temperatures and pressures in the presence of a catalyst, although direct transformation is possible.[6,7]

The first commercial products appeared in the mid-1970s as brazable blanks and indexable inserts. These early products consisted of a layer of polycrystalline CBN supported by a tungsten carbide substrate. Currently solid CBN inserts are available. The number of manufacturers of CBN cutting tools is small. De Beers produce Amborite while General Electric produce Borzan (BZN). In Japan Sumitomo

Electric produce Sumiboron BN 2000. In addition some CBN materials are being produced in Russia but these are not currently available in the West.

Amborite and BZN are the most freely available CBN materials. Each of these materials has a polycrystalline structure with extensive CBN–CBN bonding. However, additional bonding material is used in each case. Amborite has a ceramic binder while BZN uses a metallic nickel–cobalt alloy. The percentage of binder used is, however, small in both cases.[3]

Measured room temperature hardness values quoted for high concentration CBN products are considerably higher than those of carbides or ceramics, in the region of 3500–4000 HV.[3,8] At high temperature (800–1000°C) the hardness of CBN is comparable with the room temperature hardness of most carbides and alumina tooling. The use of CBN tooling has shown a steady growth since its introduction in the mid-1970s. It has replaced tungsten carbide in specific instances and is now used in areas previously considered the domain of grinding processes. To date, research into the industrial use of CBN has been mainly on turning and boring[9,10] with relatively little work done on milling.[11]

It has been suggested[3] that CBN can be used effectively on a range of hard materials, i.e. hardened tool and die steels, M2, D2, etc., Ni-hards and chilled cast iron, nickel- and cobalt-based aerospace alloys having a hardness greater than 35 RC, for example, Inconel 718 and Waspalloy. In addition some hard facing materials such as stellite 6 and calmonay have been successfully machined with CBN.[12] However, not all hard materials are capable of being machined effectively by CBN, an example being Titanium 6/4. This is due to the formation of borides and nitrides during machining. Further, CBN is not recommended for machining low carbon steels, non-hardened alloy steels, stainless steels, non-ferrous metals, and in general cast iron. However, in the case of materials on which CBN can be used extremely high cutting speeds are possible. This has the effect of producing high tool/chip/workpiece temperatures within the cutting zone. It is considered[3] that these high temperatures can induce localised softening of the work material making it easier to machine.

The work discussed in this paper involves a comparison between tungsten carbide, alumina coated carbides, sialons and De Beers Amborite CBN, in terms of tool wear, tool life and the surface finish produced. Two work materials were used, Vitallium, a cobalt-based high strength alloy, and a grey cast iron. These materials were machined using turning and milling operations on the Vitallium and turning of the cast iron. A list of the tool materials used in the tests are given at the beginning of each section dealing with a particular machining operation.

2. MACHINING OF VITALLIUM

2.1. Workpiece Material

Vitallium is a cobalt–chromium–molybdenum alloy used in the manufacture of orthopaedic implants. The chemical composition of Vitallium is given in Table 1. Vitallium bars 35 mm diameter and 300 mm long were supplied in a heat treated condition for use in the turning tests. In the case of the face milling investigations rectangular bars of Vitallium 35 mm wide by 250 mm long were used.

Vitallium possesses many features which make it a desirable orthopaedic material, i.e. high yield strength, good fracture toughness and fatigue strength, non-corrosive, etc. However, Vitallium does exhibit characteristics which affect machinability. These include:

(i) high shear strength;
(ii) high work hardening characteristics;
(iii) the presence of hard abrasive carbides in the microstructure;
(iv) low thermal conductivity and specific heat capacity leading to high temperatures during machining.

TABLE 1
Chemical Composition of Vitallium

Element	Weight %
Carbon (C)	0·2–0·35
Chromium (Cr)	27·0–30·0
Molybdenum (Mo)	5·0–7·0
Silicon (Si)	1·0–Max.
Manganese (Mn)	1·0–Max.
Nickel (Ni)	2·5–Max.
Iron (Fe)	0·75–Max.
Cobalt (Co)	Balance

2.2. Cutting Tools

For both the turning and milling tests a range of cutting tool materials were used to assess their suitability for the machining of Vitallium. The tools used included a range of K10 tungsten carbides, ceramic inserts (sialons) and CBN inserts. The tools used for the turning tests on the Vitallium are outlined in Table 2.

2.3. Turning of Vitallium

During any cutting operation complex conditions exist at the workpiece/tool/chip interfaces and these give rise to a range of wear mechanisms.[13] In the tests undertaken on the Vitallium attention was primarily paid to the progression of flank wear on the tool and its influence on tool life. From these wear tests it was established that for a given cutting tool material cutting speed was the dominant factor influencing the wear rate and hence life of the tool.

2.4. Flank Wear Curves

2.4.1. Flank wear—carbide inserts

The change in flank wear with changes in cutting speed for the Kennametal K68 and Iscar IC20 carbides are shown in Figs. 1 and 2. In general the curves exhibit three distinct regions.

(a) An initial bedding in region in which the wear rate is high during the first 2 min of cutting.
(b) A region representing a gradual increase in wear rate with time. This region covers the largest proportion of the life of the cutting tool, i.e. between 2 and 15 min.
(c) A shorter period of rapid wear culminating in tool failure.

In general it can be seen that the rate of tool wear is directly related to the cutting speed. In particular, the period of time spent by the tool in the second region is reduced as the cutting speed is increased. The K68

TABLE 2

Material	Manufacturer	Trade name	ISO grade	ISO classification
Carbides	Kennametal	K68	K10	TNMA 220408
	Iscar	IC20	K10	TNMG 220408
Sialon	Kennametal	Kyon 2000		RNGN 120400
CBN	De Beers	Amborite		RNMN 120300

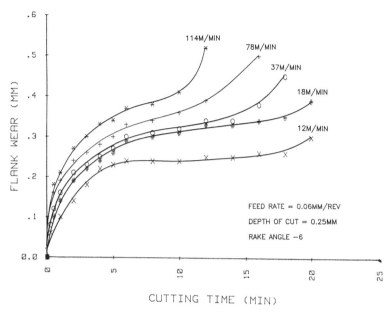

FIG. 1. Tool wear curves for turning Vitallium with tungsten carbide K68.

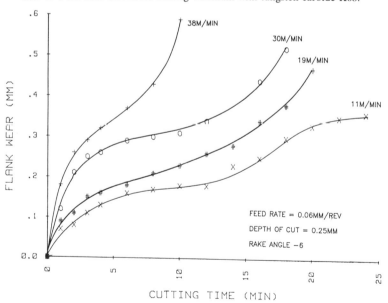

FIG. 2. Tool wear curves for turning Vitallium with tungsten carbide IC20.

carbides were found to exhibit the greatest resistance to diffusion attack. As a consequence they can be used at higher cutting speeds, i.e. up to $114\,\text{m}\,\text{min}^{-1}$ compared to $72\,\text{m}\,\text{min}^{-1}$ for IC20. The K68 carbide was also found to have the best fracture resistance. The IC20 material was slightly inferior to the K68 in terms of resistance to fracturing, grooving and cratering.

2.4.2. Flank wear—CBN

The results of flank wear against cutting time obtained when turning Vitallium using a CBN insert is shown in Fig. 3. The cutting conditions used with the CBN insert were identical to those used in the carbide tests, i.e. rake angle, feed, depth of cut. A feature of the curves shown in Fig. 3 is the absence of the three wear rate regions displayed on the curves obtained for the carbides (Figs. 1 and 2). One possible reason for this is that the cutting speeds used for the CBN tests were generally higher than those used with the carbides.

The wear mechanism on the CBN insert was also found to be influenced by the cutting speed. At a speed of $78\,\text{m}\,\text{min}^{-1}$ grooving and notching was seen on the flank of the tool. When the cutting speed

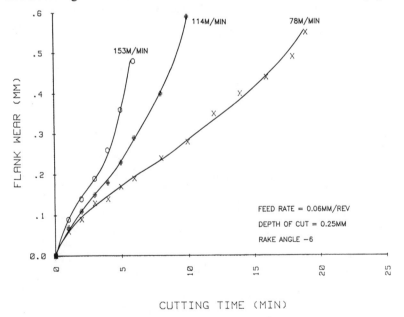

FIG. 3. Tool wear curves for turning Vitallium with CBN.

was increased to 114 m min^{-1} grooving was found to be the primary wear mechanism, whilst at the top speed of 15 m min^{-1} the flank wear showed evidence of abrasion. At each of the cutting speeds used with the CBN insert a continuous chip was formed. Evidence of high cutting temperatures was given by the production of orange coloured chips indicating a temperature in the region of 1000°C. If 0·3 mm flank wear is taken as the criterion of tool failure (BS.5423.1977), it can be seen from Figs. 1, 2 and 3 that when account is taken of the cutting speed, the CBN performs better than the carbides.

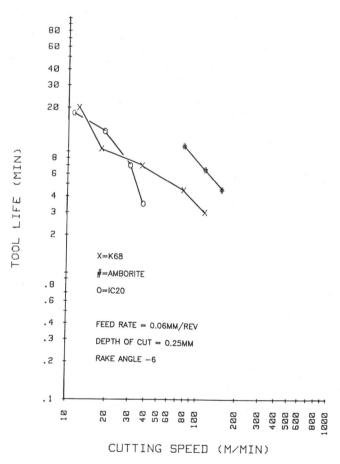

FIG. 4. V–T diagram for turning Vitallium.

2.4.3. Flank wear—silicon nitride

Attempts to produce detailed flank wear curves using silicons were unsuccessful. This was due to early failure occurring due to fracture and chipping at the cutting face. Consequently, no information was obtained on wear mechanisms. This result would suggest that tool location and rigidity are important features when attempting to machine high strength materials such as Vitallium using ceramic tools. Again, this result highlights the inferior toughness of the sialons component compared to tungsten carbides.

2.5. Taylor Tool Life Curve

A series of Taylor tool life curves obtained from the turning tests are shown in composite form in Fig. 4. These results show that at the lower cutting speeds the K68 carbide performs better than either the CBN or the sialon. However, it can be seen that at higher cutting speeds, i.e. in excess of 150 m min^{-1} the life of the carbide falls off dramatically. It would therefore appear that the CBN tools have the capability of operating at high cutting speeds and as a consequence achieve large metal removal rates. As discussed earlier the tool life results for the silicon nitride insert were poor.

3. MILLING OF VITALLIUM

In this section the results obtained when tungsten carbides, sialons and CBN were used in a face milling operation on Vitallium are discussed.

Unlike turning, milling gives rise to interrupted cutting and as a consequence the conditions at the tool/workpiece are detrimental to good tool life.[14]

During the cutting operation the cutting edge is subject to a continuous cycle of mechanical and thermal stress leading to chipping and cracking of the rake and clearance face of the tool. Cracks due to mechanical stress under certain adverse conditions lead to gross chipping and accelerated tool wear.

Tool life is also greatly influenced by the entry and exit conditions experienced by the cutting edge during the cutting operation. These are particularly important when machining high strength materials where the stresses can be 3–4 times greater than those for normal steels.

Two tungsten carbide inserts, ISO grade K10 were used. The

ISCAR IC20 and Krupp Widia G10E inserts were selected on the basis of their previously determined enhanced performance when face milling Vitallium in terms of tool life and surface finish respectively. Both the tungsten carbide and Iscanite inserts were of standard milling-insert geometry. The CBN and Syalon inserts were round on the basis of manufacturers' recommendations. A large negative rake was used to direct the cutting forces through the thicker, stronger cross-section of the insert in the hope of preventing chipping of the cutting edge.

3.1. Testing Equipment and Procedures

(i) Milling Machine—Correa Fiue.
(ii) Iscar Carbide Tool Holder (IEM 125 R8) with axial rake +6° and two triangular milling insert capacity, distance between cutting edges 32 mm.
(iii) Carbide Tool Holder with axial rake −20° and two round milling insert capacity, distance between cutting edges 45 mm.
(iv) Coolant (Hocut 711).
(v) Inserts used for purposes of tests (Table 3).

3.1.1. Tool life test procedure

Average flank wear as a function of cutting time was used as the criterion for tool life measurements. During tests the inserts were removed from the tool holder and their flank wearland measured using an inverted microscope, magnification 10, at suitable time intervals.

Maximum tool life was taken to be the time required to reach a wearland of 0·3 mm.

3.1.2. Tool life comparison

In order to compare tool life a feed of 0·1 mm per tooth and depth of cut of 0·25 mm were chosen. Inserts were then tested at a number of different speeds.

TABLE 3

Tool material	Manufacturer	Trade name	ISO grade	ISO identification
Tungsten carbide	ISCAR	IC20	K10	TPKN 1603 PPR
Iscanite	ISCAR	Iscanite		TPKN 1603 PPR
CBN	DeBeers	Amborite		RNMN 1203 DOT
Syalon	Kennametal	Kyon 2000		RNGN 1204 OOE
Tungsten carbide	Krupp Widia	G10E	K10	TPKN 1603 PDR

Using the information thus obtained and a wearland of 0·3 mm as a criterion for maximum tool life a Taylor Plot was constructed.

3.2. Tool Wear Curves

It was found that the tool wear of the CBN, Iscanite and Kyon 2000 was by continuous mechanical chipping of the cutting edge. The inherent susceptibility of these materials to such a wear mechanism would seem to render them unsuitable for face milling Vitallium.

Tool wear of the carbide insert was predominantly by abrasion. Inserts in the K10 range have a high toughness level which combined with a reasonable degree of hardness leads to a greater resistance to chipping.

The curves of flank wear against time for various cutting speeds are shown in Figs. 5–8. The two silicon nitride cutting tools, Kyon 2000 and Iscanite, perform in a similar manner. The results for the CBN tests do not indicate any marked superiority over the silicon nitride tools at low to medium cutting speeds. As in the case of the turning

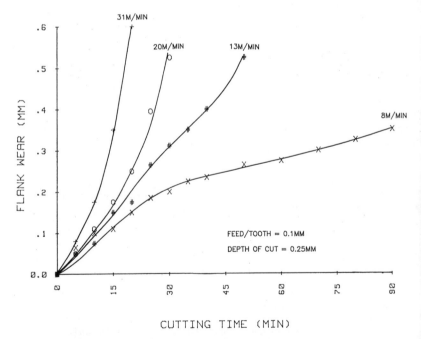

FIG. 5. Tool wear curves for milling Vitallium with tungsten carbide IC20.

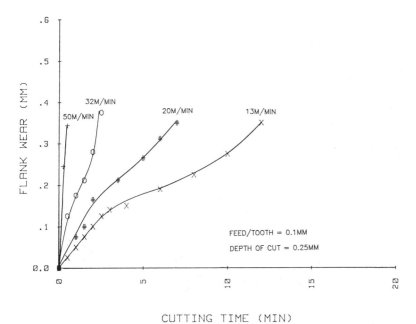

FIG. 6. Tool wear curves for milling Vitallium with Iscanite.

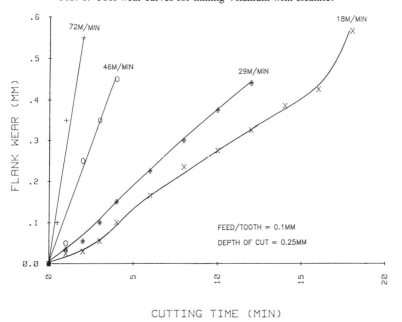

FIG. 7. Tool wear curves for milling Vitallium with sialon KY2000.

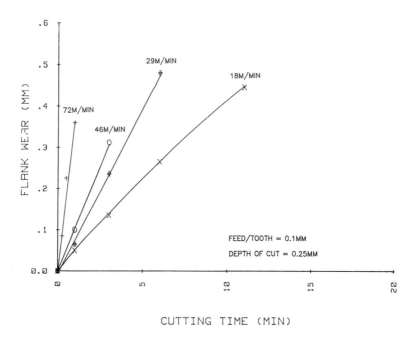

FIG. 8. Tool wear curves for milling Vitallium with CBN Amborite.

tests the K10 carbide, in this case Iscar IC20, out performs both the CBN and sialons.

3.3. Taylor Tool Life Curves

The results outlined in the discussion on tool wear are further emphasised in the Taylor tool life curves, Fig. 9. These curves show that sialons are unsuitable for milling Vitallium at normal cutting speeds. In this respect the IC20 carbide would appear to be better and cheaper than either the CBN or the sialons.

4. TURNING OF GREY CAST IRON

4.1. Work Material

Grey cast iron grade 180 was supplied as cylindrical bars of 150 mm diameter and 350 mm long. The composition and hardness of the material used is shown in Table 4.

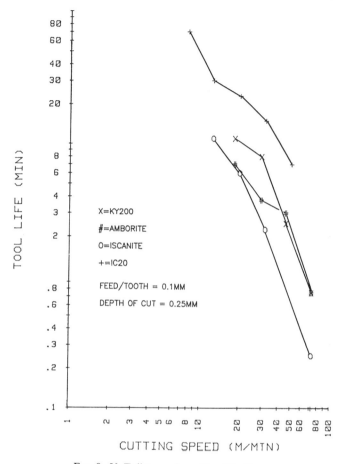

FIG. 9. V-T diagram for milling Vitallium.

4.2. Tool Materials

For these turning tests the cutting tools used included a straight carbide, an alumina coated carbide, CBN and sialon tools. The tool geometry was as follows for each cutting tool used:

Nose radius = 0·8 mm
Back rake = −6°
Side rake = −6°
Approach angle = 90°

TABLE 4
Composition and Hardness of Grey Cast Iron

Element	Weight %
Carbon	3·41–3·42
Silicon	2·23–2·33
Manganese	0·45–0·50
Sulphur	0·09–0·10
Phosphorus	0·40–0·47
Carbon equivalent	4·30–4·34

Hardness analysis results: 180 ± 28 VHN

The designation of the tools used was as given in Table 5. Because of the diameter of workpiece materials cutting speeds in the range 100–534 m min^{-1} were possible.

4.3. Flank Wear Tests

The flank wear was measured at intervals of 5 min depending on the wear rate. This procedure was repeated till the tool flank wear was 0·3 mm or greater. In the case of the alumina coated carbide the sialon and the CBN 0·3 mm flank wear was not attained even after extensive machining at high speeds. Under these circumstances a wear criterion of 0·1 mm was used when constructing the Taylor Tool Life curve.

The curves for flank wear against cutting time at various cutting speeds are shown in Figs. 10–13. A feature of the curves for the coated carbide, the Iscanite and the CBN is that a wear criteria of 0·3 mm on the flank was not attained at the lower speeds. Indeed in the case of the CBN tool the flank wear was in the region of 0·1 mm after 20 min at a speed of 534 m min^{-1}. The general performance of the various inserts can be specified as follows:

(a) CBN 'Amborite' is by far the best tool to cut cast iron. At

TABLE 5

Tool material	Trade name	Manufacturer	ISO classification
Tungsten carbide	G3 315	Sandvik	TNMA 22 0408
Alumina coated carbide	TX-10	Seco	TNMA 22 0408
Sialon	Iscanite	Iscar	TNGN 22 0408
CBN	Amborite	De Beers	TNHN 16 0408

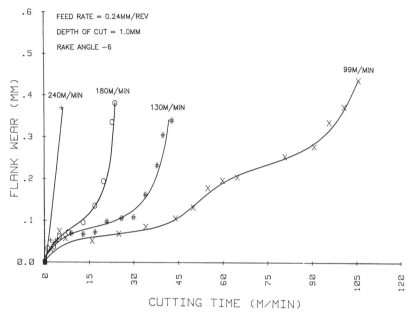

FIG. 10. Tool wear curves for turning grey iron with tungsten carbide GC315.

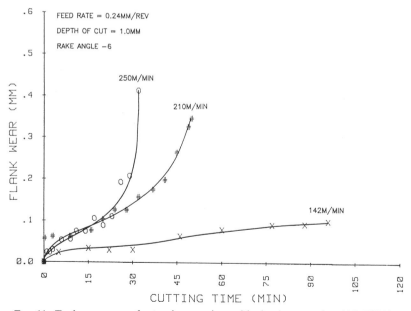

FIG. 11. Tool wear curves for turning grey iron with alumina coated carbide TX-10.

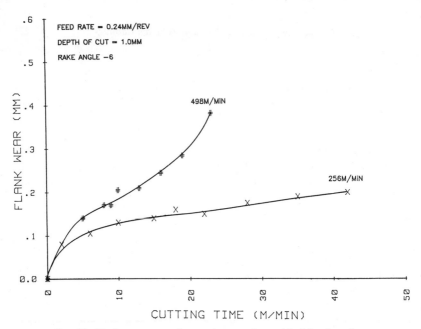

FIG. 12. Tool wear curves for turning grey iron with sialon Iscanite.

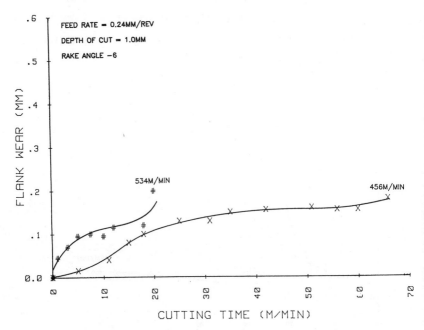

FIG. 13. Tool wear curves for turning grey iron with CBN Amborite.

speeds between 400 and 600 m min^{-1} it has at least double the tool life of Iscanite or the triple coated carbide TX-10.

(b) The triple coated carbide TX-10 has three times the life of Iscanite at speeds between 200 and 300 m min^{-1}.

(c) The tungsten carbide insert was the worst insert. At high speeds (250 m min^{-1}) it only lasted 2 min while the TX-10 insert lasted 20 min. At moderate speeds (100 m min^{-1}) the tungsten carbide lasted 45 min which is more or less to be expected with this grade of tool material.

(d) These results would seem to indicate that CBN, TX-10 and Iscanite could be used under very severe cutting conditions where high metal removal rates are required.

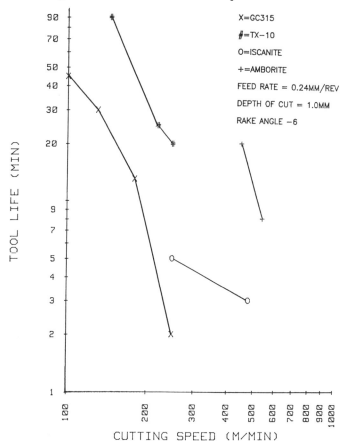

FIG. 14. V–T diagram for turning grey iron.

4.4. Taylor Tool Life Curves—Cast Iron

The V–T curves (tool failure criteria) in Fig. 14 show definite segregation between the tool materials tested, with Amborite outperforming all the other tools. The alumina coated TX-10 did not perform as well as the CBN and this is in contradiction to the results of Aspinwall et al.[3] The probable reason for this difference is due to the cast iron grade used for each series of tests. Aspinwall used a grade 260 iron while our iron was a grade 180. The grade 260 iron had a much higher shear strength (299 N mm^{-2}) compared to the grade ISO (207 N mm^{-2}) and this would probably account for difference in tool performance.

5. SURFACE FINISH TESTS

A series of tests were performed to assess the suitability of the various tool materials in terms of the surface finish produced during turning and milling of Vitallium and turning grey cast iron.

The surface finish measurements were taken using a Hommell T2-2 instrument. The meter cut-off frequency of 0·8 mm was used for each test. The sampling length was set at 5 mm for the tests on the Vitallium and 1·5 mm for the tests on the cast iron. On each of the machined surfaces an average of six readings were taken.

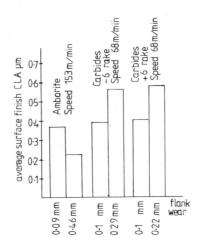

FIG. 15. Surface finish for turning Vitallium.

5.1. Turning Vitallium

The various surface finish values obtained when using the CBN and carbide inserts under a range of cutting condititions are shown in composite form in Fig. 15. The influence of flank wear on the surface finish is clearly demonstrated. The results show that the CBN tool produced a better surface finish compared to the carbide. Indeed the surface finish improved with flank wear when using CBN. This is thought to be due to the effects of work-softening as a result of the high cutting temperatures. This work-softening gives rise to a better shearing action and as a consequence the surface finish improves.

5.2. Milling Vitallium

During these tests the machining was performed in such a way as to expose the insert to the full cut per revolution. In addition the roughness measurements were made parallel to the direction of feed and along the centre of the cut surface.

Figure 16 shows the surface finish produced CLA (μm), at constant feed and depth of cut for a range of cutter speeds. Surface roughness

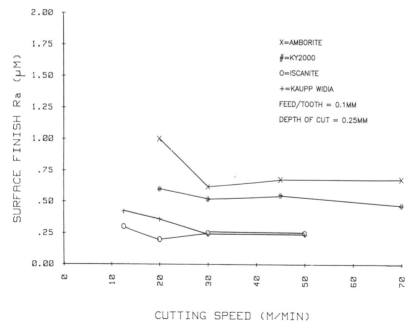

FIG. 16. Surface finish for milling Vitallium.

levels in general lie well below 1 μm for all cutting materials. The enhanced performance of the Iscanite and tungsten carbide inserts can be attributed to the use of positive rake angles and the presence of a wiper flat.[15] In contrast the Kyon 2000 and CBN inserts have a negative rake angle and lack a wiper flat.

5.3. Turning Cast Iron

This series of tests was performed at two cutting speeds, 100 m min^{-1} and 134 m min^{-1}. The depth of cut was 0·2 mm and the feed rate was varied over the range 0·06–0·24 mm per rev at each of the test speeds.

The results are shown in Figs. 17 and 18. For each of the tool materials tested the surface finish deteriorated as the feed rate was increased. By comparing Figs. 17 and 18 it can be seen that the scatter in the surface finish values decreases with increased cutting speed. At each of the cutting speeds used for the tests the best surface finish was obtained with the Iscanite with CBN producing the poorest finish, particularly at high feed rates. At the higher cutting speed the coated carbide did not perform as well as the straight carbide. However, at the lower speed of 100 m min^{-1} the coated carbide performed much better than the GC315 insert particularly at the higher feed rates. These results confirm the well established suitability of ceramic tools in the machining of cast iron.

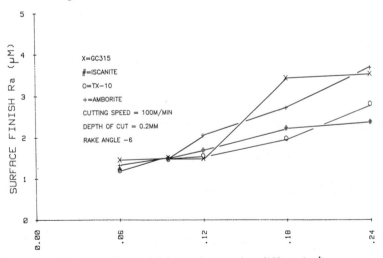

FIG. 17. Surface finish for turning grey iron (100 m min^{-1}).

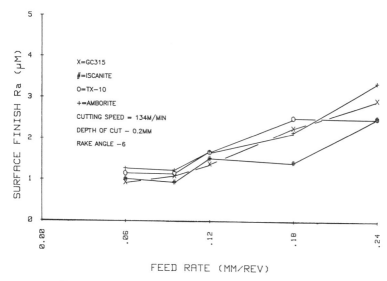

FIG. 18. Surface finish for turning grey iron (134 m min^{-1}).

6. ECONOMIC CONSIDERATIONS

In any machining operation, whether metal removal or surface finish is the prime consideration, the cost of the operation is of importance. The machining costs are associated with two major factors:

(a) the initial tool costs,
(b) the life of the tools.

In the case of the cutting materials used for the tests described earlier the cost per insert was as follows:

Carbides	K68	£1·50
	IC20	£1·00
	GC315	£2·30
KRUPP WIDIA	GI0E	£2·00
Coated Carbide	TX-10	£3·50
Sialons	Kyon 2000	£7·00
	Iscanite	£10–£16·00
CBN	Amborite	£180·00 approx.

As can be seen the carbide inserts have a considerable cost advantage, over the sialons and particularly the CBN. However, these costs must be considered in relation to tool life and metal removal rates. It was shown that for Vitallium, tungsten carbides performed very well compared to the more exotic sialons and CBN tools. CBN did have an advantage in terms of the surface finish produced. Overall, however, when account is taken of the cost of CBN compared to the carbides it would appear that the use of CBN to machine Vitallium at the cutting speeds used in the tests described here can not be justified.

In the case of the cast iron the Iscanite and alumina coated carbide were superior to either the straight carbide or the CBN, at medium to low cutting speeds. Given that the Iscanite can work at high cutting speeds over long periods of time without rapid wear would suggest that these inserts are worth considering for high volume continuous machining operations. The CBN is also capable of operating at extremely high speeds without severe wear occurring. However, the cost at approximately twelve times that of the Iscanite and fifty times that of the coated carbide would appear to restrict its use on economic grounds.

7. CONCLUSIONS

(1) For turning and milling Vitallium tungsten carbide inserts had lower wear rates compared to CBN.
(2) Sialon was not successful in turning Vitallium. This may have been due to other factors, e.g. tool holder, geometry, locating devices, etc.
(3) For the milling of Vitallium the Iscanite insert was quite successful in terms of surface finish production.
(4) The main advantage of using CBN to machine Vitallium is due to the proven ability of this material to operate at high cutting speeds without rapid tool failure. In addition, CBN was found to produce the best surface finish in the turning tests.
(5) When machining the cast iron the Iscanite insert produced good surface finish and gave a good tool life.
(6) The alumina coated carbide was found to be far superior than the straight carbide in terms of tool life and marginally better in terms of the surface finish produced.
(7) The CBN was the only material capable of operating at high cutting speeds without rapid tool wear occurring. This would

suggest that for high volume metal removal at high speeds over extended cutting times CBN has a decided advantage compared to the other materials tested.

REFERENCES

1. BOGGESS, J. *American Machinist,* January 1984, 78.
2. WOOD, D. O. *AMS Metals Congress,* St Louis Missouri, October, 1982, Paper 8201-084.
3. ASPINWALL, D. K., TUNSTALL, M. and HAMMERTON, R. In *Proceedings of the 25th International Machine Tool Design Research Conference,* p. 269, 1985.
4. ANON. *Machine and Production Engineering,* No. 4, November 1981, 31.
5. GREARSON, A. and JACK, D. H. *Proceedings of the 1st International Tool Conference,* Birmingham, p. 211, 1984.
6. WENTORF, R. H. *Journal of Chemical Physics,* **34** (1961) 809.
7. BUNDY, F. P. and WENTORF, R. H. *Journal of Chemical Physics,* **39** (1963) 1144.
8. GANE, N. and STEPHENS, L. W. *Wear,* **88** (1983) 67.
9. *Proceedings ASM International Conference on Cutting Tool Materials,* Kentucky, USA, p. 281, 1980.
10. TAKATSU, S., SHIMODA, H. and OTANI, K. *Proceedings of the International Conference on New Tool Materials, Metal Cutting and Forming,* London, 1981, Engineering Digest Technical Conference, Session 1, paper 3.
11. TONSHOFF, H. K. and BORYS, W. E. In *Proceedings of the North American Manufacturing Research Conference,* p. 278, 1983.
12. DE BEERS LTD *Production Engineer,* February 1985, 55.
13. WILSON, M. and EL-BARADIE, M. *Proceedings of the 22nd International Conference on Machine Tool Design Research,* Manchester, p. 171, 1982.
14. OPITZ, H. and BECKHAUS, H. *Annals of C.I.R.P.,* **18** (1970) 257.
15. FIELD, M. and KABLES, J. Factors influencing surface finish in turning and milling grey and ductile cast irons. In *Proceedings of the S.M.E. Creative Manufacturing Engineering Programs,* 1971, technical paper MR71-145.

20

The Wear Behaviour of Sialon and Silicon Carbide Ceramics in Sliding Contact

S. A. HORTON,† J. DENAPE,§ D. BROUSSAUD,§
D. DOWSON,‡ F. L. RILEY,† AND N. WALLBRIDGE‡

† *Department of Ceramics;* ‡ *Department of Mechanical Engineering, University of Leeds, LS2 9JT, UK*
Ecole des Mines de Paris, Centres des Matériaux, B.P. 87, 91003 Evry Cedex, France

ABSTRACT

The sialon and silicon carbide ceramics are noted for their hardness and resistance to wear in certain applications. However, there is little published information concerning the wear behaviour of these materials. We report here the results of a study of wear behaviour and associated wear mechanisms, for one material of each class.

Steady state wear factors have been determined at room temperature for homogeneous sliding couples, tested in laboratory air, using reciprocatory pin-on-plate, and tri-pin-on-disc configurations, over sliding distances of up to 900 km. Shorter term tests have been made with a plate-on-rotating cylinder configuration.

Wear rates have been interpreted in terms of the postulated mechanisms of material removal, based on microscopical and other examinations of worn surfaces and wear debris.

1. INTRODUCTION

Engineering ceramics are being considered for a variety of applications where an understanding of their tribological behaviour will be a prerequisite. Examples of areas where there is interest in their use include heat engines[1,2] and total replacement hip joints.[3] In such applications sliding wear behaviour will be of importance in governing the life of a particular component or system.

Literature on the wear of ceramics is limited. Fundamental aspects of the friction and wear of ceramics have been reviewed by Buckley[4] and by Buckley and Miyoshi.[5] Page,[6] and Czernuska and Page[7] have investigated the indentation and scratch behaviour of a variety of ceramic materials, while recent papers have described the sliding wear behaviour of oxide and non-oxide ceramic couples.[8–12]

The aim of the present study has been to investigate the sliding wear behaviour, at room temperature, of two commercial silicon-based ceramics, a β'-sialon and a sintered silicon carbide. Three types of test configuration have been used in this investigation, plate-on-rotating cylinder, reciprocatory pin-on-plate and tri-pin-on-disc. The first test configuration has been used for short term tests of up to 4 km sliding distance. The latter two test configurations have been used to determine steady state wear factors at sliding distances of up to ~900 km. Results are also available from the authors' laboratories from tri-pin-on-disc tests of alumina couples,[8] tested under similar conditions to the sialon and silicon carbide of this investigation. Direct comparisons can therefore be made of the wear behaviour of these oxide and non-oxide ceramics.

2. EXPERIMENTAL

2.1. Materials

The silicon carbide test pieces, manufactured by Ceraver, Tarbes, France, were sintered with 1 wt % boron and 0·5 wt % C to produce single phase α-SiC. The β'-sialons were produced, also by Ceraver, by sintering under nitrogen, a mixture of silicon, aluminium and 6–8 wt % yttrium oxide, to obtain a β'-sialon containing a yttrium aluminosilicate intergranular phase. Mechanical properties typical of the materials tested are listed in Table 1.

Examination of the silicon carbides and sialons by optical microscopy and by scanning electron microscopy showed that porosity was present in all of the materials tested (e.g. Fig. 3(c)). The average intercept grain sizes of silicon carbide samples etched in boiling Murakami's reagent (20 g potassium ferricyanide and 20 g potassium hydroxide dissolved in 100 ml of water) (Fig. 1) were estimated from a series of scanning electron micrographs to be ~6 µm.

TABLE 1
Mechanical Properties of the Sialon and Silicon Carbide Materials

Property	Material	
	Silicon carbide	Sialon
Young's modulus E (GPa)	460	275
Shear modulus, G (GPa)	200	110
MHV 200 g (GPa)	26·1	15·6
Modulus of rupture, σ_R (3-point bending) (MPa)	440	530
K_{IC} by indentation (MPa m$^{1/2}$)	2·7	3·5

Thin foils were prepared from the sialons and examined in a Jeol 200CX transmission electron microscope. Microstructural information was obtained in the transmission mode and compositional information in the scanning transmission mode by energy dispersive analysis of X-rays. A transmission electron micrograph of the sialon is shown in Fig. 2(a). It can be seen that the β'-grains have a prismatic morphology and that the yttrium is concentrated in the intergranular glass, Figs. 2(b) and (c). The average grain size of the β'-phase was estimated to be ~1 μm, while the largest grains attained a length of ~6 μm.

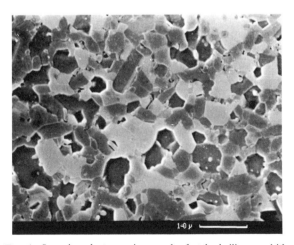

FIG. 1. Scanning electron micrograph of etched silicon carbide.

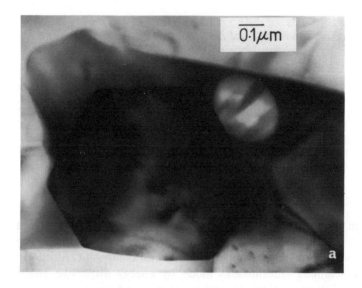

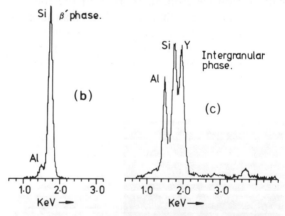

FIG. 2. Transmission electron micrograph of sialon and compositions of phases. (a) β' grains and Y–Si–Al glass; (b) composition of β' grain; (c) composition of Y–Si–Al glass.

2.2. Preparation of Surfaces

The surfaces of the pins, plates and discs were prepared by diamond grinding followed by diamond lapping to roughness average values (Ra) within the range 0·02–0·08 μm. The cylinders for the plate-on-rotating cylinder test were used in the diamond ground condition, Ra

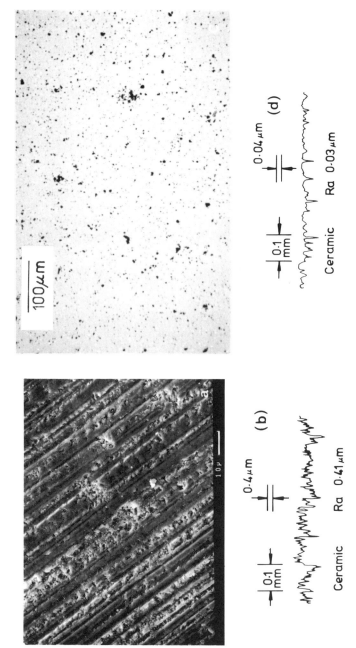

FIG. 3. Initial surfaces of wear test specimens. (a) Sialon, ground surface; (b) talysurf trace of (a); (c) sialon lapped surface; (d) talysurf trace of (c).

1–1·2 µm (Ra values determined at a cut off of 0·8 mm). Micrographs and talysurf traces representative of the surfaces tested are shown in Figs. 3(a)–(d).

2.3. Wear Tests
The three types of wear test were carried out at room temperature (~21°C), unlubricated and in air, at a relative humidity of 40–70%.

2.3.1. Reciprocatory pin-on-plate and tri-pin-on-disc tests
The reciprocatory wear machine has six stations, Fig. 4(a), each of which accommodated a reciprocating wear plate, a wear pin (rigidly held in a stainless steel collet) and a non-sliding control pin. Normal load was applied by the addition of weights to the collet shafts of the wear pin holders (low loads ~7·5–65 N) or by a cantilever arrangement (high loads ~80–240 N).

The tri-pin-on-disc machine is shown in Fig. 4(b). An annular ceramic disc was driven via a spigot at the base of a stainless steel bath against three stationary wear pins, which were rigidly held in a stainless steel holder. The weight of the pin holder was sufficient for the low load tests carried out on this machine. The truncated conical geometry of the pins used in both of the above tests is shown in Fig. 4(c).

A constant sliding speed of $0·24 \text{ m s}^{-1}$ was used for the tri-pin-on-disc test, whereas the sliding speed of the reciprocatory test was adjusted to give an average value of $0·24 \text{ m s}^{-1}$ over each cycle. The sliding speeds of the plates and discs relative to the stationary wear pins was adjusted by means of variable speed, constant torque motors. In each test a non-sliding control pin was used, to allow corrections to be made for any weight changes unrelated to wear, such as atmospheric moisture uptake.

The reciprocatory pin-on-plate and tri-pin-on-disc tests were conducted over sliding distances of up to 900 km, in order to establish steady state wear factors for the silicon carbide and sialon materials.

2.3.2. Plate-on-rotating cylinder tests
Short-term tests, where the sliding distance was limited to 4 km, were carried out on this machine. A schematic diagram of the plate-on-rotating cylinder machine is shown in Fig. 4(d) together with the test specimen geometry. The normal load (2–40 N) was applied by

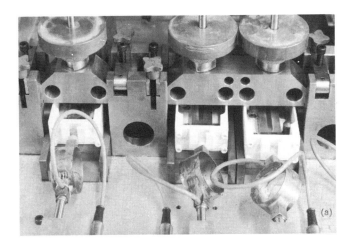

FIG. 4. Wear machines and specimen geometries. (a) Reciprocatory pin-on-plate; (b) tri-pin-on-disc.

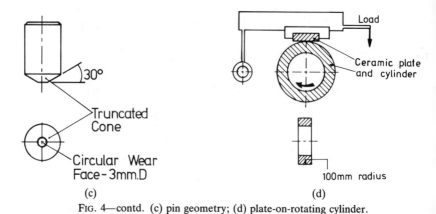

FIG. 4—contd. (c) pin geometry; (d) plate-on-rotating cylinder.

a cantilever arrangement and the sliding speed was varied over the range of $0 \cdot 1$–$4 \, \text{m s}^{-1}$.

2.4. Measurement of Wear Rates

The reciprocatory pin-on-plate and tri-pin-on-disc tests were stopped at regular intervals for the determination of the wear rates of the pins. The weight losses of the worn pins (corrected for any weight changes in the control pins) were converted to a volume loss, to an accuracy of $\pm 6 \times 10^{-4} \, \text{mm}^3$ on the basis of the sialon and silicon carbides densities of $3 \cdot 20 \, \text{Mg m}^{-3}$ and $3 \cdot 21 \, \text{Mg m}^{-3}$ respectively.

The results are presented in terms of a wear factor value, k, defined as:

$$k = \frac{V}{PD} \quad (1)$$

where V is the volume of material removed by wear (mm³), P is the normal load (N), D is the sliding distance (m).

The geometry of the plate-on-rotating cylinder test was such that ellipsoidal wear scars were formed on the plate surfaces. Measurements of the major and minor axis of the wear scar together with profilometry measurements of the depth of wear were used to calculate the volume of materials removed by wear, to an accuracy of $\pm 1\%$. A wear factor (k) was then calculated using eqn. (1).

2.5. Characterisation of the Worn Surfaces and Wear Debris

The worn surfaces were examined periodically by optical microscopy, conventional incident light and Nomarski contrast. The semiconducting properties of the silicon carbide allowed examination of the

TABLE 2
Results from the Reciprocatory Pin-on-Plate Test for Homogeneous Couples of Sialon and Silicon Carbide (Tested at an Average Sliding Speed of 0.24 m s^{-1}, in Air)

Test	Material	Normal load on pin (N)	Wear factor (k) of pins (mm^3 N^{-1} m^{-1})	Sliding distance (km)	Comments
1	Sialon	80	10^{-4}	0.2	Excessive wear of pin and plate after short sliding distance. Tested as ground Ra 0.41 μm
2	Sialon	13	2.5×10^{-6}	121	Test stopped for examination of pin and plate by SEM before steady state wear factor was confirmed. Test as lapped, Ra 0.06 μm
3	Sialon Silicon carbide	13	2.3×10^{-6} 1.9 to 2.4×10^{-6}	147–886 147–886	Steady state wear factors established after ~150 km of sliding. Tested as lapped, sialon 0.06 μm and silicon carbide 0.04 μm Ra

Ra values measured at a cut off of 0.8 mm.
Sialons of tests 1–3 were from different batches of sintered material.

worn surfaces as the tests progressed. Because of the coating requirements for the non-conducting sialon, scanning electron microscopy was not used to characterise the worn surfaces until the completion of the test.

Wear debris was collected from each of the tests and examined by scanning electron microscopy. X-ray techniques (diffractometer and powder camera) were also used to determine the structural nature of the debris. The nitrogen content of the debris from the sialon tri-pin-on-disc test was determined as ammonia by an alkaline fusion and acid titration method.

3. RESULTS

3.1. Reciprocatory-pin-on-plate and Tri-pin-on-disc Tests

The reciprocatory wear test results and the test conditions are summarised in Table 2.

At a normal load of 80 N excessive wear of the sialon pin ($\approx 10^{-4}$ mm^3 N^{-1} m^{-1}) and plate occurred after a very short sliding distance of 0·2 km. The sialon used for this test tended to fracture readily during diamond machining and the pin and plate were tested in the ground condition (Ra, 0·41 μm).

Further reciprocatory tests were carried out on sialons from two different batches of sintered material under a normal load of 13 N. Wear factor values of $2·5 \times 10^{-6}$ mm^3 N^{-1} m^{-1} after 121 km of sliding (batch 2), and a mean steady state value of $2·3 \times 10^{-6}$ mm^3 N^{-1} m^{-1} between 147 and 886 km of sliding (batch 3) were recorded. These results suggest that any microstructural and/or compositional differences in materials from different batches are not reflected in the wear factor values.

Reciprocatory pin-on-plate tests were also carried out with silicon carbide couples at a normal load of 13 N. The steady state wear factor values ranged from $1·9 \times 10^{-6}$ to $2·4 \times 10^{-6}$ mm^3 N^{-1} m^{-1}, over the sliding distance of 147–886 km. Similar values have been recorded in the early stages of testing at normal loads of 7·8, 23 and 43 N, for silicon carbide couples.

Tri-pin-on-disc tests were carried out with the silicon carbide and sialon materials at normal loads of 7·7 and 7·6 N respectively. The wear factor values obtained were $\sim 9 \times 10^{-7}$ mm^3 N^{-1} m^{-1} for the silicon carbide after 448 km of sliding and $\sim 2 \times 10^{-6}$ mm^3 N^{-1} m^{-1}

TABLE 3
Results from Plate-on-Rotating Cylinder Test, for Homogeneous Couples of Sialon and Silicon Carbide

(a) Effect of normal load, sliding speed 0.25 m s^{-1}

Load (N)	Wear factor (k) (mm^3 N^{-1} m^{-1})	
	Sialon	Silicon carbide
2	1.7×10^{-5}	2.3×10^{-6}
5	1.2×10^{-5}	3.7×10^{-6}
10	1.1×10^{-5}	4.6×10^{-6}
20	1.0×10^{-5}	4.2×10^{-6}
40	1.3×10^{-4}	5.2×10^{-6}

(b) Effect of sliding speed, normal load 5 N

Sliding speed (m s^{-1})	Wear factor (k)(mm^3 N^{-1} m^{-1})	
	Sialon	Silicon carbide
0.1	3.7×10^{-5}	6.3×10^{-6}
0.5	4.5×10^{-6}	2.8×10^{-6}
1.0	6.4×10^{-6}	3.3×10^{-6}
2.0	2.3×10^{-5}	5.2×10^{-6}
4.0	4.7×10^{-5}	6.6×10^{-6}

after 466 km for the sialon. The above values are similar to those obtained with the reciprocatory pin-on-plate tests conducted under normal loads of 13 N.

3.2. Plate-on-rotating Cylinder Test

The wear factor values for the silicon carbide couples tested under normal loads of 2–40 N, and at a sliding speed of 0.25 m s^{-1} varied from 2.3×10^{-6} to 5.2×10^{-6} mm^3 N^{-1} m^{-1}, Table 3(a). The values for the sialon couples ranged from 1.0×10^{-5} to 1.7×10^{-5} mm^3 N^{-1} m^{-1} at normal loads of 2–20 N, increasing to 1.3×10^{-4} mm^3 N^{-1} m^{-1} at 40 N, Table 3(a).

The results for tests carried out at sliding speeds of 0.1–4 m s^{-1} and a constant normal load of 5 N are listed in Table 3(b), for homogeneous couples of silicon carbide and sialon.

Over the range of loads and sliding speeds investigated with this test configuration the silicon carbide couples show only a small variation in

FIG. 5. Scanning electron micrograph of worn sialon pin, reciprocatory test 13 N. Showing: 1, flaking; 2, polished area; 3, agglomerated debris film.

their wear factor values. The sialon, however, exhibited excessive wear at a normal load of 40 N, and an order of magnitude variation in the wear factor values between sliding speeds of $0.5-4$ m s^{-1}.

3.3. Examination of Worn Surfaces and Wear Debris

A variety of features were observed on the worn surfaces of the silicon carbide and sialon test specimens, some of which were common to both materials. Macroscopic examination of the pins after testing at low loads (<13 N) showed polished surfaces. At higher loads (⩾13 N) the pin surfaces consisted of alternating polished and roughened tracks extending in the direction of sliding. In the roughened areas material had flaked from the surface. Figure 5 shows a worn sialon pin after reciprocatory testing under a normal load of 13 N. The wear debris from the latter tests was in the form of platelets and agglomerates of fine particles. At a normal load of 80 N (reciprocatory test), areas where material had flaked from the sialon pin and plate surfaces were also observed and X-ray analyses of the wear debris showed that it was structurally similar to the parent bulk material. Similar features were observed on the worn surfaces of the 40 N plate-on-rotating cylinder

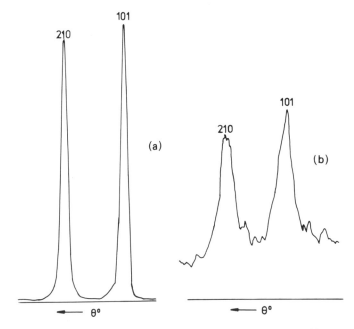

FIG. 6. X-ray diffraction line profiles of sialon. (a) Bulk material; (b) wear debris, plate-on-rotating cylinder test 40 N.

FIG. 7. Optical Nomarski contrast micrograph of worn sialon, tri-pin-on-disc test, 7·6 N.

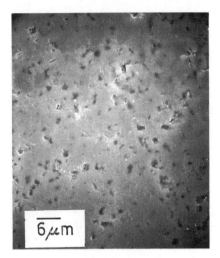

Fig. 8. Scanning electron micrograph of worn silicon carbide, plate-on-rotating cylinder test, 5 N, 0·25 m s^{-1}.

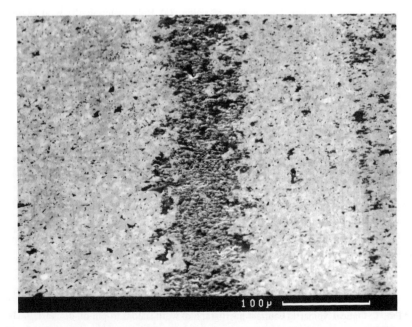

Fig. 9. Scanning electron micrograph of worn silicon carbide, reciprocatory test 23 N.

specimens (wear factor $1 \cdot 3 \times 10^{-4}\,\mathrm{mm^3\,N^{-1}\,m^{-1}}$), although the wear debris from this test was considerably finer than that of the 80 N reciprocatory pin-on-plate test. X-ray examination again showed that the debris was structurally similar to the parent bulk material (compare Figs. 6(a) and 6(b)).

Under low loads with the tri-pin-on-disc test, and at low loads and low sliding speeds with the plate-on-rotating-cylinder test, the worn surfaces of the sialon specimens were polished (Fig. 7). Films of agglomerated debris particles can also be seen in the latter micrograph; this type of feature was observed on all of the worn sialon surfaces. The wear debris from these tests was mainly in the form of agglomerates of fine particles, of $<1-2\,\mu\mathrm{m}$ in dimension. X-ray examination of the debris from the 7·6 N tri-pin-on-disc test showed that it was amorphous. Chemical analysis gave a nitrogen content of $2 \cdot 9 \pm 0 \cdot 2\,\mathrm{wt}\,\%$, compared with a theoretical value of $\sim 35\,\mathrm{wt}\,\%$ for the bulk material.

The worn surfaces of the silicon carbide specimens are shown in Figs. 8 and 9. At low loads and sliding speeds the worn surfaces were polished (Fig. 8). While at higher loads (>13 N) with the reciprocatory wear test, and at high loads and sliding speeds with the plate-on-rotating cylinder test, grooves in the surfaces and microfractured regions were observed. The latter features were most evident on the worn reciprocatory pin surfaces (Fig. 9), where rough tracks of fracture can be seen. The films of agglomerated debris particles which were present on all of the worn sialon surfaces were only observed on the worn silicon carbide at loads of $\geqslant 13\,\mathrm{N}$.

The wear debris from the 7·7 N silicon carbide tri-pin-on-disc test was largely X-ray amorphous, and EDAX analyses for oxygen gave a marked positive response. Taken together these observations indicate that the debris is probably amorphous silicon dioxide. A similar conclusion can be reached regarding the debris obtained from the sialon couples under wear conditions where polishing predominated. In this case the debris was presumably an amorphous aluminosilicate phase (the aluminium content of the sialon is presumably sufficiently low that the tendency for crystalline mullite or aluminium oxide to nucleate is small).

4. DISCUSSION

The results of this preliminary study are consistent with those of other workers using similar systems and wear conditions.[10,12] They suggest that, unlubricated, the wear behaviour of these hard non-oxide ceramics is at best equal to that of the commonly used hard oxides,[9] and that it can be markedly inferior.[8] For example Wallbridge et al.[8] have reported wear factor values of $3 \cdot 4 - 7 \cdot 3 \times 10^{-10}$ mm^3 N^{-1} m^{-1} for unlubricated homogeneous couples of alumina, where the test conditions were the same as those used in this investigation ($0 \cdot 24$ m s^{-1} and a normal load of $7 \cdot 5$ N per pin).

The microstructural studies of the worn surfaces suggests that the wear processes occurring in the polished areas of the silicon carbide and sialon are plastic in nature and associated with oxidation, either of the contacting surfaces or wear debris, as evidenced by the amorphous wear debris and the oxygen or nitrogen analyses.

At high loads and high sliding speeds fracture was observed on the worn surfaces of both materials, although this appears to vary in extent and nature, depending upon the material, test configuration, and test conditions used. The X-ray diffraction results from the higher load tests of the sialon couples (40 N and 80 N) confirmed that the wear debris was structurally similar to the parent bulk material, unlike that observed from the low load tri-pin-on-disc test.

5. CONCLUSIONS

The range of wear factor values for the sialon and silicon carbide couples tested in 'air' and at room temperature, 10^{-4} to $2 \cdot 3 \times 10^{-6}$ mm^3 N^{-1} m^{-1} and 9×10^{-7} to $6 \cdot 6 \times 10^{-6}$ mm^3 N^{-1} m^{-1} respectively, are consistent with values obtained by other workers in tests carried out under similar conditions.[10,12]

There appear to be two important sliding wear regimes.

(1) Polishing, at low loads and low sliding speeds—where local plastic deformation of the contacting surfaces plays a major role in governing wear behaviour, and is associated with the formation of amorphous silicate debris.
(2) Fracture of the parent material from the contacting surfaces, predominantly at high loads and high sliding speeds.

Significant differences in the wear factors were obtained for the sialon couples at high and low loads, 10^{-4} mm^3 N^{-1} m^{-1} and 10^{-6} mm^3 N^{-1} m^{-1} respectively. However the values for the silicon carbide couples exhibit only small variations over a similar range of loads.

The wear resistance of the silicon carbide couples was insensitive to sliding speed over the range of 0.1–4.0 m s^{-1}, whereas for the sialon the wear factor values exhibit an order of magnitude increase at sliding speeds of both 0.1 m s^{-1} and 4.0 m s^{-1}, when compared with the minimum value (4.5×10^{-6} mm^3 N^{-1} m^{-1}) shown at ~0.5 m s^{-1}. This effect is being investigated further.

REFERENCES

1. BRYZIK, W. and KAMO, R. SAE Paper 830314, 1983.
2. Heat Engine Issue, *American Ceramic Society Bulletin*, **64** (2) 1985.
3. SEMLITSCH, M., LEHMANN, M., WEBER, H., DOERRE, E. and WILLERT, H. G. *Journal of Biomedical Materials Research*, **11** (1977) 537.
4. BUCKLEY, D. H. *American Ceramic Society Bulletin*, **52** (1972) 884.
5. BUCKLEY, D. H. and MIYOSHI, K. *Wear*, **100** (1984) 333.
6. PAGE, T. F., SAWYER, G. R., ADEWOYE, O. O. and WERT, J. J. *Proceedings of the British Ceramic Society*, **26** (1978) 193.
7. CZERNUSKA, J. T. and PAGE, T. F. *Proceedings of the British Ceramic Society*, **34** (1984) 145.
8. WALLBRIDGE, N., DOWSON, D. and ROBERTS, E. W. In *Wear of Materials*, Ludema, K. C. (Ed.), American Society of Mechanical Engineers, Reston, Virginia, p. 202, 1983.
9. SCOTT, H. G. In *Wear of Materials*, Ludema, K. C. (Ed.), American Society of Mechanical Engineers, Vancouver, p. 8, 1985.
10. ISHIGAKI, H., KAWAGUCHI, I., IWASA, M. and TOIBANA, Y. In *Wear of Materials*, Ludema, K. C. (Ed.), American Society of Mechanical Engineers, Vancouver, p. 13, 1985.
11. IWASA, M., TOIBANA, Y., YOSHIMURA, S. and KOBAYASHI, E. *Journal of the Ceramic Society of Japan*, **93** (1985) 21.
12. DERBY, J., MACBETH, J. and SESHADRI, S. In *Combustion Engines— Reduction of Friction and Wear*, Institution of Mechanical Engineers, London, p. 133, 1985.

COMMENTS AND DISCUSSION

Chairman: W. G. LONG

A. J. Whickens: In your plate-on-rotating cylinder test, you said that you got an ellipsoidal scar. I believe you said you measured the wear factor directly?

S. A. Horton: This was measured at the School of Mines. I believe that what they did was to measure the axes and the depth and they have shown that by assuming a simple shape that you can calculate to 1 or 2% the actual value for that particular scar.

Whickens: Does that cross check with the assumed shape, the density and the measured weight loss?

Horton: Yes, the silicon carbide wear factor values obtained via weight loss using a microbalance or by volume loss from wear scars are quite similar.

R. T. Cundhill: Are you going to extend your research to other types of silicon nitride ceramics, i.e. materials with different hardness and toughness.

Horton: No. We are more interested in the effects of environment, load, etc. We are not really looking at differences in composition and microstructure although we have tested materials from different batches and found very little difference in the wear factor values.

W. G. Long: Will you be looking at temperature as a variable?

Horton: Only low temperatures. There has been some work done on alumina at low temperatures and it does have quite a significant effect.

M. J. Pomeroy: Do you have any idea what kind of temperature may be developed in your wear tests?

Horton: We have not attempted to measure either bulk temperatures or surface temperatures of the non-oxide materials. However, at high sliding speeds, for example 4 m s^{-1}, surface temperatures may become important. In this respect it is interesting that the material with the lower thermal conductivity, sialon, was inferior to silicon carbide.

21

Studies on Properties of Low Atomic Number Ceramics as Limiter Materials for Fusion Applications

B. A. THIELE, H. HOVEN, K. KOIZLIK, J. LINKE AND E. WALLURE

Institut für Reaktorwerkstoffe, KFA Postfach 19 13 5170, FRG

This recent R & D programme is concerned with the investigation of the potential of non-oxide ceramics for high heat flux components in nuclear fusion reactors such as limiters, divertor plates and protective tiles for the first wall. These components will be exposed to cyclic heat loads of up to 10 MW m^{-2} at operating conditions of large size plasma machines and future fusion reactors. Occasional plasma instabilities can cause heat fluxes of around 100 MW m^{-2}. Therefore, the low cycle thermal fatigue behaviour and the thermal shock resistance of candidate materials are of prime interest. A continuous flow of high energy neutral and charged particles from the plasma to the components leads to materials erosion. This eroded material contaminates the plasma which increases the heat radiation—especially in the case of high atomic number elements. Therefore low Z-material is favoured. Nowadays, graphite is successfully being used in fusion reactors. Therefore, ceramics have a promising potential as candidate materials. They may be used as bulk material or as coatings on components.

The present study deals with thermal shock and erosion–redeposition behaviour of low-Z-bulk-ceramics: SiC, SiC + Si, SiC + 3% Al, SiC + 2% AlN, AlN, Si_3N_4, BN with graphite as reference material. Also included are substrate-coating systems: TiC coated graphite, Cr_2C_3 coated graphite and TiN on Inconel. The properties are being investigated by electron beam and in-pile fusion machine tests in the KFA-Tokamak machine Textor.

Thermal shock tests have been performed in two steps by single-shot electron beam tests via electron beam welding machines. First a comparison of the thermal shock behaviour of different materials at given heat loading conditions and second determination of the critical heat flux and pulse length.

The in-pile tests were performed on a series of segments of the above mentioned ceramics, which were sandwiched between the reference limiter material Inconel 600.

The single electron-beam-shot tests showed that sublimation was the dominant damaging effect for graphite, BN and Si_3N_4. Only minor differences in the damage behaviour were found. Materials with mediocre thermo-mechanical properties, such as SiC and AlN, showed cracks. The highest energy density values were tolerated by specimens of SiC alloyed with 2% AlN.

The damage on samples of the in-pile tests was mainly caused and influenced by unipolar arcing and material redeposition. In general, the behaviour of the ceramics was comparable with the electron beam tests: BN and SiC + 2% AlN are at present regarded as the prime candidates for future irradiation tests.

COMMENTS AND DISCUSSION
Chairman: E. GUGEL

K. H. Jack: You use silicon carbide with 2% aluminium nitride in solid solution. Why was this limited to 2%?

B. A. Thiele: This was an Hitachi material. They have a patent on silicon carbide + 2% beryllium oxide. We could not work with the 2% BeO because of the toxicity problem. We knew that the material with AlN was a fine-grained material and we hoped that this would show superior behaviour. Hitachi supplied us with this material.

Jack: When you dissolve aluminium nitride in silicon carbide you get the 2H structure? Do you know if this was the 2H polytype?

Thiele: The company did some research on this material and they found out that it is not exactly what they thought it was. It was not 2% aluminium nitride but much less and apart from some aluminium oxide and a little bit of iron, it was extremely pure. This was in the grain boundaries.

Jack: And you do not know why it performed so well?

Thiele: I think it performed well because it was such a fine-grained material and the second phase in the grain boundary may be molten and then cooled down again after thermal shock and so it performed better than a pure silicon carbide.

22

Effect of Microstructural Features on the Mechanical Properties of REFEL Self-bonded Silicon Carbide

P. KENNEDY

UKAEA, Springfields Nuclear Power Development Laboratories, Preston, PR4 ORR, UK

ABSTRACT

One of the main characteristics of any ceramic material is high strength and with certain ceramics becoming increasingly important as engineering materials, and having to operate under conditions of extreme thermal and mechanical stress, strength is perhaps the key property.

However, strength and uniformity of strength depend not only on external variables such as temperature, but on the distribution of 'flaws' or microstructural features in the material and so the control of porosity, grain size and surface finish is vital.

A ceramic which has proved most successful in many high endurance applications is REFEL silicon carbide; a uniform, self-bonded material formed by infiltrating a graphite/silicon carbide body with molten silicon. As one might expect the principal microstructural features in this material are grains and uniformly dispersed regions of silicon-filled porosity and the effects of these features on strength, Young's modulus, hardness and fracture toughness have been examined. However, as temperature was found to have no effect on strength in a previous study, only room temperature measurements were made.

It was found that strength and fracture toughness decreased linearly with silicon content but that, for a given machining technique, critical flaw size remained constant. For a given silicon content the observed strength was found to increase as the grain size decreased, a plot of strength versus (grain size)$^{-1/2}$ being linear up to the point where the extrinsic flaws, introduced by machining, were larger than the grains. Beyond this point, although the slope of the strength versus (grain size)$^{-1/2}$ curve was lower, the strength continued to increase as the grain size decreased because, for a given grinding procedure, the size of the machining flaw introduced decreased with the grain size of the product.

1. INTRODUCTION

One of the main characteristics of any ceramic is high strength and with certain ceramics becoming increasingly important as engineering materials, and having to operate under conditions of extreme thermal and mechanical stress, strength is perhaps the key property. Unlike most metals, however, the strength of a ceramic is not unique but depends on a number of factors, internal and external. Furthermore the material consistency, which in general is as important if not more important than high absolute strength, depends on the same factors.

REFEL silicon carbide is a uniform, fine-grain, self-bonded silicon carbide, which, because of its high temperature resistance and its excellent performance under conditions of thermal stress and thermal shock was originally developed as a cladding for high-temperature-nuclear-reactor fuel.[1] The material has since proved most successful in many non-nuclear, high-endurance applications and is currently being exploited commercially by BNFL for a variety of applications ranging from mechanical seals and bearings to engine components. REFEL silicon carbide is a polycrystalline two phase material containing free silicon and a study has been carried out to determine the effects of grain size, free silicon content and surface finish on mechanical properties. The results of this study are presented and discussed below.

2. FACTORS AFFECTING THE STRENGTH OF A CERAMIC

The theoretical strength of a ceramic material depends on Young's modulus E, its surface energy γ and the distance between atoms in its structure. Consequently, as E and γ are high for the engineering ceramics and the interatomic spacing is generally low, the theoretical strength is very high, i.e. of the order of 100 GPa. In practice however 'flaws' or microstructural features such as pores and grains act as stress raisers and the observed strengths are much lower.

The dependence of observed strength on maximum flaw size c was first expressed by Griffith[2] in 1920 as:

$$\sigma_f = \sqrt{\frac{2E\gamma}{c}} \qquad (1)$$

and this equation has since been modified by incorporating a constant Y to allow for the geometry of the flaw and a further constant Q to allow for the geometry of the test. Thus:

$$\sigma_f = \frac{1}{Y}\sqrt{\frac{2E\gamma Q}{c}} \qquad (2)$$

and for a semicircular flaw with bend test geometry this becomes:

$$\sigma_f = \frac{1}{1\cdot 23}\sqrt{\frac{2E\gamma}{c}} \qquad (3)$$

The reason why flaws have such an influence on the strength of a ceramic is that most ceramics are brittle. Hence there is no stress relief by plastic deformation and once a flaw reaches the critical size, defined by the Griffith equation, it propagates catastrophically. A measure of the resistance of a material to crack propagation is given by the fracture toughness K_{IC} where:

$$K_{IC} = \sqrt{2E\gamma} = 1\cdot 23\sigma_f c^{1/2} \qquad (4)$$

and consequently for high strength the 'flaw size' should be minimised and the fracture toughness maximised.

3. SELF-BONDING PROCESS FOR PRODUCING DENSE SiC

Dense silicon carbide may be produced by hot pressing, sintering or by reaction bonding and REFEL silicon carbide, the subject of this paper, is made by the last of these three processes. A porous body containing graphite and silicon carbide is reacted with molten silicon. During the process the silicon moves through the material under the influence of capillary forces, reacts with the graphite converting it to silicon carbide and the silicon carbide formed in situ deposits epitaxially on the surfaces of the original grain and bonds them together. Theoretically, it is possible to calculate the green density required to yield a fully dense silicon carbide body, but in practice it is necessary to leave additional porosity to allow the silicon to move freely, and consequently the fired material contains some interconnected porosity filled with silicon. In general, there is no unreacted carbon or empty porosity and so the principal microstructural features are the original grains of silicon carbide overcoated with the silicon carbide formed in the reaction, the boundaries between the overcoated grains and the silicon filled pores.

Consider the reaction in more detail. As there is a molar volume increase when graphite is converted to silicon carbide, if the local graphite density exceeds 0.963 g cm^{-3}, porosity is sealed off and no further reaction occurs. Furthermore, the maximum packing density of silicon carbide is about 50%. Consequently, for low C:SiC ratios the silicon carbide particles form a cage which prevents the carbon being pressed to too high a density but for a high C:SiC ratio this is not the case. The relationship between carbon, silicon carbide and porosity is shown in Fig. 1 and it may be seen that for a given C:SiC ratio density can be controlled by changing the 'green' porosity by varying the pressing pressure. However, low pressures tend to give a non-uniform

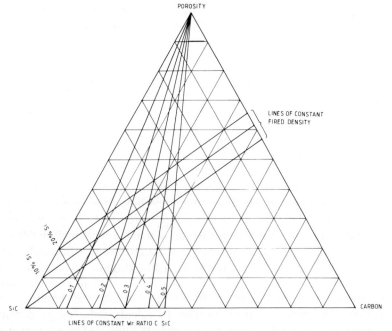

FIG. 1. The interdependence of the volume fractions of carbon, silicon carbide and porosity in the production of self-bonded silicon carbide containing different volume fractions of silicon.

product and as the pressure is increased an end-point density is reached. Consequently, fired density is best controlled by varying pressure in conjunction with C:SiC ratio. This in effect changes the thickness of the epitaxial layer but for a given fired density the extent of bonding is the same and so the material properties are the same.

In practice, the optimum REFEL silicon carbide composition from the fabrication point of view is 90% silicon carbide/10% silicon and the silicon content is usually controlled in the region 87–93%. For certain applications, however, special products with 5–25% free silicon are sometimes made.

4. EXPERIMENTAL

To cover the range of production requirements the silicon content was varied from 5 to 25%, carbon:silicon carbide ratios varying from 0·15 to 0·3 were used and all the 9 μm mean grain size materials were isostatically pressed in the region 70–120 MPa. The variable grain size materials were isostatically pressed under similar conditions using 50/50 mixtures of 600 mesh grit and grits varying in size from 50 to 800 μm, but the 4 μm mean grain size materials were slip-cast. All products were fired in the presence of silicon at 1600°C.

Modulus of rupture measurements were carried out in three-point bend using 5 mm × 3 mm × 45 mm specimens with a 27 mm span between the outer knife edges. The specimens were surface ground using a horizontal spindle machine with different grades of diamond wheel and at least 10 measurements were made for each data point. Two methods were used to measure K_{IC}: (a) the notched beam technique using specimens of similar geometry to the MOR specimens and (b) the microindentation technique described by Chantikul et al.[3] in which the reduction in strength caused by microindentation is measured. Young's modulus determinations were made using a resonant frequency technique and measurements of microhardness were carried out at a constant load of 0·5 kg using a Leitz Miniload 2.

5. RESULTS AND DISCUSSION

5.1. The Effect of Temperature on Strength and Microstructure

Before discussing the effects of microstructural features on strength and fracture toughness perhaps the most important external variable which should be considered is temperature. This was examined in some detail in 1971 by McLaren et al.[4] and the strength of material containing 10% free silicon was shown to be invariant with temperatures up to 1400°C, the melting point of silicon.

These workers also showed that, as a consequence of increased plasticity in the free silicon phase, the toughness of REFEL silicon carbide increased with temperature and in order to reconcile these results they postulated that the effective flaw size must also increase with temperature to maintain a constant γ/c ratio. Since 1971 the constancy in strength of REFEL silicon carbide up to 1400°C has been confirmed by many other workers and recently Popp and Pabst[5] at MPI Stuttgart have shown that, provided the load is not applied too rapidly, the fracture toughness does in fact increase from about 4 MN m$^{-3/2}$ at room temperature to 12 MN m$^{-3/2}$ at 1200°C. In this present study all measurements were made at room temperature.

5.2. The Effect of Maximum Grain Size and Strength

Whilst cracks can be initiated by a variety of flaws, in a dense polycrystalline ceramic the predominant 'flaw' is usually a grain and consequently, from the Griffith's relationship, a plot of observed strength versus (grain size)$^{-1/2}$ should be a straight line of slope $K_{IC}/1\cdot23$ which passes through the origin. Such was found to be the case (Fig. 2) for REFEL silicon carbide containing 10% free silicon and coarse grains ranging in size from 100 to 800 μm. As predicted, the slope of the initial part of the curve was found to be 2·91, corresponding to a K_{IC} of 3.58 MW m$^{-3/2}$, in good agreement with independent measurements, but below 100 μm a much lower slope was observed. The overall shape of the curve is in fact characteristic of that obtained for other ceramics, e.g. Al_2O_3, and originally the small grain size behaviour was interpreted by Carniglia[6] as fracture at defects produced by local plastic yielding. In 1977, however, Cranmer et al.[7] examined hot-pressed silicon carbide and showed that the low slope branch was due to surface defects becoming predominant. Thus, by improving the surface finish they were able to increase the strength at which branching occurred. The same was found to be the case for REFEL silicon carbide but before discussing the results let us consider the dependence of K_{IC} and strength on silicon content.

5.3. The Effect of Silicon Content on Fracture Toughness

Because K_{IC} is independent of surface finish, this property will be considered before strength. Measurements of K_{IC} were made using

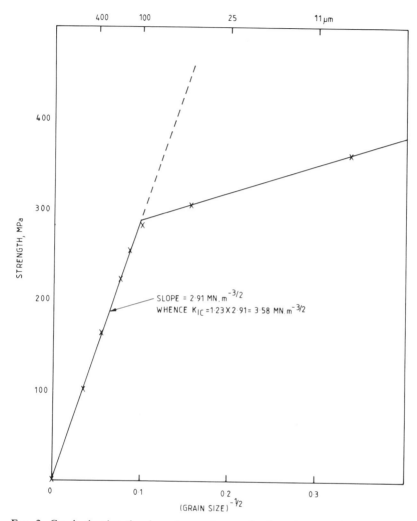

FIG. 2. Graph showing the dependence of strength of maximum grain size for SiC containing 10% free silicon.

notched beams and the microhardness indent method due to Chantikul et al.,[3] and because in the latter case Young's modulus and microhardness values are needed to calculate K_{IC}, measurements of these parameters, over a range of silicon contents, were also made. In the case of Young's modulus some data due to Carnahan[8] on hot-pressed silicon carbide were included to give low silicon values. The results are shown in Figs. 3 and 4 and both parameters may be seen to decrease as the silicon content increased.

The two sets of data for K_{IC} are given in Fig. 5 and they are consistent K_{IC} decreasing from $3 \cdot 8 \, \text{MN m}^{-3/2}$ at 0% silicon to $2 \cdot 8 \, \text{MN m}^{-3/2}$ at 25% silicon. Values of surface free energy calculated from the data are presented in Table 1 and these also decrease with

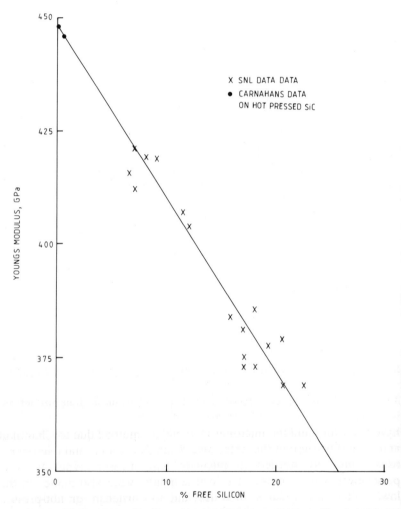

FIG. 3. Graph showing the dependence of Young's modulus on silicon content.

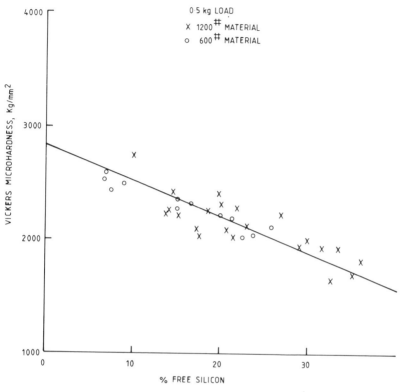

FIG. 4. Graph showing the dependence of microhardness on silicon content for a load of 500 g.

silicon content, a result which might be expected as silicon has a lower surface free energy than silicon carbide at room temperature.

5.4. The Effect of Silicon Content and Surface Finish on Strength

The variation of mean strength with silicon content for a 9 μm mean-grain-size material, finish ground with a 100 mesh diamond wheel is shown in Fig. 6 and again there is a decrease. The results, however, have been used in conjunction with the K_{IC} data to calculate flaw size and it has been found that this value is appreciably constant, at about 62 μm, and independent of silicon content, Table 2. The observed peak to valley roughness of the surface however was found to be much lower than this figure at 1·34 μm and so obviously grinding must produce damage which cannot be measured using a talysurf.

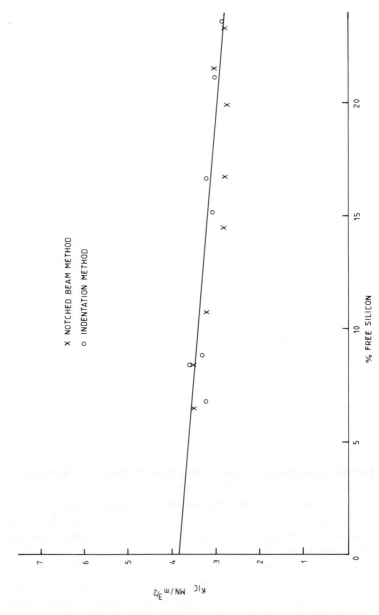

FIG. 5. Graph showing the dependence of fracture toughness on silicon content.

TABLE 1

% Silicon	K_{IC} (MN m$^{-3/2}$)	E (GPa)	γ (kJ m^{-2})
0	3·8	448	16·1
5	3·6	429	15·1
10	3·4	409	14·1
15	3·2	390	13·1
20	3·0	371	12·1
25	2·8	352	11·1

$K_{IC} = \sqrt{2E\gamma}$.

Similar specimens which had been ground with a 100 mesh wheel were finish-machined with a 220 mesh wheel. About 250 µm of material was removed and the results are shown in Fig. 7. Once more the data were used to calculate the flaw size which was found to have decreased to about 45 µm but again to be independent of silicon content, Table 2. The peak to valley talysurf measurement had also decreased to 0·88 µm but this was still considerably less than the calculated flaw size.

Finally material which had been finish-machined with a 220 mesh wheel was ground using a 600 mesh wheel. About the same amount of material was removed from the surface, i.e. about 250 µm, and although the RTM has decreased to 0·5 µm, the strength measurements were very similar to those obtained from the 220 mesh ground material, Fig. 7, and the calculated mean flaw size was the same.

For comparison with these last data the variation of mean strength with silicon content was measured for 4 µm mean-grain-size material ground using a 600 mesh wheel. The results are shown in Fig. 8. As in the case of the 9 µm material the calculated flaw size was found to be appreciably constant and independent of silicon content but considerably lower at 26 µm, Table 3.

It would thus appear that the slope of the branch curve in Fig. 2 is due to the fact that, for the same grinding procedure, the size of the flaw introduced decreases with grain size. For a given grain-size however the size of the flaw introduced on machining is independent of both the silicon content and, below some critical level, the grade of the diamond wheel used. Consequently the dynamics of the grinding machine would seem to be all important and whilst no comparative data have been obtained in this study, a mean strength of 608 MPa has

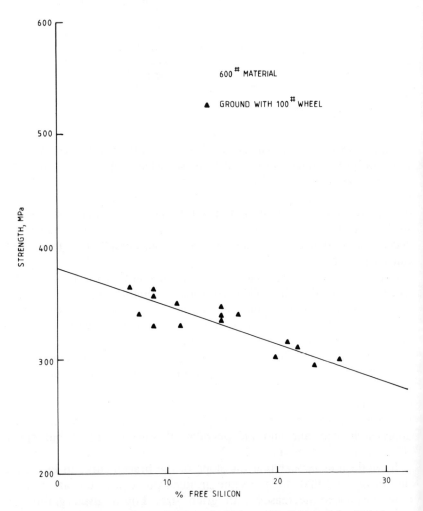

FIG. 6. Graph showing the dependence of MOR on silicon content for SiC of mean grain size 9 μm ground with a 100 mesh diamond wheel.

TABLE 2

% Silicon	K_{IC} (MN m$^{-3/2}$)	σ_{100} (MPa)	c_1 (μm)	σ_{220} (MPa)	c_2 (μm)
0	3·8	382	65·4	450	47·1
5	3·6	365	64·3	430	46·3
10	3·4	347	63·5	410	45·4
15	3·2	330	62·0	390	44·5
20	3·0	312	61·1	370	43·5
25	2·8	295	59·5	350	42·3

$K_{IC} = 1 \cdot 23 \sigma c^{1/2}$.

been obtained for 9 μm material, containing 10% free silicon, ground on a vertical spindle machine, i.e. having a flaw of 21 μm.

6. CONCLUSIONS

The principal flaws present in REFEL silicon carbide are grains and silicon filled pores.

The observed strength decreases with silicon content because the fracture toughness decreases with silicon content.

The observed strength increases as the grain size decreases and is proportional to (grain size)$^{-1/2}$ to a point where the size of the flaw introduced by machining exceeds the grain size. Thereafter the strength continues to increase as the grain size decreases because the size of the flaw introduced on machining depends on grain size. However the slope of the curve is less.

TABLE 3

% Silicon	K_{IC} (MN m$^{-3/2}$)	σ (MPa)	c (μm)	
0	3·8	604	26·2	
5	3·6	572	26·2	1 200 material
10	3·4	539	26·3	600 grind
15	3·2	506	26·4	
20	3·0	473	26·6	
25	2·8	440	26·8	

$K_{IC} = 1 \cdot 23 \sigma c^{1/2}$.

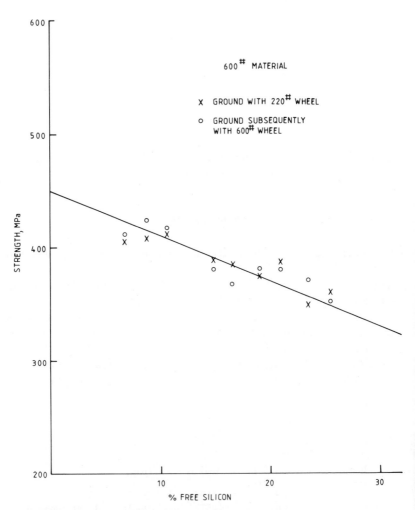

FIG. 7. Graph showing the dependence of MOR on silicon content for SiC of mean grain size 9 μm ground with a 220 mesh wheel (×) and a 600 mesh wheel (○).

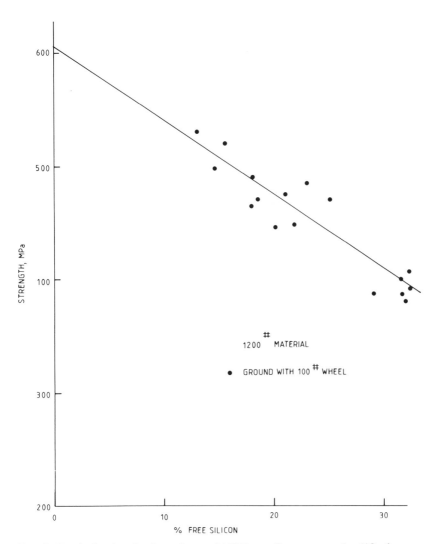

FIG. 8. Graph showing the dependence of MOR on silicon content for SiC of mean grain size 4 μm ground with a 100 mesh diamond wheel.

The point at which the discontinuity in the strength/(grain size)$^{-1/2}$ curve occurs depends in the first instance, on the grade of diamond wheel used to machine the specimens. In the limit, however, for finer grades of wheel, machine dynamics become more important.

REFERENCES

1. FORREST, C. W., KENNEDY, P. and SHENNAN, J. V. In *Special Ceramics*, Vol. 5, Popper, P. (Ed.), British Ceramic Research Association, Stoke-on-Trent, p. 99, 1972.
2. GRIFFITH, A. A. *Philosophical Transactions of the Royal Society*, **A221** (1920) 613.
3. CHANTIKUL, P., ANTIS, G. R., LAWN, B. R. and MARSHALL, D. B. *Journal of the American Ceramic Society*, **64** (1981) 541.
4. MCLAREN, J. R., TAPPIN, G. and DAVIDGE, R. W. *Proceedings of the British Ceramic Society*, **20** (1972) 259.
5. POPP, C. and PABST, R. F. *Metal Science* (1981) 130.
6. CARNIGLIA, S. C. *Journal of the American Ceramic Society*, **55** (1972) 243.
7. CRANMER, D. C., TRESSLER, R. E. and BRADT, R. C. *Journal of the American Ceramic Society*, **60** (1977) 230.
8. CARNAHAN, R. W. *Journal of the American Ceramic Society*, **51** (1968) 223.

COMMENTS AND DISCUSSION

Chairman: E. R. PETTY

P. Popper: Firstly, I think you said different customers want different silicon contents—I am intrigued by that—and, secondly, have you done any lapping for finishing?

P. Kennedy: Well, first of all, I think the lower the silicon content, the better—that seems obvious. As for higher silicon contents, well there are occasions when we have been asked to produce material containing 20–30% silicon. As far as lapping is concerned, it has a very bad effect—the strength usually drops considerably. Now, I can only suppose that all the particles in the lapping media are acting as point indentors and we are putting in a lot of sub-surface damage in the lapping operation.

D. P. Thompson: After you have carried out the machining, is there any way you can do either a heat treatment or a surface nitriding/oxidation treatment? Secondly, do you actually have silicon carbide–

silicon carbide bonds between grains or do you always have a layer of silicon?

Kennedy: Yes, surface oxidation at about 1000°C gives 10–20% increase in strength. None of these samples are oxidised before testing but it would have increased the strength. As for surface nitriding, we have had mixed results. Because the molar volume of nitride is greater than that of the silicon you can induce a lot of surface stress and the component can break up quite easily. On the bonding question, the original silicon carbide grains are over-coated by an epitaxial layer of silicon carbide, it does not precipitate out in the free silicon but actually bonds to the original grains and they grow. It is very difficult to differentiate between the original grains and the overcoat and where these epitaxial layers grow together they coalesce and a bond is formed. It is possible under certain conditions to get the silicon carbide coming out of solution, homogeneously nucleating in the free silicon phase, so forming a fine cubic silicon carbide which does not contribute to the strength of the material. When it lays down on the individual grains and bonds them together it does contribute to the strength and incidentally the overgrowth copies the morphology of the base grain so it is a hexagonal silicon carbide.

23

Failure Probability of Shouldered and Notched Ceramic Components Using Neuber Notch Theory

H. FESSLER

Department of Mechanical Engineering, University of Nottingham, UK

AND

D. C. FRICKER

Howmedica International Inc., Raheen Industrial Estate, Limerick, Ireland

ABSTRACT

Neuber's theory for hyperbolic notches and photoelastic results of edge stresses in symmetrical shouldered and notched flat bars with circular arc fillets have been used to obtain expressions for the failure probability of ceramic components. The evaluation of the failure probability or the predicted mean failure stress for any practicable shape only needs the material properties and the elastic stress concentration factor. Reference is made to previously published work concerning as-fired reaction-bonded silicon nitride (RBSN) turbine blade roots.

NOTATION

a	half minimum width of shouldered bar, dovetail and Neuber notch
b	width of shoulder
m	Weibull modulus
r	radius of circular arc fillets
s	distance along the edges from neck
t	thickness of component
u, v	elliptical curvilinear co-ordinates (see Fig. 2)
v	unit volume

x, y	Cartesian co-ordinates (see Figs. 1 and 2)
A	surface area of component
H	a step function; $H = 1$ if σ is +ve, $H > 8$ if σ is −ve
K	elastic stress concentration factor
M	bending moment per unit thickness
P	axial tension per unit thickness
P_f	failure probability
S	shear force per unit thickness
V	volume of component
ρ	minimum radius of curvature of hyperbolic notch
$\bar{\sigma}_{fv}$	mean failure stress of unit volume of material under uniform uniaxial tension
σ_{nom}	nominal stress at the neck
$\bar{\sigma}_{nom}$	mean nominal failure stress at the neck
$\sigma_{1,2,3}$	principal stresses
$\Sigma(A)$	stress-area integral
$\Sigma(E)$	edge-stress integral
$\Sigma(V)$	stress-volume integral

Subscripts

B	bending
T	tension
S	shear

1. INTRODUCTION

Because flaws of varying severity are randomly distributed throughout ceramic components, failure originates at the worst combination of stress and flaw, i.e. it need not originate at the position of greatest stress. It is therefore necessary to consider the failure probability of every part of the component. Weibull statistics have previously been developed to do this, using a four-function equation. For this (and any other rigorous) method it is necessary to know the principal stresses at every point in the component. However it has been established that in high-quality components, which have a high Weibull modulus m, most of the failures originate very near to the point of maximum stress.

This paper shows how the classical theory of elasticity solution for the stress distribution in deeply notched bars may be used to estimate the failure probability of other flat ceramic components if their stress concentration factor is known. This avoids the need for a complete stress analysis and the subsequent computation of the stress-volume integral (i.e. a significant amount of work) for every shape considered during design.

2. RIGOROUS EVALUATION OF FAILURE PROBABILITY

The failure probability of a brittle component depends on the mean strength and the variability of strength of the material, the shape and size of the component, as well as the type and magnitude of the loads. All these effects are incorporated[1] in the four-function Weibull equation

$$P_f = 1 - \exp\left\{-\left(\frac{1}{m}!\right)^m \left(\frac{\sigma_{nom}}{\bar{\sigma}_{fv}}\right)^m \frac{V}{v} \Sigma(V)\right\} \tag{1}$$

where

$$\Sigma(V) = \int_V \left[\left(\frac{\sigma_1}{\sigma_{nom} H(\sigma_1)}\right)^m + \left(\frac{\sigma_2}{\sigma_{nom} H(\sigma_2)}\right)^m + \left(\frac{\sigma_3}{\sigma_{nom} H(\sigma_3)}\right)^m\right] \frac{dV}{V} \tag{2}$$

The first function $\{(1/m)!\}^m$ is associated with strength variability only; it varies from 0·65 for $m = 5$ to 0·58 for $m = 20$. The second function $(\sigma_{nom}/\bar{\sigma}_{fv})^m$ combines the effects of material strength and applied load and is the only function affected by the magnitude of the load. The mean unit strength $\bar{\sigma}_{fv}$ allows the size-dependence of the strength of brittle components to be conveniently accounted for by the inclusion of the third function V/v. The effects of component shape and type of loading are represented in the fourth function, the stress-volume integral; the non-dimensional nature of $\Sigma(V)$, with respect to both stress and volume, is an important feature. It has previously been derived[2] that the mean nominal fracture stress $\bar{\sigma}_{nom}$ can be evaluated from $\bar{\sigma}_{nom} = \bar{\sigma}_{fv}/\{V\Sigma(V)/v\}^{1/m}$. Using the alternative assumption that all failures originate at the surface of the component leads to analogous equations in terms of areas instead of volumes. For low failure probabilities, eqn. (1) shows that P_f is linearly proportional to $\Sigma(V)$ or $\Sigma(A)$, which is the only function affected by changes to the shape of the component.

The evaluation of $\Sigma(V)$ or $\Sigma(A)$ requires significant computation. This paper shows how good estimates of the failure probability of notched and shouldered flat bars can be obtained simply, without such computation. The parts of the stress-area integrals contributed by the edges only were called $\Sigma(E)$ and were evaluated separately; $\Sigma(A)$ depends on the relative thickness a/t of the component whereas $\Sigma(V)$ and $\Sigma(E)$ do not. Experimental results for the distribution of edge stresses in shouldered bars, enabled $\Sigma(E)$ to be evaluated for them.

The proposed method depends on the observation that, although eqn. (1) assumes no threshold stress (failure may originate anywhere), failure is likely to occur very near to the greatest tensile stress, especially if this greatest tensile stress is localised, i.e. the stress gradients near it are large.

3. THE COMPONENTS

Shouldered and notched bars occur frequently in engineering; the greatest stresses normally occur in the fillets of these components, shown in Fig. 1. Because of this, much effort has been devoted to determine the elastic stress concentration factors K for these shapes by theory of elasticity, numerically using finite differences or finite elements and, experimentally, mainly using photoelasticity. Most of these results are conveniently presented by Peterson.[3]

Most of the published results only give K, i.e. the maximum tensile

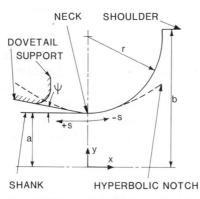

FIG. 1. Shapes, nomenclature and notation with co-ordinates.

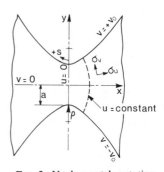

FIG. 2. Neuber notch notation.

stress divided by σ_{nom}, the nominal stress at the narrowest part. However, the stress distributions along the edges of some symmetrical shouldered plates, loaded in tension, have been published[4] and these photoelastic results are used here. Fessler *et al.*[4] give the stress distributions for 12 shapes with fillet radius ratios $r/2a = 0.036$, 0.072, 0.108 and 0.144; each of these radius ratios was tested with width ratios $b/a = 1.5$, 2.0 and 2.5. This covers the usual practical range. Fricker[5] gives results for two other edge stress distributions having $r/2a = 0.5$ and 1.0, both with $b/a = 3$ (see Fig. 1 for definitions of symbols).

Stress distributions in notched bars are available[5] from our frozen-stress photoelastic work on RBSN dovetailed turbine blade roots loaded in tension.[6] The dovetails used here had half angles $\psi = 10°$, and $30°$, fillet radius ratio $r/2a = 1$ and shoulder width ratio $b/a = 3$. The distance between the neck and start of the support was $0.4 < s/2a < 1.0$. Some of these models were loaded by static tension and these results are used here; others were subjected to centrifugal loading.

4. NEUBER NOTCHES

Neuber[7] uses curvilinear co-ordinate systems to define the geometry of a theoretical notch which best approximates the Cartesian geometry of the real notch. This doubly-symmetrical, curvilinear system shown in Fig. 2 defines boundary conditions and gives exact stress solutions for distant, uniform, static tension T, bending B and shear S. A symmetrical pair of single, deep, round (not pointed), external notches gives the best approximation to the shapes shown in Fig. 1. Because Neuber employs elliptical, orthogonal curvilinear co-ordinates, his notch profiles are hyperbolae, defined by $v = \pm v_0$ and unbounded with respect to u. The minimum radius of curvature ρ of the hyperbolic notch occurs at the neck. Figure 1 shows close agreement of profiles in the highly-stressed neck if $\rho = r$.

Appendix 1 gives the mathematical definition of the shape and size of Neuber notches, the co-ordinate axes transformation and scaling equations, followed by the complete principal stress index equations for tension. Similar derivations for bending and shear are presented in Ref. 5.

Since the boundary of the Neuber notch is simply defined as $v = \pm v_0$ in the curvilinear co-ordinate system, stress integrals are derived in Appendix 1 using this co-ordinate system. Because the extent of the notches in the x-direction is unlimited ($|u| \to \infty$), stress integrals were derived using a characteristic volume = $a^2 t$ and a characteristic edge area = $2at$. The equations for P_f and $\bar{\sigma}_{nom}$ above require the term $V\Sigma(V)/v$ for their evaluation where $\Sigma(V)$ is the integral over the component volume V divided by V. Hence $V\Sigma(V)/v$ is the value of the integral. V is usually the total volume of the component. For unlimited volumes, however, e.g. Neuber notches, shouldered bars and contact stress problems, it is necessary to refer to characteristic volumes (or areas) and evaluate $\Sigma(V)$ in terms of these volumes; the value of $V\Sigma(V)/v$ remains constant irrespective of the particular choice of characteristic volume. If the volume is unlimited, the integration must be truncated at a limit $\hat{u}$ where it is assumed that further contributions to the integral are negligible.

The evaluation of stress-volume integrals $\Sigma(V)$ and edge-stress integrals $\Sigma(E)$ was performed using double and single numerical integration procedures[5] as described in Appendix 2.

5. SHOULDERED BARS

Cubic splines were fitted to the 14 sets of edge stress values in and near the fillets because this gives good, smooth fits; the polynomial expressions are convenient for the evaluation of stress integrals, as described in Appendix 2. Figure 3(a) shows, for the shape with the highest stress concentration, the very good, smooth fit obtained. This was achieved by judicious positioning of the 'knots', the positions where the program matches the values of the functions and their first and second derivatives.

Choosing a characteristic area = $4at$ for the two free edges of these thin, flat plates, eqn. (2) reduces to

$$\Sigma(E)_T = 2 \int_{s_1}^{s_2} \left(\frac{\sigma_1}{\sigma_{nom}}\right)^m \frac{t \, ds}{4at} = \int_{(s_1/2a)}^{(s_2/2a)} (\sigma_1/\sigma_{nom})^m \, d(s/2a) \qquad (3)$$

where $\sigma_{nom} = P/2a$, the mean shank stress and where the limits of integration s_1 and s_2 along one edge are nominally chosen so that $(s_2/2a) - (s_1/2a) = 1$. The above length of edge is adequate if the limits are chosen to include the highly stressed fillet region. This was also demonstrated for the 'worst' case, i.e. for the shape with the lowest stress concentration ($r/2a = 1$, $b/a = 3$); this integration was

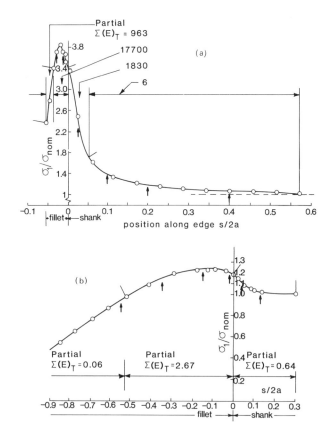

FIG. 3. Stress distribution along edges of shouldered bars in tension. Curves drawn from cubic splines fitted between arrows indicating knots. Extents of contributions to $\Sigma(E)_T$ for $m = 10$ shown by dimension lines. (a) Highest stress concentration $r/2a = 0.036$, $b/a = 2.5$; (b) lowest stress concentration $r/2a = 1.0$, $b/a = 3.0$.

carried out in three stages and the contributions to $\Sigma(E)_T$ are indicated in Fig. 3(b). Because Ref. 8 has shown that the percentage of failures initiating in any region approximately equals the proportion of the region's contribution to the total stress integral, Fig. 3 demonstrates that the likelihood of failure in the edges outside the chosen limits is insignificant. Hence their exact values are not critical and the length $(s_2 - s_1)/2a$ need not be exactly equal to 1 (it is less than unity in Fig. 3(a) but more in Fig. 3(b)).

The characteristic area and the limits of s in eqn. (3) have been specifically chosen so that $\Sigma(E)_T \to 1$ when $\sigma_1 \to \sigma_{nom}$, i.e. for a uniformly stressed uniform bar.

Edge-stress integrals $\Sigma(E)_T$ were evaluated for the 14 shapes for four m values and are presented in Fig. 4, using logarithmic scales of $\Sigma(E)_T$ and tensile stress concentration factor K_T. The equation describing the least-square fitted straight line relationships shown is

$$\Sigma(E)_T = K_T^{(0.95m-2.4)} \qquad (4)$$

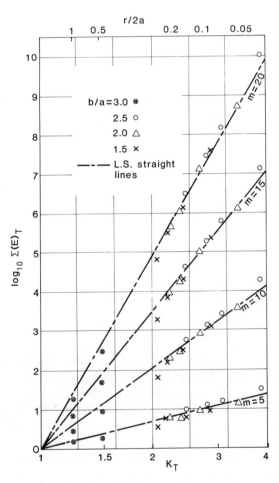

FIG. 4. Edge-stress integrals of shouldered bars in tension for four Weibull moduli m.

where it has been shown[4] that

$$K_T \simeq 1 + (0\cdot 34 + 0\cdot 1 b/a)\left(\frac{r}{2a}\right)^{-(0\cdot 28 + 0\cdot 08 b/a)} \quad (5)$$

By using eqns. (4) and (5) (or some other value for K_T) the edge-stress integral for any shouldered plate can be evaluated directly.

6. NOTCHED BARS

The differences of the principal stresses across turbine blade root models were obtained from the photoelastic fringe orders at the neck (see Fig. 1) and at several other sections. Figure 5 shows the comparison of these experimental results using a nominal stress equal to the mean neck stress, with values calculated using Neuber's deep notch solution for the same radius at the root of the hyperbolic notch ($\rho = r$). The similarity of the values at the neck for $\psi = 10°$ and $\psi = 30°$ confirms the general belief, illustrated by Peterson,[3] that the stress distribution at this neck is almost independent of the flank angles of the notches. It is more important that Fig. 5 also shows that the Neuber stress distributions for hyperbolic notches, to be used to calculate stress-volume and edge-stress integrals, are in good enough agreement with the measured values for circular-arc notches.

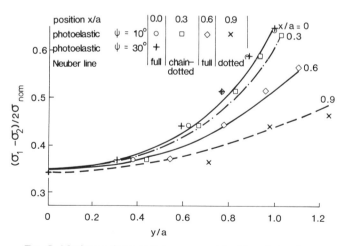

FIG. 5. Maximum shear stresses across notched bars in tension.

Figure 6 shows edge-stress integrals $\Sigma(E)_T$ and Fig. 7 shows stress-volume integrals $\Sigma(V)_T$ for four m values over the practical range of symmetrical, deep hyperbolic notches under tension. The contributions to $\Sigma(V)_T$ made by portions of the plate bounded by u = constant lines (see Fig. 2) were evaluated separately and it was found[5] that even for the largest root radius ($\rho = 2a$) and the lowest m value, 80% of failures will occur within the region bounded by $u = \pm 0.8$ (where $s = \mp 1.3a$ for $\rho = 2a$). For higher, more usual m values, practically all failures will occur in the above region, where it has been demonstrated in Fig. 5 that the stress distribution in the hyperbolic notch is very similar to that in a circular notch. For smaller root radii and typical m values, practically all failures will occur very close to the neck where the hyperbolic notches are even better approximations to circular-arc geometries (see Fig. 1) and Neuber stress distributions are even better approximations to those in circular notches. Figures 6 and 7 therefore apply to circular notches also.

Figures 6 and 7 show that $\log_{10} \Sigma(E)_T$ and $\log_{10} \Sigma(V)_T$ are approximately proportional to $\log_{10} K_T$. The least-square fitted straight lines also shown are defined by

$$\Sigma(E)_T = K_T^{(0.97m - 1.6)} \tag{6}$$

and

$$\Sigma(V)_T = K_T^{(0.84m - 3.0)}. \tag{7}$$

K_T is the tensile stress concentration factor for the notch considered, obtained from the usual sources e.g. Peterson.[3]

Pure bending of bars with deep, hyperbolic notches was also studied. From Neuber's bending theory[7] $\Sigma(E)_B$ and $\Sigma(V)_B$ were evaluated from integration over the tensile half of the bars (the contribution to failure probability of the compressive half is negligible) using the same characteristic edge area and volume as for tension. The nominal stress $\sigma_{nom} = 3M/2a^2$, the maximum elementary tensile bending stress at the neck. The results can be described by

$$\Sigma(E)_B = \tfrac{1}{2} K_B^{(0.95m - 2.1)} \tag{8}$$

and

$$\Sigma(V)_B = \frac{1}{2(m+1)} K_B^{(0.90m - 3.2)} \tag{9}$$

The least-square straight lines are as good fits for bending as those shown in Figs. 6 and 7 for tension. K_B, the stress concentration factor

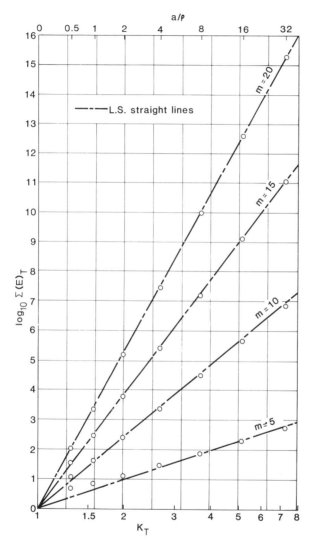

FIG. 6. Edge-stress integrals of Neuber notched bars in tension for four Weibull moduli m.

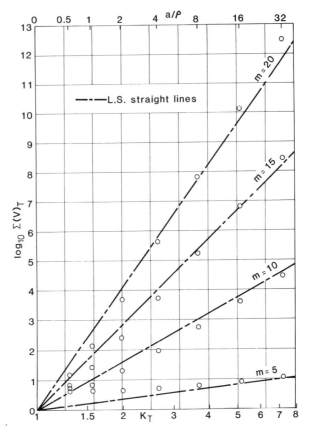

FIG. 7. Stress-volume integrals of Neuber notched bars in tension for four Weibull moduli m.

in pure bending may be obtained from Peterson.[3] The factors $1/2$ and $1/2(m+1)$ are the values of the edge-stress integral and the stress-volume integral for uniform bars (without the projections which form the notches) under pure bending.

Pure shear of bars with deep, hyperbolic notches was also studied. From Neuber's pure shear theory, $\Sigma(E)_S$ and $\Sigma(V)_S$ were evaluated[5] from integration over one quadrant of the bars (the stress distribution has anti-symmetry) using the same characteristic edge area $2at$ and characteristic volume a^2t as above. Figure 8 shows $\Sigma(E)_S$ and Fig. 9 shows $\Sigma(V)_S$ for four m values over the practical range of Neuber notches in pure shear. The maximum principal stress $\hat{\sigma}_1$ occurs at two

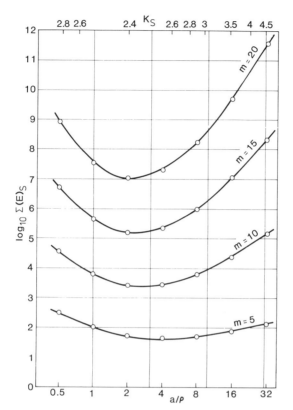

FIG. 8. Edge-stress integrals of Neuber notched bars in pure shear for four Weibull moduli m.

diagonally opposite points on opposite edges but as shown in Figs. 8 and 9, its ratio to $\sigma_{nom} = S/2a$, the mean neck shear stress gives K_S which does not increase monotonically with notch curvature a/ρ. Neuber[7] shows that the minimum value $K_S = 2\cdot 4$ occurs when $a/\rho = 2$. For this reason and also because of the overall stress distributions in shear, Figs. 8 and 9 show curved $\Sigma(E)_S$ and $\Sigma(V)_S$ variations with a/ρ, for which no attempt was made to approximate them.

7. DISCUSSION

Comparisons of eqns. (6) and (7) for tension, of eqns. (8) and (9) for bending and Figs. 8 and 9 for shear show that the edge-stress integrals

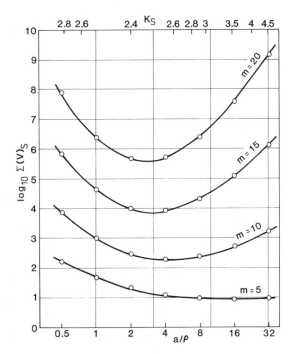

FIG. 9. Stress-volume integrals of Neuber notched bars in pure shear for four Weibull moduli m.

for notched bars are much bigger than the stress-volume integrals. For a typical value of Weibull modulus $m = 10$, $\Sigma(E)_T/\Sigma(V)_T = K_T^{2.7}$ and $\Sigma(E)_B/\Sigma(V)_B = 11 K_B^{1.6}$. This observation is primarily related to the stress gradients. The stress gradient from the surface across the neck $\partial \sigma_1/\partial y$ is much greater than $\partial \sigma_1/\partial s$, the stress gradient from the neck along the surface. (The stress gradient across the neck has to be large to satisfy equilibrium.)

The exponents in eqns. (4), (6), (7), (8) and (9) are very similar; all are less than the Weibull modulus m and all of them approach m as $m \to \infty$, i.e. in agreement with deterministic elastic stress analysis.

Although the maximum stress in shouldered bars does not occur at $s = 0$ (see Fig. 3), it is still reasonable to say that in the important (because most highly stressed) region near the neck the stress distributions in shouldered and notched bars are effectively similar for ceramic components of equivalent K_T (K_T for a shouldered bar is lower than K_T for the corresponding geometry Neuber notch). This is shown by comparing the results in Fig. 4 with those in Fig. 6: the $\Sigma(E)_T$ values in Fig. 6 are close to twice (the ratio of characteristic edge areas, viz. $4at/2at$) those in Fig. 4, which is what their ratio would be if both edge-stress distributions were identical. Because of this and the above it is suggested that, in the absence of $\Sigma(V)_T$ values (or $\Sigma(V)_B$ or $\Sigma(V)_S$ values) for shouldered bars, these may be obtained approximately from the assumption that

$$\left[\frac{\Sigma(V)_T}{\Sigma(E)_T}\right]_{\text{shouldered}} \simeq \left[\frac{\Sigma(V)_T}{\Sigma(E)_T}\right]_{\text{notched}} \quad (10)$$

Using the same value of maximum stress (e.g. K_T) not the same fillet radius,

$$\Sigma(V)_{\text{shouldered}} \simeq K_T^{(-0.13m-1.4)} K_T^{(0.95m-2.4)}$$

$$\simeq K_T^{(0.82m-3.8)} \quad (11)$$

referred to a characteristic volume $= 2a^2t$.

The use of characteristic volumes and edge areas to allow the evaluation of $\Sigma(V)$ and $\Sigma(E)$ has useful consequences, viz. the failure probability of a complex ceramic component can be calculated using eqn. (1) where the total product $V\Sigma(V)/v$ can be evaluated by simply adding together appropriate contributions from individual regions of the component, all referred to the same nominal stress. Reference 5 illustrates this for an RBSN shouldered bar of $b/a = 3$ and shank and shoulder lengths equal to $4a$. Reference 8 has used the properties of the product $V\Sigma(V)/v$ to present the value of this term evaluated from a large (linear scale factor λ) photoelastic model; the value for the ceramic prototype is then $(V\Sigma(V)/v)/\lambda^3$, i.e. $\Sigma(V)$ is the same for both the model and prototype, whose total volume is $1/\lambda^3$ times that of the model.

The disadvantage to the choice of characteristic volume and area here for Neuber notches is that the curves in Figs. 6 and 7 are not constrained to approach the origins of the axes when $a/\rho \to 0$ and $K_T \to 1$. However, such low K_T values are unusual and the least-square approximations are reasonably accurate for typical geometries and typical Weibull moduli.

The effective surface area of shouldered and notched bars consists of two faces and two edges (the stresses at the ends are too low to contribute to $\Sigma(A)$). For Neuber notches where the characteristic area = $2at$, Appendix 1 defines the stress function M and so

$$\Sigma(A) = \frac{1}{2at}\int_A M\,dA = \frac{1}{2at}\left[\begin{array}{l}\text{Integral over area of two faces}\\+ \text{Integral over area of edges}\end{array}\right]$$

Appendix 1 shows that these integral contributions give

$$\Sigma(A) = \frac{1}{2at}[2a^2\Sigma(V) + 2at\Sigma(E)]$$

Hence

$$\Sigma(A) = (a/t)\Sigma(V) + \Sigma(E) \qquad (12)$$

For shouldered bars, the characteristic volume is also $a/2$ times the characteristic area and so eqn. (12) also applies. The relevant $\Sigma(V)$ and $\Sigma(E)$ values can be used directly to simply evaluate $\Sigma(A)$ for any required (half neck width)/(bar thickness) ratio a/t and any type of loading.

Under combined tension, bending or shear loading, the individual Neuber elastic stress solutions must be superposed and stress integrals then evaluated or approximated from the total stress distribution.

8. CONCLUSIONS

It has been shown that the stress distribution in one circular-arc notch with two different flank angles is very similar in the critical neck region to that in the geometrically equivalent hyperbolic Neuber notch. It is reasonable to assume the same is true for other circular notch geometries, although more accurate approximations will be obtained from the Neuber notch whose elastic stress concentration factor is the same.

Stress integrals have been presented for Neuber notches in tension, bending and shear and these stress integrals allow the failure probability P_f or the mean failure stress $\bar{\sigma}_{\text{nom}}$ to be evaluated. Empirical equations have been given relating stress integrals to the elastic stress concentration factor K and so for shouldered and notched ceramic components with circular arc fillets, approximate values for P_f and $\bar{\sigma}_{\text{nom}}$ can easily be calculated given K and the material properties. Complex components with several fillets can also be simply treated during design.

ACKNOWLEDGEMENTS

This work was carried out with the support of the Procurement Executive, Ministry of Defence. The authors also thank the technicians at Nottingham University for their skilful assistance.

REFERENCES

1. STANLEY, P., FESSLER, H. and SIVILL, A. D. *Proceedings of the British Ceramic Society*, **22** (1973) 453.
2. STANLEY, P., SIVILL, A. D. and FESSLER, H. *Proceedings of the 5th International Conference on Experimental Stress Analysis*, CISM, Udine, 2.42, 1974.
3. PETERSON, R. E. *Stress Concentration Factors*, John Wiley, New York, 1974.
4. FESSLER, H., ROGERS, C. C. and STANLEY, P. *Journal of Strain Analysis*, **4** (1969) 169.
5. FRICKER, D. C. Ph.D. thesis, University of Nottingham, 1979.
6. FESSLER, H. and FRICKER, D. C. *Proceedings of the British Ceramic Society*, **26** (1978) 81.
7. NEUBER, H. *Theory of Notch Stresses*, J. W. Edwards Co., Ann Arbor, pp. 38–47, 1946.
8. STANLEY, P., SIVILL, A. D. and FESSLER, H. *Journal of Strain Analysis*, **13** (1978) 103.
9. FESSLER, H. and FRICKER, D. C. *Journal of the American Ceramic Society*, **67** (1984) 582.
10. Nottingham Algorithms Group Library, University of Nottingham, 1973.

APPENDIX 1: STRESS INTEGRALS FOR A SYMMETRICAL PAIR OF UNIFORMLY 'THIN', SINGLE, EXTERNAL, ROUNDED, DEEP NEUBER NOTCHES SUBJECTED TO TENSION

Neuber[7] shows that the shape and size of any notch are defined in terms of the neck geometry by $\tan^2 v_0 = a/\rho$ and $a = \sin v_0$; the transformation from elliptical, orthogonal curvilinear co-ordinates to normalised Cartesian co-ordinates is given by $x/a = \sinh u \cos v / \sin v_0$ and $y/a = \cosh u \sin v / \sin v_0$. Furthermore for this co-ordinate system transformation, the 'distortion' factors h_u and h_v are

$$h_u^2 = \left(\frac{\partial x}{\partial u}\right)^2 + \left(\frac{\partial y}{\partial u}\right)^2 = h_v^2 = \left(\frac{\partial x}{\partial v}\right)^2 + \left(\frac{\partial y}{\partial v}\right)^2 = h^2$$

So $h^2 = \sinh^2 u + \cos^2 v$ and the Jacobian

$$J = \begin{vmatrix} \dfrac{\partial x}{\partial u} & \dfrac{\partial x}{\partial v} \\ \dfrac{\partial y}{\partial u} & \dfrac{\partial y}{\partial v} \end{vmatrix} = h^2$$

Neuber gives the elastic plane stress solution as

$$\frac{\sigma_u}{\sigma_{\text{nom}}} = \frac{B}{h^2} \cosh u \cos v \left(2 + \frac{\cos^2 v_0 - \cos^2 v}{h^2}\right)$$

$$\frac{\sigma_v}{\sigma_{\text{nom}}} = \frac{B}{h^4} \cosh u \cos v (\cos^2 v - \cos^2 v_0)$$

$$\frac{\tau_{uv}}{\sigma_{\text{nom}}} = \frac{B}{h^4} \sinh u \sin v (\cos^2 v_0 - \cos^2 v)$$

where $B = \sin v_0/(v_0 + \sin v_0 \cos v_0)$.

The principal stress indices are then

$$\frac{\sigma_1}{\sigma_{\text{nom}}}, \frac{\sigma_2}{\sigma_{\text{nom}}} = \frac{\dfrac{\sigma_u}{\sigma_{\text{nom}}} + \dfrac{\sigma_v}{\sigma_{\text{nom}}}}{2} \pm \left[\frac{\left(\dfrac{\sigma_u}{\sigma_{\text{nom}}} - \dfrac{\sigma_v}{\sigma_{\text{nom}}}\right)^2}{4} + \left(\frac{\tau_{uv}}{\sigma_{\text{nom}}}\right)^2\right]^{1/2}$$

$$\frac{\sigma_3}{\sigma_{\text{nom}}} = 0$$

The stress concentration factor K_T is the value of $\sigma_1/\sigma_{\text{nom}}$ at the points $u = 0$, $v = \pm v_0$ and Neuber has shown that

$$K_T = \frac{2(a/\rho + 1)\sqrt{a/\rho}}{(a/\rho + 1)(\arctan \sqrt{a/\rho}) + \sqrt{a/\rho}}$$

Using the independent-action failure criterion,[1] the multiaxial failure criterion function[9] M is given by

$$M(u, v) = (\sigma_1/\sigma_{\text{nom}})^m + (\sigma_2/\sigma_{\text{nom}})^m$$

since the biaxial stress state is always tension–tension for tensile loading.

Using a characteristic volume = $a^2 t$ and the symmetry of stresses in each quadrant, the stress-volume integral is given by

$$\Sigma(V)_T = \frac{4}{a^2 t} \int_0^{\hat{a}} \int_0^{v_0} M(u, v) J \, dv \, du \, t$$

so

$$\Sigma(V)_T = \frac{4}{\sin^2 v_0} \int_0^{\hat{u}} \int_0^{v_0} M(u, v) J \, dv \, du$$

Using a characteristic area $= 2at$, the edge-stress integral is given by

$$\Sigma(E)_T = \frac{1}{2at} \int_s M(u, \pm v_0) \, ds \, t$$

Now

$$ds = -\left[\left(\frac{dx}{du}\right)^2 + \left(\frac{dy}{du}\right)^2\right]^{1/2} du = -h \, du$$

and the symmetry of edge stresses gives

$$\Sigma(E)_T = \frac{2}{\sin v_0} \int_0^{\hat{u}} M(u, v_0) h \, du$$

APPENDIX 2: NUMERICAL EVALUATION OF INTEGRALS

Single Integrals

An available Fortran subroutine D01ACF[10] was used to evaluate the definite line integral of the form

$$I_1 = \int_{s_1}^{s_2} f(s) \, ds$$

where the user-supplied function $f(s)$ is given in eqn. (3) for shouldered bars and in Appendix 1 for a Neuber notch in tension. Besides the function and the limits of integration, a required relative accuracy is an input to the subroutine. This then successively applies a series (maximum number eight) of Gaussian quadrature formulae, using the Patterson method of optimum addition of points to Gauss formulae (the maximum number of points here is 255) until two successive results differ by less than the required relative accuracy. The latest value of the integral, the number of points used, the relative accuracy and a success or failure message are then the subroutine output. This subroutine works well with 'difficult' quadratures with rapidly varying integrals, e.g. $(\sigma_1/\sigma_{nom})^{20}$, although the integration range was subdivided to forestall problems and to give the distribution of contributions to the edge-stress integral. Accidental premature convergence never occurred (see Fig. 6).

For shouldered bars, Fig. 3 shows the typical subdivision of the range $s_2 - s_1 = 2a$ into three or four regions where the required relative accuracy for each was 0·1%; $\Sigma(E)_T = I_1$ as given by eqn. (3). For Neuber notches, the subdivided integration range was in steps of constant Δu (geometrically smaller near the neck than away from it) using a constant required relative accuracy of 0·01% until the latest contribution was less than 0·1% of the total integral up to $\hat{u}$. For $m = 5$, 10, 15 and 20, the step sizes were respectively $\Delta u = 0.4$, 0·3, 0·2 and 0·2 for all a/ρ values. $\Sigma(E)_T = 2I_1/\sin v_0$ as shown in Appendix 1.

Double Integrals

Two different subroutines D01ACF and D01AAF[10] were used to evaluate the integral

$$I_2 = \int_0^{\hat{u}} \int_0^{v_0} M(u, v) J \, dv \, du = \int_0^{\hat{u}} \left[\int_0^{v_0} M(u, v) J \, dv \right]_u du$$

the output of the 'difficult' inner quadrature given by D01ACF, being the input for D01AAF which specified the local value of u. The subroutine D01AAF operates in a similar way to D01ACF in that it successively applies a series of Chebyshev's quadrature formulae (the maximum number of points being 128) until the required relative accuracy is achieved. It is best suited to 'well-behaved' quadratures as here.

For Neuber notches, the inner quadrature range from 0 to v_0 was not subdivided but the outer range of u was subdivided into the same step sizes as above. The required relative accuracies were both 0·01% as above and the limit for $\hat{u}$ was again defined as above. $\Sigma(V)_T = 4I_2/\sin^2 v_0$ as shown in Appendix 1.

COMMENTS AND DISCUSSION

Chairman: E. R. PETTY

S. Hampshire: I know that some of your previous work was concerned with reaction-bonded silicon nitride. What implications does your present paper have for design of engineering components in this type of material?
D. Fricker: The design of a component is simply trying to minimise the maximum tensile stress in the component and still allowing it to

function in the way that it is required to. Better material will give more safety for a specific load or you can increase the load and get better performance with the same reliability. You can design irrespective of the material that you are going to use.

W. Grellner: Did you make any comparisons between your solutions and possible finite element analysis calculations?

Fricker: No. This has been done previously. In fact, manufacturers would put into the calculations the parts of the analysis to calculate failure probabilities after getting principal stresses from finite element analysis. They simply take the elements and do some integration over the volume.

24

Characterisation and Mechanical Properties of Hot-pressed and Reaction-bonded Silicon Nitride

R. C. PILLER, K. P. BALKWILL, A. BRIGGS AND R. W. DAVIDGE

AERE Harwell, Didcot, Oxon, OX11 0RA, UK

ABSTRACT

Samples of RBSN and HPSN provided by British Companies are characterised in terms of their chemical compositions and microstructures. Details of ceramic machining procedures are given and mechanical properties at ambient and temperatures up to 1500°C measured using Modulus of Rupture and Stepped Temperature Stress Rupture tests.

1. INTRODUCTION

The work described constitutes the initial stages of a programme whose objectives are to generate high temperature mechanical properties data for a wide range of ceramics and to understand failure mechanisms in order to be able to predict service life. The programme will eventually produce data for various commercial silicon nitrides, silicon carbides, alumina, zirconia and sialon samples using techniques such as stress rupture, stepped temperature stress rupture, 3 and 4 point bend, creep and dynamic fatigue tests. This paper concentrates on the characterisation and properties of reaction-bonded silicon nitride (RBSN) and hot-pressed silicon nitride (HPSN).

2. MATERIALS

A.E. Ltd provided six $125 \times 125 \times 5$ mm plates of RBSN (Nitrasil) produced in May 1984. A.M.E. Ltd supplied two slabs of HPSN produced in July, 1984; one 155 × 110 × 15 mm and the other cut into two pieces 110 × 80 × 18 mm. Chemical analysis using optical emission spectroscopy gave the results in Table 1 which show that the main impurity elements found in the RBSN were aluminium, calcium and iron whilst the HPSN contained these and a high level of magnesium. The presence of all these elements can be accounted for either as intentional additions or by pick-up during processing, e.g. the amount of magnesium in the HPSN indicates that MgO was used as a sintering aid during hot-pressing. Iron and calcium are common impurities in silicon powder and iron has also been shown to have beneficial effects on sintering.[1] It may also be used as a catalyst in the nitriding process for RBSN.[2]

All the samples required cutting and machining to give suitably sized specimens for mechanical testing. The cross-sectional dimensions of all specimens were 4 × 4 mm, and two lengths, 20 mm and 40 mm, were used; the shorter lengths in 3-point bend Modulus of Rupture (MOR) tests and the longer lengths in the Stepped Temperature Stress Rupture tests. Special care was taken in machining the specimens to size, including directing all grinding to the longitudinal direction, chamfering the edges to remove damage (due to a breakdown in communications only the HPSN specimens were chamfered), and light cuts only for removal of the last 0·5 mm of material. Finishing was done using a resin-bonded diamond grit impregnated wheel (100 μm grit size) and a maximum depth of cut of 0·025 mm on each pass.

The densities of the individual specimens of both materials were measured using a water immersion technique. The average value obtained from 62 RBSN specimens was $2·529 \pm 0·013 \text{ g cm}^{-3}$ and for 20 HPSN specimens $3·170 \pm 0·015 \text{ g cm}^{-3}$. Both values vary slightly from those quoted in the manufacturers' brochures ($2·35-2·45 \text{ g cm}^{-3}$ for RBSN and $3·19 \text{ g cm}^{-3}$ for HPSN).

Scanning electron micrographs of machined specimens of RBSN and HPSN are shown in Figs. 1 and 2 respectively. The predominant

TABLE 1
Chemical Composition of A.E. RBSN and A.M.E. HPSN

	Al	B	Ca	Cr	Cu	Fe	Mg	Mn	Na	Ni	Y
AE RBSN	0·7%	—	0·1%	20	30	0·5%	300	100	500	70	30
AME HPSN	0·7%	200	0·1%	300	70	3%	0·7%	200	—	100	5

All values, except where noted, are in parts per million.

Mechanical Properties of Hot-pressed and Reaction-bonded Silicon Nitride   343

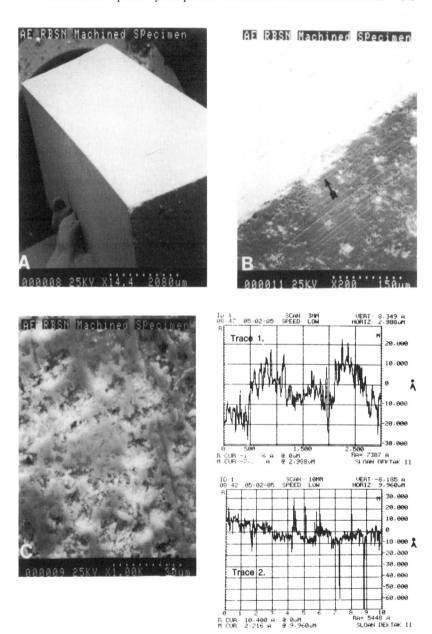

FIG. 1. Scanning electron micrographs and surface profile traces for as-machined specimen of reaction-bonded silicon nitride.

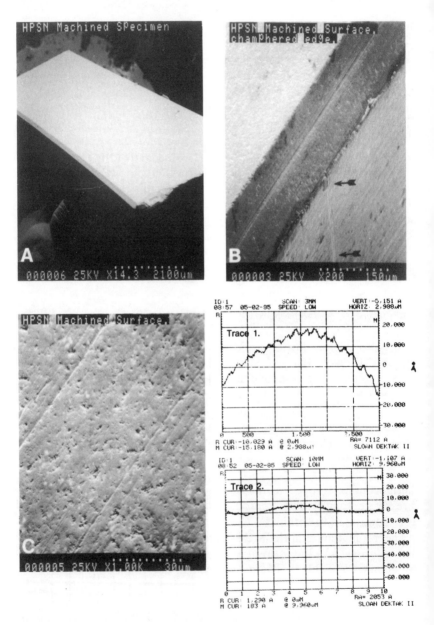

FIG. 2. Scanning electron micrographs and surface profile traces for as-machined specimen of hot-pressed silicon nitride.

feature of RBSN is its high porosity, much of which is inherent, but some of which may be caused by pull-out of grains. Other machining damage, in the form of long striations and cracks, might be negligible in comparison with the porosity. In unchamfered specimens edge damage such as that shown in Fig. 1(B) occurred as a result of chipping during processing or careless handling. The defect shown in Fig. 1(B) is of the order of 100 μm long. The surface of machined HPSN, Fig. 2(C), is less porous than that of RBSN, but has more predominant machining striations. The traces shown in Figs. 1 and 2 are surface profiles in the transverse (trace 1) and longitudinal (trace 2) directions of both materials after machining, obtained with a DECTAK surface profiler. The traces corroborate the SEM observations on the respective qualities of the machined surfaces.

Although the HPSN specimens were chamfered occasional minor edge damage still occurred, as shown in Fig. 2(B), and associated cracks could serve as fracture origins.

3. EXPERIMENTAL AND RESULTS

Two types of data are reported, Modulus of Rupture (MOR) under constant displacement rate and Stepped Temperature Stress Rupture (STSR) under static loading conditions. Room temperature MOR data were obtained from 3-point bend-tests, with a span of 18 mm, using an Instron Universal testing machine with a constant cross head displacement speed of 0·5 mm min^{-1}. For high temperature tests a furnace was fitted to the machine, and a span of 15 mm was used. A cross-check, using a 15 mm span at room temperature, gave consistent results. The furnace was calibrated at each temperature using a Pt/Pt + 13% Rh thermocouple strapped to the tensile face of the specimen. Temperatures were found to be reproducible to ±5°C and, once stabilised, were controlled to within ±1°C.

The room temperature MOR data for the RBSN and HPSN are shown in Fig. 3 in the form of Weibull failure probability plots,[3,4] and are detailed in Table 2. The mean strength determined for the RBSN, 217 ± 17 MPa, is greater than that quoted by the manufacturer, 190 MPa, while the Weibull modulus of 15 is good although values greater than 20 have been obtained with other RBSNs.[7] A selection of photographs of the fracture surfaces is shown in Fig. 4. The specimens are positioned so that the tensile surfaces of the two halves are

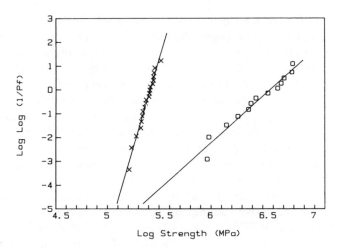

FIG. 3. Weibull plots of Modulus of Rupture strengths against failure probability (Pf) for RBSN (×) and HPSN (○). Tested at room temperature.

together, and fracture initiation sites are marked with arrows. Most fracture sources were identified and they were generally regions where large pores intersected the surface; a typical example is shown in Fig. 5.

The HPSN gave a mean strength of 648 ± 175 MPa compared to the manufacturer's value of 800 MPa. The spread of the results is large, as indicated by the very low Weibull modulus of 4. Fracture surfaces of

TABLE 2
Summary of MOR Results

Temperature (°C)	A.E. RBSN MOR (MPa)	Weibull modulus	A.M.E. HPSN MOR (MPa)	Weibull modulus
25	217 ± 17	15·2(15)	648 ± 175	4(20)
627	220 ± 15	16·1(11)	581 ± 137	—(5)
834	230 ± 15	16·7(7)	708 ± 72	—(5)
1 066	214 ± 23	—(5)	525 ± 29	—(5)
1 242	213 ± 30	—(5)	346 ± 57	—(5)
1 410	174 ± 8	—(5)	128 ± 9	—(5)
1 455	—	—	120	—(1)
1 505	173	—(3)	98	—(1)

Numbers in parentheses indicate number of specimens tested at that temperature.

FIG. 4. Fracture surfaces of specimens of RBSN broken in 3-point bend tests at room temperature. Magnification ×10.

some of the specimens are shown in Fig. 6, and again arrows indicate the fracture initiation sources, where identified. In contrast to the case of RBSN, fracture initiation in HPSN occurred at various sites; in some cases, such as specimens 3, 10 and 11, a surface flaw which could be associated with the machining damage was found but in others, for example specimen 6, the fracture was initiated at an internal flaw. Specimen 9, the strongest of the batch, with a MOR strength of 888 MPa, shows no identifiable fracture source. A high magnification

FIG. 5. SEM high magnification of typical fracture initiation source in RBSN specimen.

micrograph of the fracture origin for specimen 3 (Fig. 7) shows sub-surface damage caused by two adjacent machining marks. In at least eight of the fifteen specimens tested at room temperature fracture was initiated at such flaws created by machining.

Sets of five specimens of each material were tested at various temperatures between 25 and 1500°C. Typical Instron load/deflection curves are shown in Fig. 8 and strength results are detailed in Table 2. No reduction in the strength of the RBSN was observed up to 1240°C; in fact, as Fig. 9 shows, a slight increase occurred. Above 1240°C the strength decreased but even at 1500°C 80% of the room temperature strength was retained. The MOR strength of the HPSN remained constant, within the experimental scatter, up to 850°C but then fell rapidly until at 1410°C it was only 20% of the room temperature value. At 1410°C the strength of the HPSN was less than that of the RBSN. Both load/deflection graphs for RBSN are linear up to the point of

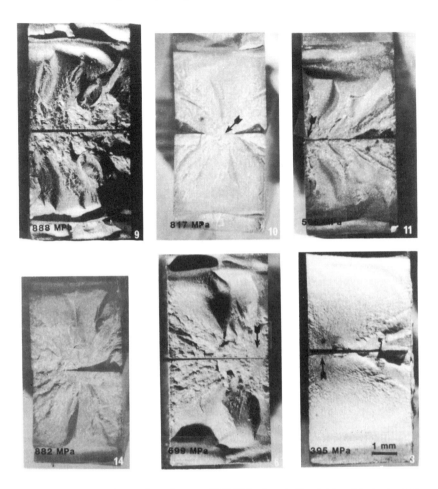

FIG. 6. Fracture surfaces of specimens of HBSN broken in 3-point bend tests at room temperature. Magnification ×10.

failure whereas those for HPSN show curvature at high loads. Normally, curvature in such graphs would indicate plastic deformation but in the HPSN viscous flow of a glassy phase at grain boundaries is probably responsible for this observation. The glass transition temperature for HPSN containing MgO as a sintering aid is in the range 850–900°C,[5] this is also the region in which curvature appears in the load/displacement graphs.

The Stepped Temperature Stress Rupture (STSR) test under static

FIG. 7. SEM high magnification of fracture initiation source in HPSN specimen showing sub-surface damage associated with machining striations.

loading was developed and used extensively by Quinn and co-workers[6,7] at the American Army Materials and Mechanics Research Centre, and is a quick method of assessing a material's stability over a wide range of temperature and stress. The specimen is heated to 1000°C in air, loaded to the required stress level and left for 24 h. If at the end of this period it is still intact the temperature is increased to 1100°C and every 24 h thereafter raised a further 100°C until failure occurs. The results, shown in Fig. 10, are presented on a graph of temperature against time; the arrows indicating at what stage of the cycle failure occurred under the indicated applied stress. Both the RBSN and HPSN failed at stresses lower than those found in the dynamic MOR tests. For example, under a stress of 121 MPa two RBSN specimens failed on heating from 1100°C to 1200°C whereas dynamic tests at 1100°C and 1200°C gave strengths of ~214 MPa.

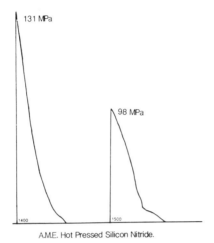

FIG. 8. Load/displacement graphs from 3-point bend tests at elevated temperatures.

Similarly, the HPSN failed on heating from 1100 to 1200°C under a static stress of 166 MPa (one specimen failed even earlier in the cycle) whilst the dynamic tests gave strengths of 480 MPa at 1100°C and 390 MPa at 1200°C respectively. Exposure to oxidising conditions for long periods under static stress obviously had an adverse effect on the strength.

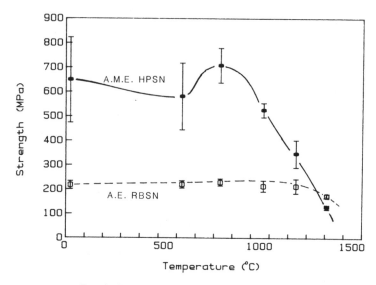

FIG. 9. Temperature dependence of MOR strength.

Figure 11 shows SEM micrographs, at various magnifications, of the surface of a RBSN specimen which was stressed to 82 MPa and survived until 2 h into the 1300°C stage of the STSR cycle. The surface was oxidised to a depth of 5–10 μm and the oxide layer contained an apparently random distribution of large pits; diameter approximately 50 μm. In the regions of greatest stress the oxide had cracked and in places was detached from the bulk of the specimen as revealed in Fig. 11(A) by the contrasting area near the fracture surface (bottom of micrograph). No obvious source of fracture initiation could be found (Fig. 11(B)), although there were several regions of high porosity or large oxidation pits which could have served as fracture origins. There was no evidence for slow crack growth on the fracture surface.

Figure 12 shows SEM micrographs of a HPSN specimen which failed 1 min into the 1300°C stage of the STSR cycle under a stress of 132 MPa. As in the case of RBSN there was a 5–10 μm thick oxide layer and pit formation, although the morphology of the pits, evident in Fig. 12(A), was very different from those found with the RBSN, they were flat-bottomed and penetrated only slightly into the bulk of the material. The fracture origin on this specimen was identifiable, being in the upper right-hand corner of the fracture face shown in Fig.

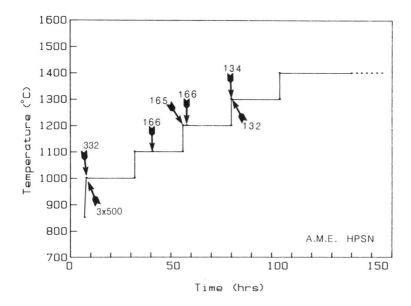
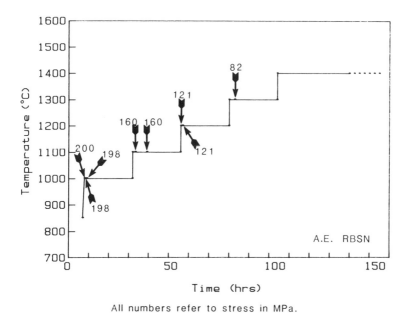

All numbers refer to stress in MPa.

FIG. 10. Stepped temperature stress rupture results. (Top) HPSN; (bottom) RBSN.

FIG. 11. SEM micrographs of RBSN specimen after STSR test. Failed 2 h 1300°C.

Mechanical Properties of Hot-pressed and Reaction-bonded Silicon Nitride   355

FIG. 12. SEM micrographs of HPSN specimen after STSR test. Failed 1 min 1300°C.

12(B). The fracture face can be divided into two regions macroscopically, one rough and one smooth, which probably indicates initial slow crack growth followed by fast growth to failure. One other feature of this specimen, shown in Fig. 12(C), is a continuation of the oxide film around the corner from the tensile surface on to the fracture surface. This might indicate that the initial stages of crack growth occurred early enough in the test cycle to allow significant oxidation within the growing crack, or alternatively, that the oxide film had a sufficiently low viscosity to be drawn into the crack by capillary action during the slow growth period.

4. DISCUSSION

The MOR strength results at room temperature for the RBSN are consistent with those obtained in similar work on other RBSN materials reported in the literature.[7-9] Quinn,[7] for example, gave results for three reaction-bonded silicon nitrides; Norton NC350, Kawecki and Ford 2.7. With the Norton NC350 he obtained a mean strength of 294 MPa and a Weibull modulus of 7; the Kawecki a mean strength of 206 MPa and a Weibull modulus of 15, and the Ford 2.7 a mean strength of 288 MPa and a Weibull modulus of 14. In the two stronger materials the fracture origins were either machining defects or large inherent pores, whilst in the Kawecki material they were either large inherent pores or pores associated with unreacted silicon. In the RBSN tested in the present work 12 of the 20 specimens had fracture initiation sites at large single pores, or concentrations of smaller pores, which intersected the surface. Two had fracture initiation at large pores well away from the surface and two more at large defects created by machining on the unchamfered edges of the specimens. Four specimens had fracture origins which were unidentifiable. In none of the RBSN specimens could fracture initiation be connected with sub-surface damage associated with machining striations. The two specimens in which fracture initiation was connected with machining damage had edge defects similar to that shown in Fig. 1(B). These observations support the accepted hypothesis[10] that in RBSN, the distribution and size of the inherent pore population determine the strength and it is only in the stronger RBSNs, where the pores are smaller, that defects created during machining become important in controlling the MOR strength.

The discrepancy in strength of HPSN between measurements made in the present work and the value quoted by the manufacturer can only be explained in terms of relative machining damage. Unlike the RBSN there is no large pore distribution and the MOR strength is very sensitive to surface damage. The fracture surfaces of the specimens were carefully examined to determine fracture origins, and in the strongest four specimens no source of any type could be identified. In a few specimens fracture initiation occurred at 'white spots' resulting from contamination by inclusions but in the majority of cases fracture initiation could only be attributed to sub-surface damage arising from machining. Such damage is commonly quoted as affecting the mechanical properties of ceramics and has been theoretically modelled by Heckel and Heigel,[11] Marshall[12] and Hakulinon[13] amongst many, but quoted values of mean strength and Weibull modulus[7,14,15] are still higher than those found in this work. Great care was taken in the preparation of our specimens and their surface quality, as shown in Fig. 2(C), was comparable to that of others shown in the literature.[7] We can only conclude that some part of our machining procedure induced severe sub-surface damage which persisted through to the finishing stage; indeed, it is possible that fine machining as a final processing step was itself deleterious to the mechanical properties. Further work is planned to investigate the matter.

The variation of MOR with temperature illustrates the differences between the two different forms of silicon nitride; RBSN was weaker than HPSN at temperatures up to 1300°C but at higher temperatures was the stronger material. HPSN exhibited a strong temperature dependence of MOR above 800°C whereas RBSN was relatively temperature insensitive, even increasing slightly around 1000°C. HPSN showed some plastic deformation at the higher test temperatures whereas RBSN remained elastic, macroscopically, even at 1500°C. Oxidation, in the 0·5 h the specimens spent at that temperature, was not excessive except in the case of HPSN tested at 1410°C or above where the oxidation became severe. In the RBSN specimens fracture origins were consistently regions of high porosity on or close to the tensile surface at all test temperatures. With the HPSN there was an increase in the occurrence of fracture initiation at the corners of the specimen as the temperature increased. At, and above, 1410°C oxidation of the HPSN obscured most fracture origins but in one case at 1410°C fracture initiation was obviously connected with a large oxidation pit. We infer that in the RBSN material the original defect size and distribution was the strength controlling feature at all temperatures. Slight variations in strength with temperature occurred because of slight modifications to this flaw population by oxidation,[8] or removal of residual stresses introduced during processing;[12] such forces would act to supplement the opening force on cracks during the application of tensile stresses in the bending test and weaken the material. The situation with HPSN was different. At room temperature the strength was determined by the flaws introduced during machining, and even at higher temperatures such flaws appeared to induce failure even though corner defects became the more dominant ones. Jakus et al.[16] showed that short-term exposure of HPSN to temperatures greater than 800°C in an oxidising environment gave an increase in the room temperature strength as a result of the formation of a glassy surface phase which smoothed out machining flaws. It would be expected that a similar mechanism might be equally important for tests performed at high temperatures but does not help to explain the increasing importance of edge defects.

Short-term dynamic tests are useful in assessing materials but long-term tests such as static fatigue and creep are often more relevant to engineering applications. Future work in this programme will extend the STSR experiments and emphasise creep and static fatigue tests. The immediate value of the STSR results shown in Fig. 10 is in the definition of parameters for the creep and stress rupture tests; for example there would be little point in setting up a creep experiment with HPSN at 1300°C and applying stresses in excess of 120 MPa. The oxidation of the specimens used in determining the STSR diagrams also indicates an area where extensive investigation will have to be done to separate oxidation effects from the true mechanical property variations.

REFERENCES

1. BOYER, S. M. and MOULSON, A. J. *Journal of Materials Science,* **13** (1978) 1637.
2. MITOMO, M. *Journal of Materials Science,* **12** (1977) 273.
3. WEIBULL, W. *Journal of Applied Mechanics,* **18** (1951) 293.
4. BRAIDEN, P. M. AERE-R7166, 1975.
5. MOSHER, D. R., RAJ, R. and KOSSOWSKY, R. *Journal of Materials Science,* **11** (1976) 49.

6. QUINN, G. and KATZ, R. N. *American Ceramic Society Bulletin*, **57** (1978) 1057.
7. QUINN, G. D. Characterisation of Turbine Ceramics after Long Term Environmental Exposure, Army Materials and Mechanics Research Center, TR80-15, Massachusetts, USA, April, 1980.
8. DAVIDGE, R. W., EVANS, A. G., GILLING, D. and WILLYMAN, P. R. In *Special Ceramics*, Vol. 5, Popper, P. (Ed.), British Ceramic Research Association, Stoke-on-Trent, p. 329, 1972.
9. GUGEL, E., LINDNER, H. A. and LEIMER, G. *Ceramic Forum International*, **62** (1984) 127.
10. EVANS, A. G. and DAVIDGE, R. W. *Journal of Materials Science*, **5** (1970) 314.
11. HECKEL, K. and HEIGEL, H. *ICM 3*, **3** (1979) 27.
12. MARSHALL, D. B. In *Progress in Nitrogen Ceramics*, Riley, F. L. (Ed.), Martinus Nijhoff, The Hague, p. 635, 1983.
13. HAKULINON, R. *Journal of Materials Science*, **20** (1985) 1049.
14. LANGE, F. F., DAVIS, B. I. and METCALF, M. G. *Journal of Materials Science*, **18** (1983) 1947.
15. EASTER, T. E., BRADT, R. C. and TRESSLER, T. R. *Journal of the American Ceramic Society*, **65** (1982) 317.
16. JAKUS, K., RITTER, J. E. and ROGERS, W. P. *Journal of the American Ceramic Society*, **67** (1984) 471.

COMMENTS AND DISCUSSION

Chairman: M. FARMER

E. A. Belfield: Have you made any attempt to improve the surface finish of the HPSN? Our own results on similar pieces would indicate modulus of rupture as quoted in A.M.E. literature of approximately 800 MPa. You have shown that in weaker specimens, fracture initiation occurs at surface flaws possibly caused by machining.

R. C. Piller: We tried to put very strict conditions down on our machine and attempted to make it as good as possible. This shows the difficulty of trying to produce a 'well-machined' specimen. Strengths of pieces where machining flaws were not observed were in the range you have quoted.

25

Tensile Strength of a Sintered Silicon Nitride

T. SOMA, M. MATSUI AND I. ODA

Materials Research Laboratory, NGK Insulators Ltd, 2–56 Suda-cho Mizuho-ku, Nagoya 467, Japan

ABSTRACT

To clarify the size effect on strength of ceramics, tensile and bend strengths were measured for a sintered silicon nitride. The tensile strengths for specimens of 6, 10, 14 and 20 mm diameters were 523, 460, 461 and 400 MPa. The strength decreased with increasing effective volume. The Weibull modulus estimated from the size effect on tensile strength was 15. The Weibull modulus estimated from the distribution of bend and tensile strengths ranged from 11 to 27. The size effect on tensile strength was proved to be essentially interpreted by Weibull statistics. Fracture surfaces were examined by SEM, and the flaw size of the fracture origin is discussed in terms of Griffith theory.

1. INTRODUCTION

High-performance ceramics are currently being studied for use as structural components such as turbo-charger rotors, gas turbine rotor blades and diesel engine insulators. The size effect on the strength of ceramics is an important concept in designing ceramic components. In previous work, the size effect on the strength of ceramics has been studied comparing 4-point bend data and 3-point bend data[1,2] or bend test and tensile test data.[3,4] However, there have been few studies which discussed the size effect on tensile strength by means of changing the stressed volume of the tensile specimen. In the present work, the size effect on tensile strength of ceramics was examined using tensile specimens of various diameters and 4-point bend specimens.

2. EXPERIMENTAL PROCEDURE

Sintered Si_3N_4 specimens with densities of $3 \cdot 2$ g cm^{-3} were used in this study. The gauge diameters of the tensile specimens are 6, 10, 14 and 20 mm. The gauge length is four times as long as the diameter of the specimen. The geometry and dimensions of the tensile specimen are shown in Fig. 1. The fracture strengths of ceramics are usually evaluated by means of bend tests instead of tensile tests. The reason is

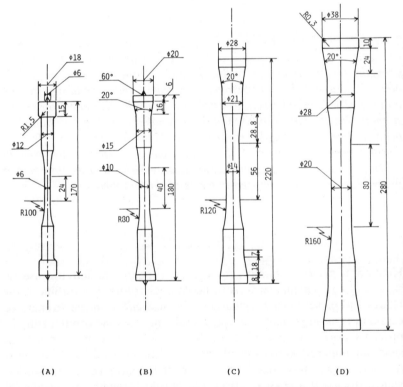

FIG. 1. Geometry and dimensions of tensile specimen. (A) 6ϕ buttonhead type; (B) 10ϕ taper type; (C) 14ϕ taper type; (D) 20ϕ taper type.

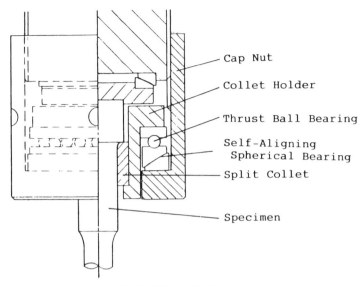

FIG. 2. Schematic of the grip.

that, owing to the brittleness of ceramics, the tensile test needs specifically designed jigs and specimens to achieve a good alignment and it needs special techniques and takes time for the sample preparation and measurement. Many techniques have been used to fix tensile specimens, such as collet chuck,[5] BN-powder cushion,[6] gas bearing,[7] pin grip[8] and hanging grip.[4] In ceramic tensile testing, it is important to align the axis accurately to conduct the test easily. Considering these conditions, the split collet chuck with spherical bearings was used in this study. Figure 2 shows the schematic diagram of the grip, which is designed to give self-axis alignment. The accuracy of the alignment for this technique was verified by measuring the bending stress component, which was detected by the three or six strain gauges at the centre of the gauge length. The cross-head speed of the testing machine was 0·05 mm min^{-1}.

The bend strength was measured by a 4-point bend test with 30 mm outer span and 10 mm inner span, using a test bar of $3 \times 4 \times 40$ mm according to JIS R-1601. The bend test was conducted for two different batches of specimens. The cross-head speed of the testing machine was 0·5 mm min^{-1}.

Fracture surfaces of tested specimens were examined by scanning electron microscopy (SEM).

3. RESULTS

The measured strength and the fracture origin for tensile specimens are given in Tables 1–4.

The bending stress component during tensile testing was given by eqn. (1), using the strain value measured by a strain gauge bridge on the specimen.

$$\% \text{ bending} = \frac{(\varepsilon_{max} - \varepsilon_{min})}{2 \times \varepsilon_{av}} \times 100 \quad (1)$$

where ε_{max}, ε_{min} and ε_{av} are maximum, minimum and average strains, respectively. Figure 3 shows the percentage of bending for 20 mm diameter tensile specimens as a function of applied load. The percentage of bending decreased with increasing applied load, and it was finally below 5% for all specimens.

The tensile strength data measured for specimens of 6, 10, 14 and 20 mm diameters were 523, 460, 461 and 400 MPa, respectively. This experimental result indicates the existence of the size effect on the strength of ceramic materials. Table 5 summarises the 4-point bend strength data. Average strength and the Weibull modulus were 811 MPa and 15 for batch 1, respectively. For batch 2, they were 794 MPa and 27, respectively. When the total data of batches 1 and 2 were used, the average strength and Weibull modulus were estimated

TABLE 1
Experimental Results for Tensile Specimens of 6 mm Diameter

Sample no.	Tensile strength (MPa)	Fracture origin
6-1	469	Surface
6-2	568	Internal
6-3	518	Internal
6-4	459	Surface
6-5	542	Internal
6-6	562	Internal
6-7	533	Internal
6-8	568	Internal
6-9	464	Surface
6-10	547	Surface

TABLE 2
Experimental Results for Tensile Specimens of 10 mm Diameter

Sample no.	Tensile strength (MPa)	Fracture origin
10-1	450	Internal
10-2	515	Surface
10-3	513	Internal
10-4	424	Internal
10-5	378	Internal
10-6	500	Internal
10-7	443	Internal

TABLE 3
Experimental Results for Tensile Specimens of 14 mm Diameter

Sample no.	Tensile strength (MPa)	Fracture origin	Sample no.	Tensile strength (MPa)	Fracture origin
14-1	458	Surface	14-26	461	Internal
14-2	430	Surface	14-27	472	Surface
14-3	367	Internal	14-28	516	Surface
14-4	445	Internal	14-29	508	Internal
14-5	489	Surface	14-30	410	Internal
14-6	387	Internal	14-31	564	Surface
14-7	464	Surface	14-32	514	Surface
14-8	462	Surface	14-33	501	Internal
14-9	458	Surface	14-34	382	Internal
14-10	461	Surface	14-35	481	Internal
14-11	473	Surface	14-36	526	Surface
14-12	474	Surface	14-37	462	Surface
14-13	426	Internal	14-38	453	Internal
14-14	447	Surface	14-39	537	Internal
14-15	448	Internal	14-40	437	Internal
14-16	513	Internal	14-41	444	Internal
14-17	544	Surface	14-42	441	Internal
14-18	487	Internal	14-43	381	Internal
14-19	492	Internal	14-44	402	Internal
14-20	473	Surface	14-45	374	Internal
14-21	513	Internal	14-46	379	Internal
14-22	354	Internal	14-47	414	Internal
14-23	513	Surface	14-48	415	Internal
14-24	491	Internal	14-49	522	Internal
14-25	492	Surface	14-50	481	Internal

TABLE 4
Experimental Results for Tensile Specimens of 20 mm Diameter

Sample no.	Tensile strength (MPa)	Fracture origin
20-1	425	Internal
20-2	451	Surface
20-3	376	Surface
20-4	374	Surface
20-5	425	Surface
20-6	403	Surface
20-7	387	Surface
20-8	371	Surface
20-9	413	Surface
20-10	380	Surface

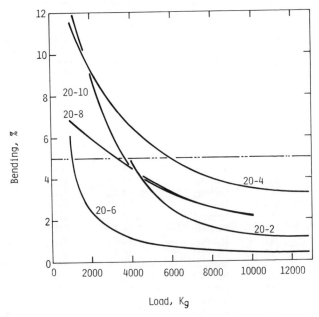

FIG. 3. Percentage of bending imposed on the tensile specimen of 20 mm diameter as a function of applied load.

TABLE 5
Four Point Bending Strength Data for Batches 1 and 2

Batch	Strength (MPa)								
1	874	698	860	866	812	719	852	900	889	766
	855	886	880	902	812	860	796	702	799	852
	715	737	823	754	822	789	884	844	705	818
	843	779	870	725	735	753	679	850	710	857
	823	764	779	824	836	826	559	851	697	854
	801	803	901	902	832	821	823	882	908	817
	821	860	892	856	750	836	821	870	802	821
	786	879	783	808	817	778	832	807	678	828
	719	858	848	786	767	837	872	870	870	838
	797	844	852	779	894	738	709	740	771	825
2	860	801	827	769	808	858	814	799	852	864
	789	850	807	792	726	775	809	844	761	802
	800	827	700	769	834	763	834	766	802	757
	818	850	823	779	818	798	798	801	803	827
	785	687	789	809	779	713	822	778	779	785
	733	789	794	830	796	761	803	723	816	789
	811	808	773	817	789	792	796	848	753	753
	818	836	753	822	775	688	841	796	797	807
	779	774	795	810	814	778	803	767	813	737
	775	812	751	759	768	834	787	818	807	836

at 802 MPa and 19, respectively. Figure 4 shows the Weibull plots for tensile strength and bend strength data.

Typical microstructures of fracture surfaces for tensile and bending specimens are shown in Fig. 5. These fracture surfaces have mirror, mist and hackle regions which are typical features of brittle fracture.

4. DISCUSSION

4.1. Size Effects on the Strength of Ceramic Materials

Size effects on the strength of ceramic materials can be related to the flaw size distribution, which cause a strength variation of ceramic materials. This strength variation and size effect can be characterised by the weakest link theory of Weibull.

$$F = 1 - \exp\left\{-\int_v \left(\frac{\sigma}{\sigma_o}\right)^m dV\right\} \qquad (2)$$

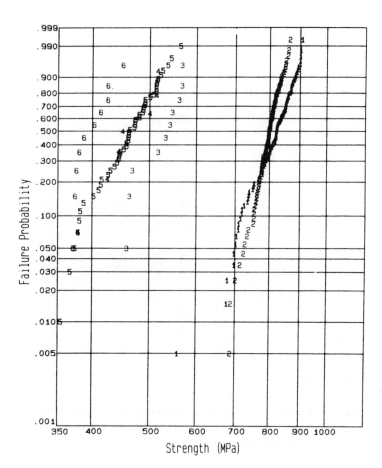

Plot symbol	Specimen	Lot number	Meas. temp.	Average strength	Weibull modulus	S.D.	No. of specimens
1	SN-Bending	batch 1	RT	811	15	63	100
2	SN-Bending	batch 2	RT	794	27	36	100
3	SN-Tensile	6 mm dia.	RT	523	13	44	10
4	SN-Tensile	10 mm dia.	RT	460	10	51	7
5	SN-Tensile	14 mm dia.	RT	461	11	50	50
6	SN-Tensile	20 mm dia.	RT	400	17	27	10

FIG. 4. Weibull plots of tensile and bending strength data.

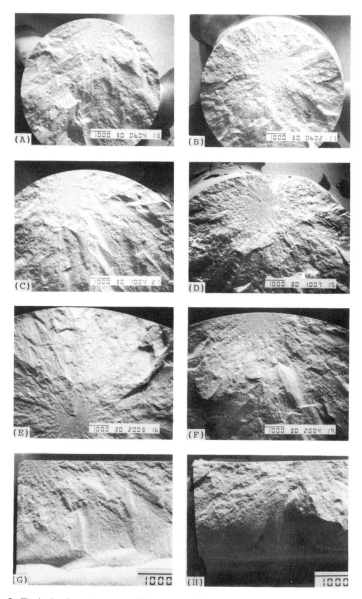

FIG. 5. Typical microstructures of fracture surfaces of tensile and bending specimens. (A) No. 6-4, $\sigma_f = 459$ MPa; (B) No. 6-8, $\sigma_f = 568$ MPa; (C) No. 10-2, $\sigma_f = 515$ MPa; (D) No. 10-4, $\sigma_f = 424$ MPa; (E) No. 20-1, $\sigma_f = 425$ MPa; (F) No. 20-8, $\sigma_f = 371$ MPa; (G) No. 4P-68, $\sigma_f = 710$ MPa; (H) No. 4P-99, $\sigma_f = 894$ MPa.

where F is failure probability, σ is applied stress, V is volume of a specimen, m is Weibull modulus and σ_o is characteristic strength. Using maximum principal stress in the stressed solid, eqn. (2) is reduced to

$$F = 1 - \exp\left\{-V_e\left(\frac{\sigma_{max}}{\sigma_o}\right)^m\right\} \quad (3)$$

where V_e is the effective volume defined by

$$V_e = \int_v \left(\frac{\sigma}{\sigma_{max}}\right)^m dV \quad (4)$$

Using a Gamma Function, average strength, $\bar{\sigma}$, is given by

$$\bar{\sigma} = \sigma_o V_e^{-1/m} \Gamma\left(\frac{m+1}{m}\right) \quad (5)$$

The ratio of average strength of two specimens, whose effective volume is V_{e1} and V_{e2}, is given by

$$\frac{\bar{\sigma}_1}{\bar{\sigma}_2} = \left(\frac{V_{e2}}{V_{e1}}\right)^{1/m} \quad (6)$$

In eqn. (6), it is assumed that m and σ_o are independent of specimen shape and testing method.

The effective volume, the average strength and Weibull modulus are shown in Fig. 6 and Table 6. The effective volume of a 4-point bend specimen with the ratio of inner span, S_1, to outer span, S_2, $S_1/S_2 = 1/3$, is obtained by integrating eqn. (4) to give

$$V_{e4p} = \frac{bhS_2(m+3)}{6(m+1)^2} \quad (7)$$

where b is width and h is thickness of the specimen. The effective volume of the 4-point bend specimen in this study is estimated at 3.30 mm^3 by eqn. (7) using the Weibull modulus, $m = 19$. In Fig. 6, the slope of the straight line is $1/m$ as indicated by eqn. (6). When both bend and tensile strength data were used for estimating size effect on strength, the Weibull modulus was estimated at 13, by the method of least squares. When only tensile strength data were used, the Weibull modulus was estimated at 15.

As shown in Table 6, the Weibull modulus calculated from the

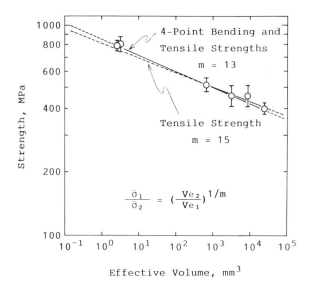

FIG. 6. Size effect on strength.

distribution in bend and tensile strengths ranged from 11 to 27. The Weibull modulus, $m = 15$, estimated from the size effect on tensile strength was in this range. The size effect on tensile strength can be essentially interpreted by Weibull statistics.

It is noteworthy that the minimum strength of bend specimens almost coincides with the average strength of 6 mm diameter tensile

TABLE 6
Effective Volume, Average Strength and Weibull Modulus for Various Test Specimens

Specimen	Effective volume (mm³)	Average strength (MPa)	Specimen number	Weibull modulus	
				(by strength distribution)	(by size effect)
6φ Tensile	6.79×10^2	523	10	13	15
10φ Tensile	3.14×10^3	460	17	10	
14φ Tensile	8.26×10^3	461	50	11	13
20φ Tensile	2.15×10^4	400	10	17	
4-point bend					
Batch 1	3.30	811	100	15	
Batch 2	3.30	794	100	27	
Batch 1 + Batch 2	3.30	802	200	19	

specimens, as shown in Fig. 4. By the weakest link theory, this indicates that the total effective volume for a population of 200 bend specimens is almost equivalent to the effective volume of one tensile specimen with 6 mm diameter. The calculated effective volume of a 6 mm diameter tensile specimen is $6 \cdot 79 \times 10^3$ mm^3 and near to the total of the effective volume of the tested bend specimens. This means that 200 bend specimens are required to find the typical Griffith flaw in the 6 mm diameter tensile specimen. For example, in the case of a rotating disc with a dimension of 200 mm in diameter and 10 mm thick, the effective volume is about 30 000 mm^3. If we try to find the typical

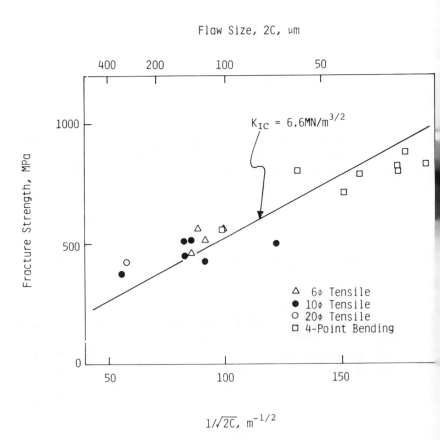

FIG. 7. Fracture strength versus flaw size.

Griffith flaw of this rotating disc by bend test, about 9000 individual bend specimens are required. Instead of bend specimens, if we use the 10 mm diameter tensile specimen for this purpose we require only one piece, because its effective volume is almost equivalent to the rotating disc. Thus, the tensile test has the advantage of reducing the total cost and time to examine the fracture behaviour of a ceramic structural component with large effective volume.

4.2. Fracture Origin

Figure 7 shows the relationship between the flaw size in fracture origin and the fracture strength. Most of the fracture origins were pores or porous regions which are formed during the fabrication process. According to Griffith theory, the relationship between Griffith flaw size, C, and fracture strength, σ_f, is given by

$$\sigma_f = \frac{K_{IC}}{\sqrt{\pi C}} \quad (8)$$

where K_{IC} is fracture toughness. The K_{IC} was $6 \cdot 6$ MN m$^{-3/2}$ which was measured by the chevron-notched beam method. The straight-line relationship in Fig. 7 was predicted by substituting the K_{IC} value in eqn. (8). Although there is considerable scatter in the data, the measured strengths show good agreement with those predicted from the Griffith criterion.

5. SUMMARY

(1) The tensile strengths for specimens with 6, 10, 14 and 20 mm diameters of a sintered silicon nitride were successfully measured with the small bending percentage below 5%. The tensile strength decreased with increasing effective volume.

(2) The Weibull modulus, calculated from the size effect on tensile strength was $m = 15$. The Weibull modulus calculated from distribution of bend and tensile strengths ranged from 11 to 27. The size effect on tensile strength was proved to be essentially interpreted by Weibull statistics.

(3) The fracture strength was related to the flaw size of the fracture origin in terms of Griffith theory.

ACKNOWLEDGEMENTS

This work was performed under the contract between the Agency of Industrial Science and Technology of MITI and the Engineering Research Association for High Performance Ceramics, as a part of R & D Project of Basic Technology for Future Industries.

REFERENCES

1. DAVIES, D. G. S. *Proceedings of the British Ceramic Society*, **22** (1973) 429.
2. KATAYAMA, Y. and HATTORI, Y. *Journal of the American Ceramic Society*, **65** (1982) C-164.
3. ASHCROFT, W. In *Special Ceramics*, Vol. 6, Popper, P. (Ed.), British Ceramic Research Association, Stoke-on-Trent, p. 245, 1975.
4. MATSUSUE, K., FUJISAWA, Y. and TAKAHARA, K. *Yogyo-Kyokai-shi*, **9** (1983) 325.
5. BORTS, S. A. and WADE, T. B. In *Structural Ceramics and Testing of Brittle Materials*, Acquaviva, S. J. and Borts, S. A. (Eds), Gordon and Breach, London, p. 47, 1968.
6. LANGE, F. F., DIAZ, E. S. and ANDERSON, C. A. *American Ceramic Society Bulletin*, **58** (1979) 845.
7. PEARS, C. D. In *Structural Ceramics and Testing of Brittle Materials*, Acquaviva, S. J. and Borts, S. A. (Eds), Gordon and Breach, London, p. 139, 1968.
8. GOVILA, R. K. *Journal of the American Ceramic Society*, **65** (1982) 15.

COMMENTS AND DISCUSSION

Chairman: E. R. PETTY

W. Grellner: Have these specimens been tested by NDT in advance and, if so, which defects have an effect in comparison with the defects you saw in the fracture surfaces.

T. Soma: We have used NDT (X-ray) but it is very difficult to detect defects.

26

Selecting Ceramics for High Temperature Components in IC Engines

M. H. FARMER, M. S. LACEY AND J. N. MULCAHY

Mechanical Engineering Department, University College Dublin, Upper Merrion Street, Dublin 2, Ireland

ABSTRACT

Refractory ceramics range from domestic ovenware and bulky furnace refractories to precisely graded high purity ceramic components. The cost reflects the quality and complexity of production. Selection of optimum material involves economic compromise based on acceptable performance. Appraisal of ceramics on offer demands reliable application of technical data to component design. This paper describes preliminary research undertaken in the Mechanical Engineering Department of University College Dublin to determine the bursting strength of hollow cylinders of silicon nitride and silicon carbide and to investigate the correlation of tensile properties as determined on thin bend test specimens to specimens of a thickness more appropriate to cylinders and pistons for an opposed piston, two-stroke diesel engine. Tests on silicate based clay specimens indicated that application of Weibull statistics do not accurately correlate the tensile strengths determined experimentally on specimens of differing thicknesses.

1. INTRODUCTION

Motivation is a pre-condition for change. The tremendous increase of effort to introduce ceramic components in the hot zones of reciprocating engines is no exception; profit being an important motivating force. Every manufacturer, involved from processing raw minerals through to competitive customer satisfaction, must achieve commercial viability to ensure an unbroken line of enthusiasm without which potential participants in research are reluctant to provide practical support for projected programmes. Such fluctuations of effort affecting application of ceramics to heat engines in the United States, England, Germany and Japan are explained by Lenoe and Meglen,[1] Johnson et al.[2] and Harmon and Beardsley.[4]

In order to engender significant involvement in the evolution of a 'ceramic diesel engine', two main factors emerge. Firstly, the feasibility of ceramic components to withstand the aggressive conditions imposed by operation must be proved and secondly, the change from traditional materials to ceramics must offer commercial benefit and/or strategic advantage.

If, before the advent of internal combustion engines, ceramics with adequate properties had been available to the innovative engineer, the concept of the popular cooling arrangement with labyrinth castings, pumps, plumbing, radiators and fans would never have reached the design office. Ceramics for the hot zones would have been the automatic choice. It was the inadequacy of cast iron and other materials at that time which stimulated ingenious but complicated solutions from which the automotive power unit of today evolved. The present question—why try to use ceramics?—may arise from inertia to challenge long proven, familiar systems. Perhaps the motivation to change may await vastly improved engine efficiency and reduced NO_x emission when oxygen enrichment of combustion air by membrane technology becomes a reality and Hi-tech Ceramics become mandatory.

2. ENGINE RESEARCH AT UCD

University College Dublin (UCD) became involved in ceramics for diesel engines seven years ago using a 500 ml opposed piston, two-stroke engine. The design and operation of the engine used for the initial research are described by S. G. Timoney in the EEC Monograph No. EUR 7660 EN1981.[5] This reference also reviews the application of ceramics in internal combustion engines from 1970,[6] when a reaction bonded silicon nitride (RBSN) piston was fitted to a 900 W Villiers four-stroke air cooled single-cylinder petrol engine. This was followed in 1973 by testing a RBSN piston in a Gardner single-cylinder water cooled diesel engine at the Royal Naval Engineering College.[7]

The work carried out at UCD under EEC Contract No. 143-76-EE-EIR[8] indicated that the opposed piston two-stroke engine incorporating the Timoney Variable Compression Ratio mechanism offered considerable advantages for the application of ceramics to diesel engines especially for cylinder liners and pistons. The final report[9] issued in 1981 illustrates the manner in which, after 35 h of motoring (being driven at 1000/1500 rev min^{-1} without fuel injection) the glass ceramic liners and pistons disintegrated within 3 min after starting fuel injection. Examination of the broken parts (Fig. 1) by the manufacturers, Owen Illinois, gave no clue as to the cause of failure, but it was supposed that the material failed under the pressure loading generated by the combustion.

Next, two silicon carbide pistons were assembled into a steel liner, but after 23 min motoring the pistons seized into the liner (Fig. 2). A set of silicon carbide components which started out as shown in Fig. 3 shattered into fragments shown in Fig. 4. Such debris presents a real challenge to the fracture analyst, especially in consideration of the complexity of parameters which could interact to complicate the failure. Obviously, precise identification of the cause would be of immense value to the design engineer.

FIG. 1. The disintegrated glass/ceramic liner and pistons after motoring for 35 h and combustion cycles for 3 min.

FIG. 2. Showing the fragmented conditions of the silicon carbide pistons after motoring in the steel liner for 23 min. The crowns of the pistons had seized into the liner.

After these somewhat disappointing experiences, a cylinder of α-silicon carbide (SiC) and a pair of solid α-SiC pistons supplied by the Carborundum Resistant Materials Company were installed in July, 1982. This research event, which is described by Timoney and Flynn,[10] proved the feasibility of successfully operating such components without lubrication and cooling.

FIG. 3. A silicon carbide cylinder and pistons before engine trials.

FIG. 4. The debris after failure during engine trials.

The 500 cc engine was found to be too small for realistic evaluation of technical ceramic materials and subsequently, Timoney was awarded an EEC contract[11] No. SUT-127-EIR to design a 1000 cc engine which would also provide greater ease for exchange of cylinders and pistons. At present the new engine has been constructed and proved to be satisfactory by operating with conventional cylinders and pistons.

3. UCD SEMINAR, OCTOBER 1984

The new engine had been designed and was at an advanced stage of construction while the EEC Basic Technological Research programme was being formulated. Eventually, the programme was renamed 'Basic Research for Industrial Technology for Europe' (BRITE) and within nine major categories the one entitled Reliability, Wear & Deterioration included 'Performance reliability and design of advanced engines, components and systems, in particular, fully ceramic and ceramic surfaced components including piston and liner systems'. A seminar was held in UCD on 4th and 5th October, 1984 to promote discussion in a European context for a research and development programme on engineering ceramics suitable for application in reciprocating internal combustion engines. Research workers in Ireland, UK, France, Italy, Germany, Belgium and Holland and members of the relevant EEC Commission were invited. The expertise of the delegates covered the range from raw material suppliers, ceramic technologists, engine development engineers and vehicle manufacturers. The proceedings were summarised in an Action Report.[12]

Preliminary exploratory research being undertaken by UCD with support by the NBST and co-operation by Associated Engineering Developments and British Nuclear Fuels Limited was described. This effort was to establish, by cold bench tests, whether cylinders of AED's Nitrasil and BNFL's REFEL silicon carbide would be able to withstand the gaseous pressure generated in the diesel engine cycle.

The first stage of this programme consisted of bursting short ceramic cylinders by compressing a rubber disc between two pistons (Figs. 5, 6 and 11). Strain gauges were fixed to the circumference of the ceramic cylinder to measure the hoop strain and strain was plotted against load on the pistons. The load was gradually increased until the cylinder fractured. The results are presented in Table 1.

By this method of testing the bursting pressure is applied as a narrow band to simulate the way combustion gases apply radial pressure in the diesel cylinder when the pistons are near to top dead centre. Typically, the ceramic cylinders fragmented as shown in Fig. 7 in contrast to the failure of a cast iron cylinder which broke into three

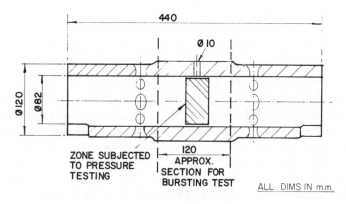

FIG. 5. Ceramic liner.

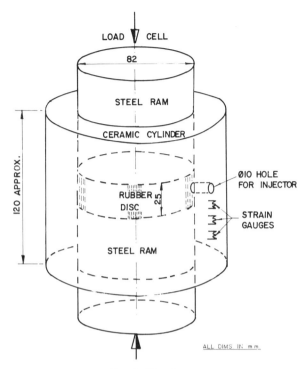

FIG. 6. Bursting test.

pieces with essentially longitudinal fractures (Fig. 8). By comparison the fractures of the ceramics tended to radiate from crack initiation centres probably due to generation of longitudinal components of stress combined with hoop stresses.

In an operational cylinder it is necessary to provide a 10 mm hole in the wall to accommodate a fuel injection nozzle. Associated Engineering Developments supplied five short cylinders of Nitrasil silicon nitride, one with no hole and the others with holes ranging from 4 to 12 mm diameter in order to explore the effect of holes on the bursting strength. Whereas it was possible to fix strain gauges at the position of maximum strain on cylinders without holes, the strain at the extreme edges of the holes could not be measured by the same technique. However, comparing the forces on the ram recorded for tests No. 1 (no hole—222 kN) and No. 2B (4 mm hole—225 kN), it seemed that the variation in material quality or latitude in experimental method

TABLE 1
Static Bursting Tests on Short Cylinders

Sample and test no.	Description of sample and test	Injector hole diameter (mm)	Force on ram (kN)	Internal pressure (note 1) (MPa)	Strain (note 2) ($\mu\varepsilon$)	Hoop stress (MPa)
1	Cylinders of Nitrasil supplied by AED Tensile strength = 136 MPa, E = 168 GPa O.D. = 118·4 mm, I.D. = 82·0 mm, area of bore = 5281 mm²	Nil	222	42	600	77
2B		4·0	225	43	>473	78
3B	Tested with injector pressing on cylinder (270 N)	12	Not valid	37	>440	>74
4C	With saddle and injector block torqued to 200 lb in (24 N m). The fracture did not involve the injector hole	10	236	(from ε) 45	>402	83
5B	Torqued saddle and injector block to 200 lb in and cylinder block to 100 lb in. The cylinder did not fracture	8	0/56 360	0/10 68	−ve. >331	125
5C	As 'B' but after changing the gasket. The cylinder fractured along the edge of the injector block		136	26	>205	47
6	Cylinders of Refel silicon carbide supplied by BNFL. Tensile strength = 298 MPa, E = 420 GPa O.D. = 100 mm, I.D. = 74·2 mm, area of bore = 4324 mm². The injector hole was bored at UCD with an inferior drill	10	133	31	>104	75
7	Cast iron cylinder liner from a diesel engine. Tensile strength = 200/300 MPa. E = 107 GPa. I.D. = 82 mm, wall thickness varied from 6·8 to 14 mm		333	63	>1747	320

Note 1. The peak internal pressure inside the cylinders of the diesel engine is not expected to exceed 15 MPa acting over a width of 12 mm. These static tests apply the bursting force over a width of 25 mm.
Note 2. As it is improbable that the strain gauges could be located at the source of the initial cracks the strains would be less than the

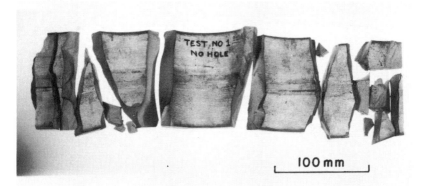

FIG. 7. A silicon nitride cylinder showing typical fragmentation after being burst by internal pressure from the rubber disc.

was more significant than the stress concentration introduced by a 4 mm hole.

The fuel injection nozzle requires a 10 mm diameter hole and it is necessary to provide a gas tight joint to prevent leakage at pressures approaching 15 MPa. Accordingly, the Nitrasil cylinders with 12 mm, 10 mm and 8 mm holes were used to test methods of fixing injection nozzles, for example Fig. 9 illustrates the system used for cylinder No. 3. The importance of resolving this problem is demonstrated by two

FIG. 8. The cast iron cylinder after bursting in the same manner as the silicon nitride cylinder shown in Fig. 7.

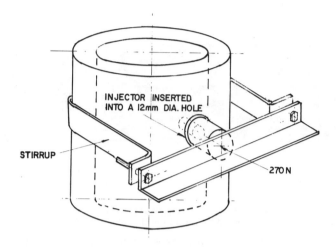

FIG. 9. AED Nitrasil. Sample No. 3: Test B. Fractured through hole, internal pressure: 37 MPa, Hoop Stress 74 MPa.

tests undertaken on cylinder No. 5. By using the arrangement of a saddle and simulated cylinder block (Fig. 10) an injection nozzle was fitted to the cylinder. The ram was loaded to 360 kN without the cylinder fracturing (compared with unsupported cylinders which fractured at 225 kN). However, when the assembly was dismantled and refitted with a different gasket the cylinder fractured along one edge of the injector block when the ram force had reached only 136 kN (Fig. 11).

British Nuclear Fuels supplied four short cylinders of Refel Silicon Carbide. Only one was tested and burst at an internal pressure of 31 MPa, whereas the cast iron cylinder tested on the engine sustained up to 63 MPa. However, all of the tests performed indicated a pressure capability of more than twice the expected pressure generated in the engine and furthermore because the objective of utilising ceramics is to retain the heat within the combustion chamber, the thickness of ceramic cylinders could be increased if necessary to cater for the gaseous pressures.

The delegates at the seminar recommended several areas of research for positive action. UCD agreed to co-ordinate effort in the following topics:

1. *Model material study* of both solid ceramic components and coated metal components for application in diesel engines.

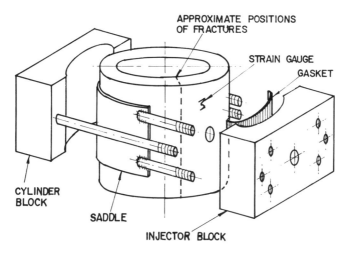

SAMPLE : 4 TEST C
FRACTURED REMOTE FROM HOLE
INTERNAL PRESSURE : 45 MPa ∴ HOOP STRESS 83 MPa
HOOP STRESS 1 (68 MPa from ϵ)

SAMPLE : 5 TEST B (without fracture)
INTERNAL PRESSURE 0–10MPa : HOOP STRESS–Ve to ZERO
INTERNAL PRESSURE 68MPa : HOOP STRESS 56 MPa from ϵ

SAMPLE : 5 TEST C
FRACTURED ALONG EDGE OF INJECTOR BLOCK AFTER
ADJUSTING GASKET
INTERNAL . PRESSURE 26 MPa : HOOP STRESS 47 MPa 34 MPa from ϵ

FIG. 10. AED Nitrasil.

2. *Mathematical modelling* for stress analysis and heat flow analysis in ceramic engine components.
3. *Laboratory assessment* of ceramic and coated components for:
   — wear and friction;
   — influence of combustion gases;
   — response to combustion phenomena;
   — correlation of mechanical properties derived from test specimens with behaviour of engine components.

4. MECHANICAL TESTS ON FLAT SPECIMENS

In pursuance of the recommendation to correlate results of cold bench tests on simple specimens with the behaviour of engine size components, it was decided firstly to correlate the fracture strength of flat

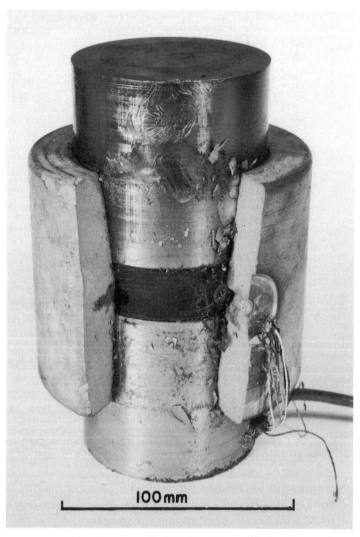

Fig. 11. The bursting test assembly showing the longitudinal fracture of the silicon nitride cylinder. The fracture on the left coincided with the edge of a saddle.

specimens with the resistance exhibited by cylinders to typical gaseous pressure generated during the diesel engine cycle. It was presumed that the forces causing a cylinder to rupture by internal pressure would produce tensile hoop stress which would be simulated more closely by tensile tests than by three point or four point bend tests. Nevertheless, reported values for strength of proprietary ceramics are often determined from bend tests, albeit of smaller dimensions than would seem to be appropriate to the wall thicknesses of cylinders for testing in the diesel engine.

Consequently, the initial programme included appraisal of bend testing in order to investigate the feasibility of utilising such published data for component design.

This research was started by D. Dwyer as an undergraduate BE project at UCD,[13] and continued by M. Lacey and J. N. Mulcahy in the Mechanical Engineering Department.

4.1. Tensile Tests

The geometry of the tensile specimen was programmed into a Kitamura S15 computer controlled machining centre. This facilitated machining photoelastic specimens, ceramic moulds and tensile testing machine grips, to closely similar profiles. Nevertheless, the photoelastic specimen proved that the stress concentration was significantly greater where the specimen touched the grips than in the parallel portion which was intended to provide a useful 'gauge length' uniformly stressed to enable reliable assessment of the tensile properties and the effect of a hole in the mid-point of the 'gauge length'.

Tensile specimens were initially made of plaster and tested to fracture. The design was modified until the specimens broke regularly in the 'waisted' zone, whether or not there was a hole through this zone of the specimen. The final design is shown in Fig. 12.

Fifteen specimens were also moulded in 'pot-clay' fired in a pottery kiln. Two of these specimens were coated with reflective photoelastic plastic films, but the value of strain to fracture of the ceramic was too low for this technique to give quantifiable results. Strain gauges were also attached to several specimens to determine the mechanical properties and to appraise the experimental procedure. A proportion of the specimens, after firing to the biscuit stage, were drilled with a 10 mm diamond hollow mill and then fired to full temperatures of 1260°C. Some other specimens were similarly drilled after fully firing and the remainder were tested without holes in the fully fired

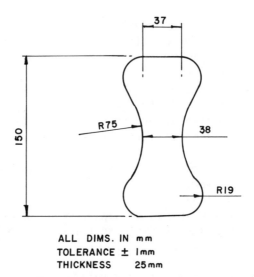

FIG. 12. Ceramic tensile test specimen.

condition. In an effort to reduce stress concentration at the points of contact the grips were lined with sheet rubber. The results of the tests are presented in Table 2.

4.2. Bend Tests

Bend testing using 3-point or 4-point loading systems is a well established method of evaluating the mechanical strength of brittle

TABLE 2
Tensile Tests on Dogbone Shapes

	Without hole	With 10 mm hole drilled after firing	With 10 mm hole drilled before firing
No. of samples	6	5	4
Average area of fracture (mm²)	900	600	600
Average stress to failure (MPa)	13·0	13·8	14·4

1. Holes all drilled with 10 mm diamond core bit.
2. The 10 mm holes drilled in the unfired samples all shrank to 9·3 mm average after firing.
3. The improved strength from both types of samples with holes compared to samples without holes could be explained by the drill removing gross flaws from the centre of the sample.

materials. The specimens are simple in shape and the test equipment uncomplicated. Consequently, tensile strength and Weibull moduli are often reported from bend test data for proprietary ceramics. However, the standardised tests[14] specify specimens that are considerably thinner than sections required for diesel engine components and therefore the primary objective of the first phase of the bend test programme was to establish the reliability of current theories in estimating the strength of large specimens from results obtained from fracture strength of small ones. For this exercise specimens were made by two local craft potters[15,16] from fully fired 'oven-ware pot-clay'. The specimens were approximately 114 mm long, 25·4 mm wide and 3, 6, 12 and 25 mm in thickness. The specimens were broken by 4-point bending on an Instron 5000 kgf tensile/compression test machine. Any broken pieces that were long enough were retested by adjusting the beam length as appropriate. The combined results are presented in Tables 3 and 4. The Weibull moduli were calculated by two different systems, viz. Weibull[17] and Fulmer Institute.[19] Each method gave different values. The results for the 5·8 mm thick specimens were used to estimate the

TABLE 3
Summary of Bend Test Results

Bar width (mm)	24	24	24	24·5
Bar depth average (mm)	4·4	5·8	12·2	23·2
Stress to fail (MPa)	26·4	31·0	24·7	21·3
Standard deviation	3·1	5·4	3·7	2·7
Strain to fail ($\times 10^{-6}$)	—	480	465	445
Young's modulus (GPa)	—	48	56	55
Poisson ratio	—	0·17	0·17	0·17
Weibull modulus (method of Davidge) Ref. 18	6·9	6·4	6·9	7·9
Weibull modulus (method of Fulmer) Ref. 19	10·2	7·0	8·2	9·5
Weibull modulus $\left(\text{solve } \dfrac{\sigma_1}{\sigma_2} = \left(\dfrac{V_2}{V_1}\right)^{1/m}\right)$ Ref. 18	−ve	3·3	4·2	3·8

TABLE 4
Nominal Bar Size

24 × 12 × 3		24 × 12 × 6		24 × 12 × 12		24 × 12 × 24	
Rank	Stress	Rank	Stress	Rank	Stress	Rank	Stress
1	18·1	1	22·3	1	15·5	1	15·0
2	19·3	2	22·5	2	16·7	2	15·4
3	19·7	3	24·2	3	18·4	3	18·7
4	20·2	4	25·8	4	20·5	4	19·0
5	22·4	5	26·0	5	20·8	5	19·0
6	22·9	6	26·1	6	22·0	6	19·3
7	23·0	7	26·2	7	22·0	7	19·4
8	23·6	8	26·3	8	22·5	8	20·5
9	24·2	9	26·5	9	22·5	9	20·6
10	24·2	10	27·3	10	23·0	10	20·6
11	24·2	11	27·4	11	23·1	11	21·0
12	24·3	12	28·1	12	23·3	12	21·0
13	24·5	13	28·9	13	24·2	13	21·4
14	24·5	14	29·2	14	24·3	14	21·8
15	25·7	15	29·4	15	25·1	15	21·8
16	25·7	16	30·2	16	26·4	16	21·9
17	26·4	17	30·6	17	26·5	17	21·9
18	26·8	18	30·7	18	26·5	18	23·2
19	26·8	19	30·8	19	26·6	19	23·8
20	27·3	20	31·4	20	27·0	20	23·8
21	27·4	21	32·7	21	27·0	21	24·2
22	27·7	22	32·9	22	27·0	22	24·0
23	28·2	23	33·1	23	27·0	23	24·9
24	28·4	24	33·3	24	27·3	24	25·1
25	28·5	25	33·6	25	27·5	25	25·5
26	28·7	26	33·8	26	27·9		
27	28·9	27	35·7	27	27·9		
28	29·1	28	36·1	28	28·3		
29	29·1	29	37·0	29	28·5		
30	29·3	30	37·2	30	28·8		
31	29·5	31	37·7	31	29·8		
32	29·7	32	38·7	32	29·9		
33	30·0	33	43·6	33			
34	30·2	34	44·5				
35	31·6						
36	31·8						
37	32·0						
38	41·2						

Nominal bar size	24 × 12 × 3	24 × 12 × 6	24 × 12 × 12	24 × 12 × 24
Mean stress (MPa)	26·4	30·7	24·7	21·2
Standard deviation	3·1	5·4	3·7	2·7
Weibull (Davidge)	6·9	6·4	6·9	7·9
Weibull (Fulmer)	10·2	7·0	8·2	9·5

strength of the 12·2 mm and 23·2 mm specimens from the equation:

$$\frac{\sigma_{5\cdot 8}}{\sigma_x} = \left(\frac{V_x}{V}\right)^{1/m} \quad \text{after Davidge}[18]$$

The predictive values of stress to failure are plotted versus volume from $10^1\,\text{cm}^3$ to $10^{-2}\,\text{cm}^3$ in Fig. 13. (The appropriate specimen thicknesses are indicated along the abscissa.) The experimental result obtained for the 5·8 mm thick specimens has been taken as the basic value.

5. DISCUSSION

By discussion with commercial manufacturers of ceramic shapes, it appears that the 'dog bone' tensile specimen would be more trouble to manufacture than either rectangular specimens, cylinders or rings;

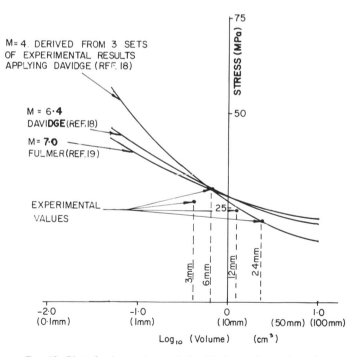

FIG. 13. Plot of volume–stress relationship for various values of m.

depending on the main line of production. Also because of the problems of minimising inadvertent stress concentration effects and avoiding difficulties in accurate load alignment, mentioned by Fessler,[20] work with axial tensile tests was discontinued in favour of bend tests. However, from the limited number of tensile tests performed it was noticed that the stress to fracture of specimens with holes was more than for specimens without holes.

As stated previously, the principal objective of performing bend tests on sets of specimens of different thickness was to investigate the validity of methods for estimating the fracture strength of ceramics, thick enough for engineering components, from values determined on relatively small specimens. The relationships of breaking stress and specimen thickness plotted in Fig. 13 show the deviation from the experimental results by curves generated using Weibull moduli of 7·0, 6·4 and 4·0. Clearly, the value of 4·0 determined by substituting the experimental results of the three sets of specimens in the formula:

$$\frac{\sigma_1}{\sigma_2} = \left(\frac{V_2}{V_1}\right)^{1/m}$$

gives a better fit than either of the other two. It is logical to conclude that the most reliable data for design purposes requires testing of full size specimens.

The limitations of predicting the design strength of large components by extrapolation based on Weibull statistics are further demonstrated by Fig. 14, which plots volume versus strength to fracture for two values of m included in Fig. 13 and also a curve for $m = 17$, which would be appropriate for consistent quality ceramic.

With increasing section thickness the three curves merge and fracture strength approaches zero, which is contrary to practical experience. With reducing specimen thickness the predicted strength increases exponentially to infinity, irrespective of the practical reality that as the specimen thickness is reduced to the same order of magnitude as the inherent defects, the stress to fracture must be impaired. It is interesting to speculate that by performing tests on successively thinner specimens it may be possible to derive information about the nature of critical defects.

In conclusion, this preliminary investigation into testing ceramics indicates that much more basic research is required to establish reliable methodology of generating data suitable for design of components for internal combustion engines.

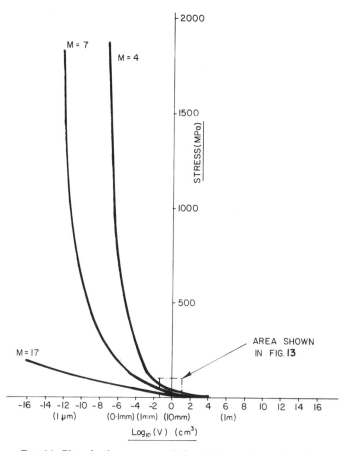

FIG. 14. Plot of volume–stress relationship for various values of m.

ACKNOWLEDGEMENTS

The authors wish to express their appreciation of the support by NBST, to AED for supplying silicon nitride cylinders and to BNFL for the silicon carbide cylinders. Thanks are also due to Mona Parkes and to Mr and Mrs Wynne, who adapted their potter's art to make mundane technical shapes. Assistance by Margaret Bell for typing, Paul Bright for photography, S. Flynn for the illustrations and several colleagues for assistance and advice is also acknowledged.

REFERENCES

1. LENOE, E. M. and MEGLEN. J. L. *American Ceramic Society Bulletin,* **64** (1985) 271.
2. JOHNSON, L. R., TEOTIA, A. P. S. and HILL, L. G. A Structural Ceramic Research Programme—A Preliminary Economic Analysis, Argonne National Laboratories, ANL/CNSU, 38, 1983.
3. JOHNSON, D. R., SCHAFFHAUSER, A. C., TENNERY, V. J., LONG, E. L. and SCHULZ, R. B. *American Ceramic Society Bulletin,* **64** (1985) 276.
4. HARMON, R. A. and BEARDSLEY, C. W. *Mechanical Engineering,* **106** (5) (1984) 22.
5. TIMONEY, S. G. Survey of the Technological Requirements for High Temperature Materials R & D Diesel Engines, EEC Monograph No. EUR 7660 EN, 1981.
6. GODFREY, D. J. and MAY, E. R. W. In *Ceramics in Severe Environments,* Kriegel, W. W. and Palmour, H. III (Eds), Plenum Press, New York, p. 149, 1971.
7. GODFREY, D. J. *Transactions SAE,* **83** (1974) 1036.
8. TIMONEY, S. G. No Coolant Diesel Engine, EEC Report No. EUR 8358 EN, 1983.
9. TIMONEY, S. G. No Coolant Diesel Engine, Final report on Contract No. 143–77 EIR, September, 1981.
10. TIMONEY, S. G. and FLYNN, G. SAE Technical Paper No. 830313—International Congress and Exposition, Detroit, Michigan, U.S.A., March 1983.
11. TIMONEY, S. G. *Trans. IEI,* **107** (1983) 29.
12. Ceramic Components in Diesel Engines, Action Report on Seminar held at UCD 4th/5th October, 1984.
13. DWYER, D. BE thesis, UCD, Mechanical Engineering, 1985.
14. American Society for Testing and Materials, ASTM—B406/76, 1976.
15. WYNNE, R. and WYNNE, P. Springfield, St Clair's Villas, Sandyford, Co. Dublin, Ireland.
16. PARKES, M. Enniskerry Pottery, Enniskerry, Co. Wicklow, Ireland.
17. WEIBULL, W. *Journal of Applied Mechanics,* **18** (3) (1951) 293–297.
18. DAVIDGE, R. *Mechanical Behaviour of Ceramics,* Oxford University Press, Oxford, p. 132, 1982.
19. Fulmer Research Institute Ltd, Stoke-Poges, *Materials Optimiser,* Section III, C, Appendix, 1985.
20. FESSLER, H. Paper presented at *Introduction to Carbide and Nitride Ceramics for Engineering Applications,* University of Leeds, September, 1983.

APPENDIX: CALCULATION OF WEIBULL MODULUS

All bars were broken on an Instron machine using a 4-point bending rig.

The maximum stress to fail was calculated from

$$\sigma = 3\frac{Wa}{bd^2}$$

A. Weibull Modulus
Calculated from Davidge[18]

$$\ln \ln \frac{1}{P_s} = m \ln(\sigma) + K$$

where

$$P_s = 1 - \frac{r}{N+1}$$

σ = stress to fail

K = constant

B. Weibull Modulus
Calculated from Fulmer[19]

$$m = \frac{1 \cdot 2 \times \sigma_{mean}}{\text{Standard deviation}}$$

COMMENTS AND DISCUSSION

Chairman: W. G. LONG

P. Popper: This is a comment about your ceramic breaking into a lot of bits and pieces. It reminded me of the experiments on ceramic armour where with a projectile, you also get ceramics breaking up into a lot of small pieces and this is supposed to stop the projectile because there is a lot of energy absorbed as a result. The ceramic usually has a backing of glass-fibre reinforced plastic and the pieces remain attached. You might consider this approach.

M. Farmer: Well, in our future programmes we are considering extending our bursting tests into shock-loading. We are also considering, very seriously, actually undertaking manufacture of composites—composite components rather than composite materials. The approach you have suggested may be the direction in which we should go.

27

Oxidative Removal of Organic Binders from Injection-molded Ceramics

B. C. MUTSUDDY

Battelle Columbus Laboratories, 505 King Avenue, Columbus, Ohio 43201, USA

ABSTRACT

Binder removal is a key operation in successful ceramic injection-molding. After investigating some alternative approaches to binder removal, we have concluded that oxidative degradation is still the most viable approach. From a series of experiments, we have been able to establish the temperature- and time-dependent degradation behavior of a polyethylene and silicon nitride mix. Also, we have determined the amount of volatiles as a function of temperature. Combining this information, we have put forward a binder removal scheme for this mix. It is apparent from our investigation that the lower removal time and better control over flaws during binder removal are critically dependent on some understanding of the mechanism involved. It is essential to know, on the one hand, the amount and composition of the volatile and nonvolatile products of decomposition and, on the other hand, the rates and orders of reactions involved in the oxidative degradation process. Reactions resulting in oxidative degradation are not identical for all polymers; they vary with chemical structure of the polymers and with the conditions at which oxidation occurs. Therefore, the scheme presented in this paper should be treated as case-specific, and the scheme should be modified with changing binder formulation as well as with shape and size of the component.

1. INTRODUCTION

Complex ceramic shapes are in increasing demand for many industrial products as well as for high-temperature applications in aerospace, in advanced heat engines, and in electronics. Production of these shapes by injection-molding can open the door to improved productivity and/or product performance. For some years, ceramics have been injection-molded on a limited basis to produce thread guides, welding nozzles, small electronics components, and spark plug insulators. Several patents exist for various improved methods of molding and for a wide range of ceramic/plastic compositions of molding mixes. Increased use of ceramic injection-molding has been restricted primarily due to problems associated with binder removal prior to densification. Defects like cracks, lamination, distortion, voids, blisters, etc., arise during the binder-removal process. These defects are more severe with increasing section thickness. Furthermore, the binder removal is a slow process that could take as long as 7 days. Several approaches[1] are being developed to reduce the binder-removal time and to minimize internal flaws. But, none of these approaches attempts to provide a basic understanding of the binder-removal process.

This paper is designed to investigate the oxidative degradation of a specific organic binder system and to analyze the degradation process.

2. EXPERIMENTAL PROCEDURE

Typical properties of the pure polymer, plasticizer, and processing aid are given in Table 1. A composition of a silicon nitride mix is given in Table 2.

Loss of weight was measured with the pure polymer, plasticizer, processing aid, and the silicon nitride mix in an Airsworth thermal balance capable of supporting a maximum load of 4·0 g and selecting weight changes to 0·0005 g. A 2-g sample, in a small high-purity alumina cup, was suspended in the furnace by a platinum wire. The furnace, capable of maintaining a maximum temperature of 1000°C, was programmed to heat the sample to 500°C at a uniform rate of 2°C min^{-1}. All tests were performed in air. A Pt–Pt/10% Rh thermocouple was used to measure temperature. The temperature was maintained constant within ±1°C. The weight loss was recorded on a

TABLE 1
Typical Properties of the Binder Ingredients

Binder ingredients	Specific gravity	Softening point (°C)	Melt flow index (g per 10 min)	Molecular weight	Thermal stability (°C)	Heat capacity (cal deg^{-1} g^{-1})
Polyethylene (MN711-20)	0·915	83	22	12 000	375–436	0·475
Butyl oleate	0·865	—	—	340	140–200	—
Castor oil	0·945	—	—	—	140–200	—

continuous strip chart. As organic binders are expected to degrade to some extent during the heating-up period, the results represent rates on a rising temperature curve. For MN711-20 polyethylene, the temperature range investigated was between 300 and 500°C; for the butyl oleate and castor oil, between 150 and 300°C, and for the silicon nitride mix, between 200 and 500°C. In general, we would expect the materials to degrade within the ranges tested. At temperatures above 600°C, the Si_3N_4 may be susceptible to oxidation.

A Consolidated Electrodynamics Model 21-620 mass spectrometer was used to determine the quantity of the volatile products and their compositions. The tests were determined at 150, 300 and 600°C, in a known volume of air and the products accumulated were measured and analyzed. A 0·3 g sample was used for each experiment.

3. RESULTS AND DISCUSSION

3.1. Thermogravimetric Analysis

Cumulative percentage weight loss for the polyethylene and butyl oleate as a function of time are shown in Figs. 1 and 2. The percentage

TABLE 2
Si_3N_4 Mix Formulation

Ingredient	Volume %	Weight %
Si_3N_4[a]	57	82
MN711-20	31	13
Butyl oleate	9	3·5
Castor oil	3	1·5

[a] GTE SN502 Si_3N_4 plus 8 wt % Y_2O_3 and 4 wt % Al_2O_3.

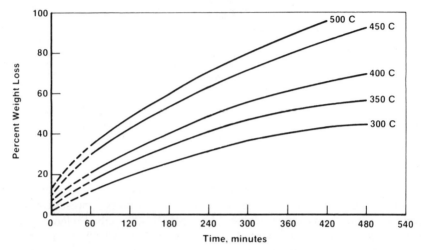

FIG. 1. Thermal degradation of MN711-20 polyethylene.

weight loss for castor oil is very similar to that for butyl oleate. Time was recorded from the moment the thermocouple indicated a constant temperature. This usually occurred about 5–10 min after the heating started. Therefore, the curves in Figs. 1 and 2 start at 3–15% volatilization. It is seen that degradation begins at an initial high rate and then moves down gradually as the reaction proceeds. The rate

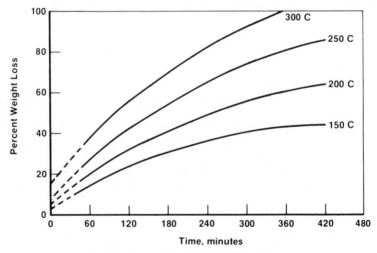

FIG. 2. Thermal degradation of butyl oleate.

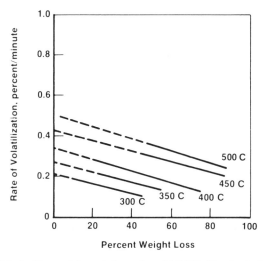

FIG. 3. Thermal degradation rates of MN711-20 polyethylene.

curves shown in Figs. 3 and 4 are straight lines beyond 10–35% degradation. The initial rate constants (K) obtained by extrapolating the straight part of the rate curves to the ordinate axis are given in Table 3. The experimental initial rates (K) obtained in this work may be explained on the ground that different reactions, or combination of reactions, prevail in the initial stages of thermal degradation. The rate

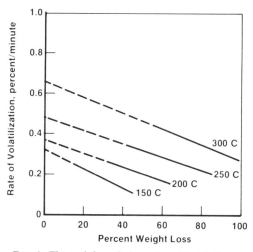

FIG. 4. Thermal degradation rates of butyl oleate.

TABLE 3
Initial Rate Constants by Extrapolation

Binder component	Temperature (°C)	Rate constant (% min^{-1})
MN711-20 polyethylene	300	0·21
	350	0·27
	400	0·34
	450	0·42
	500	0·50
Butyl oleate	150	0·31
	200	0·37
	250	0·48
	300	0·65
Si$_3$N$_4$ mix	250	0·24
	300	0·31
	400	0·41
	500	0·51

curves do not begin at the ordinate because some material volatilizes before the crucible attains the operating temperature.

A comparison of cumulative percentage weight loss of the polyethylene and butyl oleate suggests that the overall degradation process for both materials may be somewhat similar. One significant difference emerges clearly: the initial rate constants of butyl oleate is much higher. A partial explanation may be found in the fact that butyl oleate has a low average molecular weight. At higher temperatures, the shorter chains escape in the initial stage of degradation either unbroken or after a few random breaks.

The weight loss data on the individual binder component prompted a series of experiments with silicon nitride mix (Table 2) between 200 and 500°C. The sample weight was 2 g. The weight loss as a function of time is shown in Fig. 5 and the rate constants are given in Table 3 and Fig. 6. The initial rates for the silicon nitride mix versus those for pure binder ingredients may be explained by a breaking hypothesis. While random breaking of the chain is the principal mechanism of the process, other mechanisms, such as preferred breaking at the ends of monomer or monomer-like fragments, either followed by propagation or not, may take place concurrently to some extent. Furthermore, degradation may be initially retarded by groups such as H_2O, O_2, CO_2 clinging to the free-radical ends of the chains.

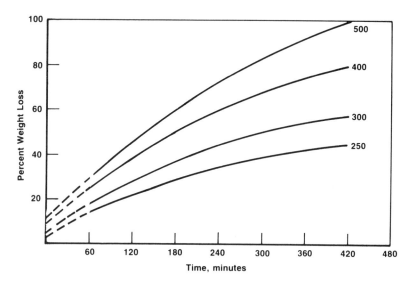

FIG. 5. Thermal degradation of silicon nitride mix.

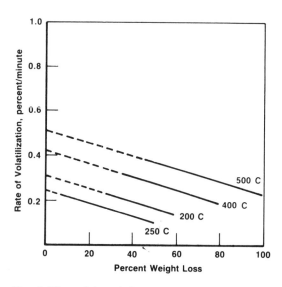

FIG. 6. Thermal degradation rates of silicon nitride mix.

The order of reaction involved in thermal degradation of polymeric binders is of great interest since it may throw light on the mechanism of the binder-removal process. In reality, one cannot speak of any given order of reaction when several reactions are involved. Nonetheless, one can visualize a composite reaction approximating a given order. An indication of the order of reaction involved in the thermal degradation of polyethylene, butyl oleate, and silicon nitride mix can be obtained from Fig. 7. Here, log of initial rate constants (K) at 400°C is plotted against log W, the weight loss at a given time. The order of reaction is obtained from the slope. The reaction order is about 1 in all three cases.

3.2. Analysis of Volatile Products

Information on the amount and composition of the volatile products is as important as the volatilization rate. Although extensive information is available on the oxidation and pyrolysis of organic materials, this study focuses on the oxidative degradation of a binder composition used with silicon nitride.

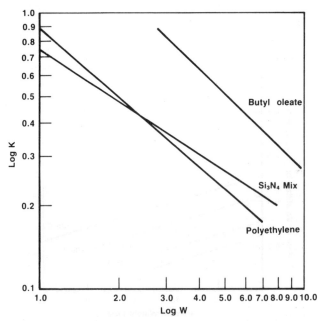

FIG. 7. Order of reaction for 400°C derived from initial rates (K) and weight loss (W) at given time.

Table 4 presents the volatile components detected after heating to 150, 300 and 600°C for 30 min. The blank run consisted of a mullite furnace tube and a zirconium boat. Milled silicon nitride was tested to provide a reference point.

At 150°C, the gas composition is very close to normal air with a slight increase in carbon dioxide (0·1% versus 0·03% for normal air). The difference in water values of <0·1% and 0·1% are not significant since 0·1 is about the lower limit of detection for water. On the whole, heating this mix at 150°C clearly produces little chemical change in the binder components either by oxidation or pyrolysis.

At 300°C, the blank and the Si_3N_4 reference sample showed no changes, but the formulation showed a definite oxidation. Depletion of oxygen with corresponding increase in CO plus CO_2 level, and the start of pyrolysis (as shown by the appearance of organics) is a clear indication of oxidation. Also, note that the nitrogen content increases proportionately as the oxygen is consumed. Water formation is probably higher than indicated in Table 4, because the water formed condensing in the cold zone would have very little vapor pressure at atmospheric pressure or above. However, most of this water would vaporize at 600°C when the pressure is reduced by expanding the total gases into the calibrated volume of the mass spectrometer inlet system.

At 300°C, the principal products of combustion are CO and CO_2 which may arise from direct oxidation of binder components or oxidation of pyrolytic fragments. Excess unconsumed oxygen at 300°C no doubt explains the failure of any organic species produced to survive at that temperature.

At 600°C, most of the oxidation has been completed and pyrolysis is well under way as indicated by the presence of increased organic species. Table 4 shows these and the total volume of gas left after 600°C for 30 min. If the volume of gas produced by the reference sample is subtracted from the total volume of gas produced by the formulation, and correlated with total organics formed, it should provide a clue to the oxidative degradation of the formulation.

In the final analysis, at 600°C the kinetics of the reactions are markedly accelerated and pyrolysis degradation of a sort not likely to occur at 300°C takes place. With time, essentially all the oxygen is consumed. The ratio of CO/CO_2 increases due to limited oxygen supply, and the anticipated high concentration of water is observed. In addition, significant amounts of hydrocarbon products survive since the oxygen required to oxidize them has been exhausted.

TABLE 4
Chemical Components Detected at Different Temperatures

	Volume %								
	150°C			300°C			600°C		
Component	Blank run mullite tube	Si_3N_4 wet milled	Si_3N_4 mix	Blank run mullite tube	Si_3N_4 wet milled	Si_3N_4 mix	Blank run mullite tube	Si_3N_4 wet milled	Si_3N_4 mix
Oxygen	20.9	20.8	20.8	20.9	20.8	7.9	17.4	18.6	0.5
Nitrogen	78.0	78.0	78.1	77.9	78.0	84.1	80.4	78.3	22.5
Argon	0.94	0.95	0.94	0.96	0.94	1.0	1.03	0.96	0.3
Carbon dioxide	0.1	0.1	0.1	0.2	0.1	3.0	1.1	0.8	2.3
Water	<0.1	0.1	<0.1	<0.1	0.1	0.2		1.3	61.0
Carbon monoxide				<0.1		3.1			5.2
Methane						0.01			2.1
Ethane						0.13			1.1
Ethylene						0.01			1.5
Propylene						0.08			1.0
Methanol and ethanol						0.2			0.5
Toluene						—			0.2
Benzene						—			0.4
Propionic and acetic acid						0.2			0.6
Miscellaneous hydrocarbons						—			0.1
Hydrogen						—			0.4
Propane									0.3
Total organic						0.63			0.8
Total gas[a]						—			152

[a] Cubic centimeters at standard pressure and ambient temperature (approximately 25°C).

The production of light hydrocarbons can be explained by the pyrolytic degradation of larger molecules and radical-induced hydrogen transfer among the molecular species present. Thus, methane, ethane, propane, and the dehydrogenated species ethylene and propylene are found along with a small amount of hydrogen. Benzene and toluene can be formed (as they are in petroleum reforming) by the cyclization and dehydrogenation of six- and seven-carbon-atom aliphatic compounds. Organic acids can arise from the partial oxidation of small hydrocarbon molecules or oxidative fragmentation of larger ones.

There is little doubt that the more oxidized species are produced as the temperature is advanced from 300 to 600°C (and very early in the hold period at 600°C), either by direct oxidation or by oxidation of the pyrolysis fragments. Also, as soon as the oxygen is exhausted, unoxidized hydrocarbons and hydrogen can remain in the gas. If the supply of oxygen were not as limited there is no doubt that the products would be almost entirely CO, CO_2, and water (at 600°C) and that even CO would be produced only in small amounts.

3.3. A Conceptual Binder-removal Profile

Using the rates of thermal degradation data, the amount of volatiles and their compositions one can now design a binder-removal profile as a function of temperature and time. Other factors that would influence the profiles are the atmosphere and the sample shape and size. A conceptualized profile is shown in Fig. 8. In this approach, a sample

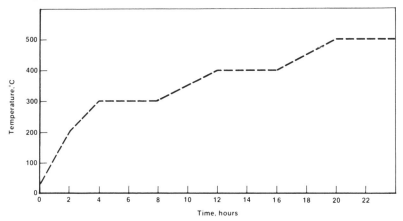

FIG. 8. Conceptual binder-removal profile.

can be rapidly heated up to the first temperature range where a noticeable weight loss is recorded. The sample should be held at that temperature (around 300°C) until the loss reaches a minimum. If the rate of loss is too high (above, say 0·65% per min), the temperature should be lowered to reduce the rate of volatilization. Then, the temperature should be raised until another noticeable weight loss is recorded and held to attain a minimum. The procedure should be repeated up to the peak temperature (i.e. 500°C). The air-flow rate should be calculated on the basis of the composition of the volatile products. The air-flow rate should be reduced during the cooling cycle. In general, the sample can be oven cooled by just turning off the heat.

4. SUMMARY

After investigating alternative approaches to the binder-removal process, we have concluded that thermal degradation at ambient air or gas pressure is still the most viable approach. From a series of experiments, we have been able to establish the following points.

We have attempted to explain the initial rate constants in the light of binder ingredients breaking at random as well as preferred breaking at the ends of monomer or monomer-like fragments. Slow degradation of silicon nitride mix may be associated with groups such as H_2O, O_2, CO_2, etc., clinging to the free-radical ends of the chains. Although no specific order of reaction was established, it is evident that an understanding of the rate constants can serve a useful purpose in designing a binder-removal schedule.

Degradation proceeds with temperature 300–600°C either by direct oxidation or by oxidation of the pyrolysis fragments of the binder components.

Finally, combining this information we have put forward a binder-removal scheme for a specific mix. However, this scheme should be treated as case-specific and should be modified with shape and size of the part.

REFERENCE

1. PASTO, A. L., NEIL, J. T. and QUACKENBUSH, C. L. In *Proceedings of the International Conference on Ultrastructure Processing of Ceramics, Glasses and Composites,* Hench, L. L. and Ulrich, D. R. (Eds), John Wiley, New York, p. 476, 1984.

28

Silicon Nitride Ceramics and Composites: A View of Reliability Enhancement

S. T. BULJAN, J. T. NEIL, A. E. PASTO, J. T. SMITH AND G. ZILBERSTEIN

Ceramics and Metallurgy Technology Centre, GTE Laboratories, Waltham, Massachusetts 02254, USA

ABSTRACT

Research to enhance the reliability of silicon nitride monolithic and composite ceramics is traced from fundamental laboratory studies through component implementation. The results indicate that an optimum green body microstructure is the key to improved reliability, as measured by Weibull statistics. A discussion of metallic impurities, and the requirement to minimize such inclusions, is presented. Again, enhanced reliability accompanies the limitation of these impurities in ceramic components.

1. INTRODUCTION

Silicon nitride ceramics and composites have been developed in various laboratories throughout the world. As a basic understanding of consolidation,[1-7] strengthening mechanisms,[8-10] oxidation resistance[10-12] and other characteristics emerged, the development of an embryonic, high technology ceramics business began. The research achievements were to be translated into articles of commerce with properties[1] goals established by the laboratory results. The ambitious aim of achieving such properties in components fabricated by several powder consolidation practices has been successful. This paper will discuss the role of microstructure, purity and consolidation practice in achieving high reliability in silicon nitride ceramics and ceramic composites. Reliability, in turn, depends on the ability of a ceramic manufacturer to reproducibly fabricate parts of low flaw content, since the brittle nature of these materials implies high sensitivity to flaws.

At temperatures in which fast fracture occurs, the strength behavior of ceramics and ceramic composites is characterized by brittle failure. That is, flaws inherent in the microstructure interact with the applied stress field so as to increase the local effective stress. The result of this stress intensification is that the flaw rapidly propagates across the specimen, causing fracture to occur. The stress intensity is directly related to the size and shape of these flaws, with the result that fracture strength depends inversely on flaw size. Thus, to increase the reliability of ceramic parts, one must either:

1. Increase the fracture toughness of the material;
2. non-destructively examine and eliminate parts with critical flaws;
3. control the flaw population, in terms of size, shape and distribution.

The first option is not open for monolithic compositions, since the toughness is basically a material constant. Toughening of a particular body can be achieved by either microstructure or compositional changes, e.g. dispersoids, which are reviewed later in this paper. The second is currently beyond the NDE state-of-the-art for the small, critical flaws of concern. This leaves the third option as the only meaningful choice, in the short term, for reliability enhancement. This paper discusses a reliability improvement program focused to determine:

A. Those microstructural factors influencing strength and toughness;
B. processing parameters leading to these microstructural factors, and
C. a means to increase the reliability of silicon nitride ceramic and ceramic composite components.

At higher temperatures, the brittle material failure becomes dependent on slow crack propagation. This complex topic, while key to the ceramics industry reliability concerns, will not be reviewed in this paper.

2. SILICON NITRIDE CERAMICS

2.1. Microstructure
Without the use of densification additives, silicon nitride powder (Si_3N_4) shows a remarkable resistance to sintering, e.g. a density of

about 85% theoretical was achieved using diamond forming conditions (1750°C, 60 kbars pressure). Sintering of Si_3N_4 has been successful only by the addition of densification aids, which allow liquid phase sintering. In sintering Si_3N_4, up to 20 wt% of a densification aid is used: MgO, CeO_2, Y_2O_3 and ZrO_2 are typical. The additive combines, at the sintering temperature, with the SiO_2 oxidation layer on the Si_3N_4 powder raw material to form a liquid which wets the Si_3N_4. This liquid penetrates between the Si_3N_4 grains providing a high diffusivity path for mass transport. Material transport and reprecipitation on the Si_3N_4 surfaces adjacent to the pores cause the grain reshaping necessary for sintering. There are two hexagonal polymorphs of silicon nitride, the α- and β-phases, with the latter stable at higher temperatures. A basic consequence of the consolidation is the irreversible conversion from α- to β-phase by a solution–reprecipitation process. On cooling from the sintering temperature, the liquid phase solidifies into a glass or partially crystalline phase at the grain boundaries.

By this mechanism, Si_3N_4-based ceramics are sintered to a minimum 98% of theoretical density. The same densification mechanism prevails in hot-pressing and hot isostatic pressing. The only difference is that the applied, external pressure forces the grains together, and increases the rate of Si_3N_4 solution into the liquid. In addition, the applied pressure enhances particle rearrangement in the early stages of sintering.

Based on years of research, GTE has developed two sinterable compositions for structural components. These are $Si_3N_4 + 6$ wt% Y_2O_3, designated PY6, and $Si_3N_4 + 6$ wt% $Y_2O_3 + 2$ wt% Al_2O_3, designated AY6. The added alumina improves sinterability. For background to the study presented here and to understand the nature of the defects causing failure in consolidated silicon nitride, a comprehensive fracture strength-fractography study[13] will be reviewed. A set of 103 cold-pressed and sintered AY6 strength bars was broken in 4-point flexure and subjected to detailed optical and scanning electron microscopy (SEM) examination. Definite flaws were located at fracture origins on 42 of these specimens, and a correlation determined between flaw type, frequency and severity. As shown in Table 1, the most frequent flaw type, was the pit/white spot. It was named for its appearance as a hole or light scattering area observed by low-power stereomicroscopy. The pits are illustrated in Fig. 1. In Fig. 2, the pit is largely filled with Si_3N_4 grains, which scatters light, creating contrast with the darker matrix and appearing as a white spot. As noted in Table 1, these flaws are by far the most frequent, though not as severe

TABLE 1
Correlation of Flaw Type, Frequency and Severity for Sintered Silicon Nitride Strength Specimens

Flaw type	Frequency (%)	Fracture stress (% of average)
Pit/white spot	88	85–91
Metallic inclusion	10	77

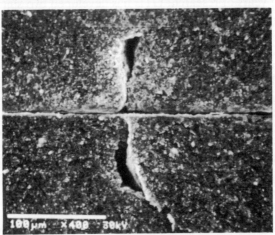

FIG. 1. Fractographic examination of a strength bar identified a pit as the fracture source (top). At higher magnification (bottom) the critical flaw is seen to be a void at the tensile surface of the fractured sample.

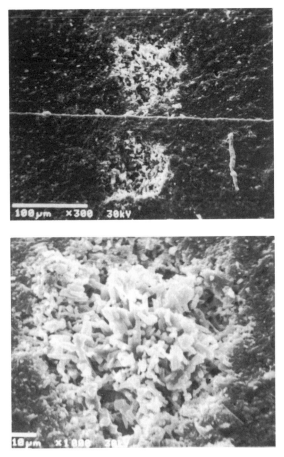

FIG. 2. Fractographic examination of this strength bar reveals a white flaw in bright field examination (top). At higher magnification, the white spot is shown to be a void into which beta silicon nitride grains grew during consolidation (bottom).

in terms of strength degradation as the other major type. The latter is found to be a metallic area in the specimen and will be discussed later in this paper.

Having determined the pit/white spot defect to be a major factor in controlling strength, the origin of these flaws was studied by preparing identical isopressed specimens and heating them to different points on the time–temperature sintering schedule. They were removed from the furnace, fractured, and examined in the SEM. A study of several specimens from each sintering was sufficient to locate several defects.

Tracing the pit/white spot-type defect morphology backwards through the sintering cycle showed its characteristic form to be present in the green-body microstructure as a density inhomogeneity. This inhomogeneity is likely caused by local differences in density among the agglomerates which typically occur in dry-processed silicon nitride and which are not completely eliminated during pressing. Increased forming pressure was found not to eliminate these defects. These density inhomogeneities took the form of sudden jogs in the path of the fracture front as it proceeded across the body. These fracture path jogs manifest themselves in several ways as shown in Fig. 3 (top). At left center is a smooth surface at an angle to the major fracture surface. Just above this is a more vertical path change with an incipient crack at its juncture with the major fracture surface. In the center of this micrograph (and at higher magnification in Fig. 3(bottom)) is shown the third form, a pullout of material. Figure 4 illustrates the changes that occur to this type of green body defect on sintering. As densification proceeds ((a)–(d)), the differential shrinkage caused by the density variations causes the opening of a void, which may become filled with β-Si_3N_4 grains (c) or not (d). These ultimately end up as flaws in the dense microstructure and thus, potential failure origins.

The above study identified inhomogeneities in the isopressed green microstructure as a major source of the lowered reliability of individual strength bars. A ceramic processing study was initiated aimed at eliminating or, at least, minimizing these green-body flaws. A single lot of GTE's SN-502 Si_3N_4 powder was used to prepare a single lot of the AY6 composition and a second of the PY6 material. Since all strength samples originated from a single powder lot processed to a single composition batch, trends and data interpretation can be reliably correlated to the forming and consolidation processes. The Si_3N_4, Y_2O_3 and/or Al_2O_3 components were milled with silicon nitride media to reduce tramp impurities. The shaping techniques utilized included isostatic dry-pressing, slip-casting and injection-molding.[13] Subsequently, certain samples were hot-pressed in BN-lined graphite dies, while others were formed into simple rectangular shapes and sintered in a graphite furnace under nitrogen. Yet other samples were shaped by injection-molding and hot isostatically pressed in evacuated glass ampules. Strength testing was completed by the 4-point flexure method. The resultant data was analyzed with Weibull statistics using a computer program which calculates the Weibull parameter (m) by least-squares fit.

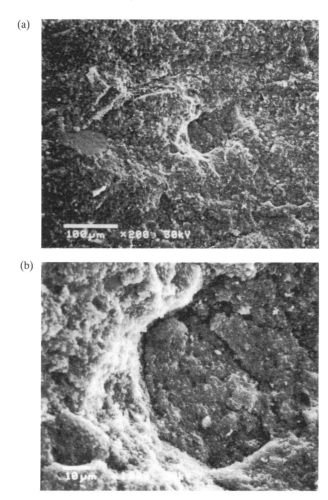

FIG. 3. (a) The fracture surface of an isopressed strength bar, prior to consolidation, is shown (above), with several incipient crack sites visible. The center of this photo has a void which is unlikely to close during sintering, see (b).

The strength and strength distribution for AY6 MOR bars fabricated by these several processing and consolidation methods, with their inherent defects are shown in Fig. 5. This is a Weibull plot for sintered AY6 specimens representing examples in which the green bodies were isopressed, slip-cast, and injection-molded. The isopressed data are bimodal. Two specimens are anomalously low strength,

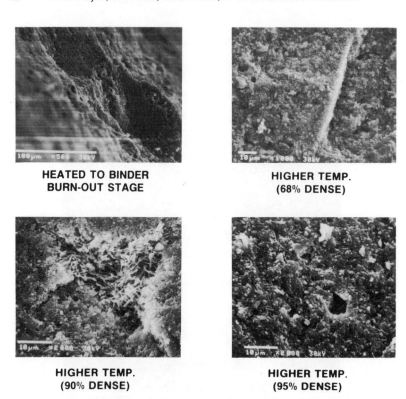

HEATED TO BINDER
BURN-OUT STAGE

HIGHER TEMP.
(68% DENSE)

HIGHER TEMP.
(90% DENSE)

HIGHER TEMP.
(95% DENSE)

FIG. 4. These photos follow the retention of isopressed flaws, at upper left, through densification with pits and white spots, Figs. 1 and 2, evident.

while the remaining eight would have yielded a much higher Weibull modulus and average strength than the population of 10.

Having established a correlation between the isopressed green microstructure homogeneity and sintered body strength, the green fracture surfaces of materials processed by 'dispersed particle' methods were examined. The photomicrographs of Fig. 6 illustrate that both the slip-cast (SC) and injection-molded (IM) green-body microstructures are free of the sharp edged density inhomogeneities shown in Figs. 3 and 4. The IM body microstructure is not as featureless as the SC body, containing several rounded features which do not cause difficulty during sintering. The exact nature of these features is yet unknown, but is undergoing further study. The beneficial effect of this homogeneous green microstructure on sintered body strength is

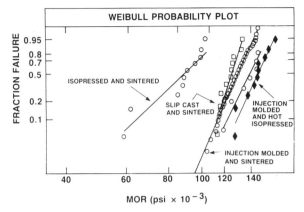

FIG. 5. Weibull probability plot for the strength of samples produced by the identified routes, see the text for discussion. The average strength and Weibull modulus are: IP—101 000 psi and 9; SC—121 000 and 20; IM—130 000 and 14; IM and HIP—143 000 and 20.

illustrated in Fig. 5. Both processes yield sintered materials of superior average strength and Weibull modulus as compared to the baseline isopressed material. In fact, the injection-molded material is essentially equivalent to the hot-pressed AY6.

The best properties were attained with injection-molded and HIP'd AY6, which yielded a mean strength of 143 000 psi and a Weibull modulus of 20. Fractography of these latter materials failed to disclose definite defects at fracture origins. This result is as expected, since flaw sizes are small at these strength levels.

The same property improvement can also be obtained for the alumina-free PY6 materials. The average strength and Weibull modulus increased from 80 000 psi and 11 for isopressed material to 100 000 psi and 21, respectively, for slip cast materials.

2.2. Chemistry

The effect of impurities on strength and reliability was presented in Table 1, where samples with metallic inclusions showed reduced strength to about 75% of that measured for the total population. Such metallic impurities typically originate as a distinct particle from the equipment used for handling and processing the powders to a finished shape. Energy dispersive X-ray analysis of these metallic areas usually shows high Si plus any of several metallic species, principally iron, nickel and chromium, suggesting their pickup from steel based

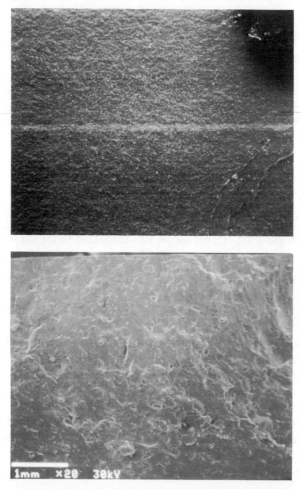

FIG. 6. The fracture surfaces of slip cast (top) and injection molded (bottom) strength bars as formed, showing a nearly flaw free green microstructure leading to higher reliability in consolidated components.

process equipment. Evidence for the particulate inclusion fracture of origin is presented in Fig. 7, which depicts an Fe-bearing inclusion present on the fracture surface of an unfired body. Its presence as a particle in this stage of processing implies its presence as a particle before cold-pressing of the shape.

After sintering, optical microscopy examination of sintered AY6 or

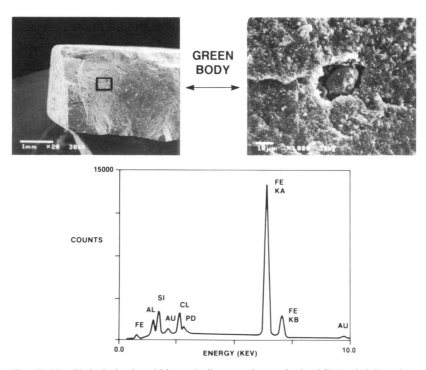

FIG. 7. Metallic inclusion in cold isostatically pressed green body of Si_3N_4. (A) Scanning electron micrograph of fracture surface; (B) scanning electron micrograph of inclusion on fracture surface; (C) energy-dispersive X-ray spectrum of inclusion in (B).

PY6 bodies with such metallic impurities shows the presence of a widely scattered, highly reflecting phase. This phase has been identified as metallic and these metallic areas consist of a mixture of metal intergrown with β-Si_3N_4 grains. It is hypothesized that at the sintering temperature the metal was a liquid saturated with nitrogen. As the body cooled, Si_3N_4 grains precipitated, dividing the liquid metal into regions which look like droplets in a two-dimensional microscope field. Further cooling caused the metal to solidify.

Examples of these inclusions found at fracture origins are shown in Fig. 8 with the metal residing in a pore, accompanied by a void area. The inclusion in Fig. 8 contained major amounts of Fe and Si, as determined by energy dispersive X-ray analysis, Fig. 8(C). Other metallic elements are commonly found associated with these inclusions, e.g. the Cr and Al shown (the Au and Pd are present in the

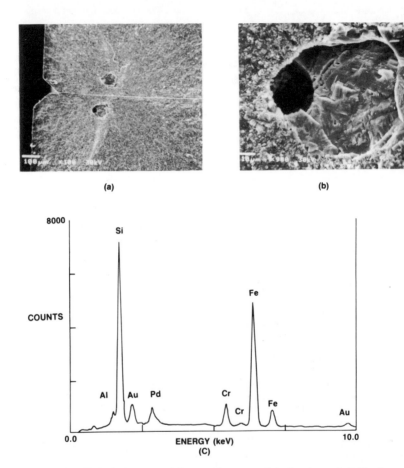

FIG. 8. Metallic-inclusion-type defect at fracture origin in sintered Si_3N_4 shown in scanning electron micrographs of fracture surfaces at (A) and (B). (C) Energy-dispersive X-ray spectrum of inclusion in (B).

conductive SEM coating applied). In several such analyses performed at GTE Laboratories, it was never proved that brightly reflecting areas, noted in the optical microscope, were pure Si; rather they were metal silicides as shown in Fig. 8.

The size and quantity of these inclusions are highly dependent on the consolidation temperature. As the sintering temperature increases, as necessary for PY6 versus AY6 bodies, the quantity and size of these inclusions similarly increases. Based on thermodynamic principles, it has been shown that the activity of silicon is lowered in the presence of iron and other metals.[15] In turn, this leads to Si_3N_4 decomposition at typical nitrogen sintering pressures. The two or more metals alloy at temperature creating the metallic inclusion observed at room temperature with the associated void resulting from solidification shrinkage and the decomposition reaction.

The obvious corrective action is to minimize the metallic contamination sources present in the processing equipment, a corrective measure in place within GTE to enhance the reliability of Si_3N_4 ceramic components.

3. SILICON NITRIDE CERAMIC COMPOSITES

3.1. Microstructure

This section discusses a silicon nitride ceramic composite material developed by combining the AY6 alloy discussed above with titanium carbide (TiC) particulate dispersoids, typically 2 μm average diameter. These silicon nitride–titanium carbide particulate composites form a new class of wear resistant materials which combines high hardness, fracture toughness, abrasion and corrosion resistance. Design and property tailoring of particulate composites require an in-depth understanding of chemical and mechanical interactions between the material's components.[2] This section concentrates on the chemical interactions between the Si_3N_4-based matrix and the TiC dispersoid, during densification, in the presence of the liquid sintering aids formed by the Y_2O_3, Al_2O_3, SiO_2 interaction. The liquid may also contain some TiO_2 originating from the TiC powder. Liquid formation, considered an event marking the beginning of densification, also gives rise to a number of chemical reactions between the components. Thermodynamic evaluation of the reactions proceeding during consolidation has indicated a high probability of TiN and SiC formation as stable products of Si_3N_4 + TiC reaction. It was anticipated that the products of any Si_3N_4 and TiC reaction would either be located in the intergranular glass phase or, more likely, be concentrated on TiC particle surfaces as reaction zones.

For standard processing, composites consolidated from the GTE SN-502 silicon nitride exhibit a reaction zone of about 0·1 μm (Fig. 9). To study this zone, the composite sintering time was extended which enlarged the reaction zone for experimental evaluation. Optical

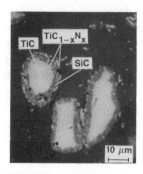 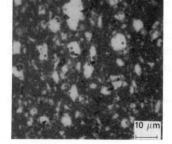

EXTENDED SINTERING TIME   STANDARD PROCESSING

FIG. 9. The original TiC particles and the resultant interfacial reaction products are shown for a sample sintered for an extended time to enlarge these phases (left) and the same composite with the preferred microstructure (right).

examination of a polished composite cross-section revealed the presence of well defined reaction zones, uniform in width, around every TiC particle (Fig. 9). The zones appeared to have a polycrystalline morphology and to consist of fine grains of at least two phases. The crystals of both phases constitute a mixture with well defined grain boundaries. The reaction zone has a distinctly different color than the TiC grain interior and is separated from it by a clearly defined interface.

Concluding that these reaction zones were a general case of the composite microstructure, electron probe microanalysis was employed to examine the elemental distribution within the reaction products. Morphologically, the TiC dispersoid particles appeared to be composed of three phases. In the phase at the interface between AY6 and TiC, the elements titanium, carbon and nitrogen are detected, suggesting titanium carbonitride Ti(C, N) formation. Isolated concentrations of silicon and carbon were also detected indicating that SiC is an interface reaction product also. An inner phase, similar in appearance to the TiC particle core, indeed shows the presence of only titanium and carbon, corresponding to the unreacted portion of the original TiC grain. Similar analysis of other TiC particles indicates the occasional presence of oxygen in the interface zone suggesting that it may be present with the carbon and nitrogen in the Ti(C, N).

The original TiC particles, when less than 1 μm diameter, may be fully converted to SiC and Ti(C, N) reaction products. The SiC crystals were chemically extracted from the composite samples and examined by scanning transmission electron microscopy which indicated the presence of β-SiC and a variety of α-SiC polytypes. This data confirms the reaction of TiC to SiC. To consolidate a reliable composite with reproducible properties, close control should be exercised during consolidation to maintain a minimal reaction zone, similar to the 0·1 μm shown in Fig. 9.

3.2. Chemistry

In addition to the matrix-dispersoid reaction control, the role of chemical impurities, such as iron or carbon, was studied. Such impurities increase the reaction kinetics and limit the resultant composite properties. For experimental purposes, the silicon nitride–titanium carbide composite was seeded with iron powder, at low levels to determine the consequence of such metallic impurities. In the iron-contaminated material, the microstructure which developed adjacent to the iron inclusions consisted, for the most part, of low melting iron silicides and iron–aluminium–silicide phases, as well as the Ti(C, N) interface, SiC particles and TiC core structure (Fig. 10). Around the iron inclusion, in the areas with somewhat lower level of iron contamination, a pattern indicative of catalyzed matrix-dispersoid reaction is observed. The cores of initial TiC grains are now fully occupied by SiC, whereas TiC$_x$N$_y$ forms only peripheral layers, in contrast to what was observed in contamination-free material. Optical examination of the areas located at some distance from the cavity

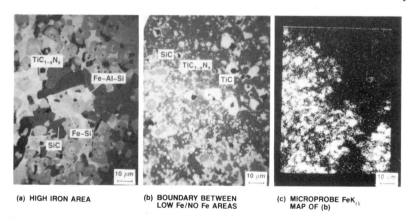

FIG. 10. The TiC and Si$_3$N$_4$ composite microstructure in a sample seeded with iron showing the accelerated reaction ((a) and (b)) and the normal microstructure in the area without iron, top right of (b) and (c).

reveals an unaltered microstructure similar to the contamination-free microstructure. This observation, together with the absence of iron in these areas, points out that the catalytic effect of iron on the matrix-dispersoid reaction was restricted to a relatively narrow zone around the iron particle.

Hence, iron segregations in Si_3N_4–TiC composites appear to act as sources of substantial microstructural inhomogeneities. Iron silicides and/or iron–aluminium silicides, which usually form in such areas and which are characterized by extreme brittleness and relatively low melting points, can make a composite unsuitable for high temperature, high stress applications.

Since impurities, such as iron, lead to substantial alterations of Si_3N_4–TiC composite microstructure, it was anticipated that the mechanical properties would be affected. The hardness, fracture toughness[16] and strength of the AY6–TiC composite (SN-I) and the SN-II composite, seeded with iron, are summarized in Table 2. They appear to indicate close microstructure-property interdependence.

As can be seen from this table, the Si_3N_4–TiC composites containing iron impurities (SN-II) are characterized by properties consistently lower than the material prepared from high purity powder (SN-I). The loss of hardness can most certainly be related to the greater amount of reaction products, silicon carbide and titanium carbonitride, whose hardness is lower than titanium carbide. For example, the Si_3N_4-based hot-pressed composites, containing the same volume fraction of SiC and TiN dispersoid as the materials under study, exhibited hardness of

TABLE 2
Mechanical Properties of Si_3N_4–TiC Composite Materials

Si_3N_4–TiC composite	Microhardness Knoop ($GN\,m^{-2}$)	Indentation fracture toughness, K_{IC} ($MN\,m^{-3/2}$)
SN-I	15·08(±0·35)	4·45(±0·22)
SN-II	13·95(±0·15)	3·88(±0·14)

	Modulus of rupture strength (MPa)		
	RT	1 000°C	1 200°C
SN-I	552[80 000]	827[120 000]	441[64 000]
SN-II	559[81 000]	531[77 000]	241[35 000]

13·65 GN m^{-2} and 12·99 GN m^{-2}, respectively. The change in microstructure associated with the increased matrix-dispersoid reactions caused by the iron impurity also decreases the fracture toughness. The observed changes in K_{IC} translates into an apparent reduction of critical flaw size by 25%. The room temperature strength of the two composites, SN-I and SN-II, are unaffected by the impurities. The elevated temperature strength is largely controlled by the properties of the intergranular glass, such as its composition and viscosity. The strength of the SN-I composite prepared from high purity powders without iron increases at 1000°C and is reduced at 1200°C. It has been postulated that the strength maximum at 1000°C is associated with stress relaxation between the Si_3N_4 matrix and the TiC dispersoid or that localized crack blunting occurs by viscous flow of the second phase. Above this temperature, the viscosity of the Y_2O_3–Al_2O_3–SiO_2 glass is reduced and the strength of the material decreases rapidly. The softening point and the viscosity of this glass, with iron present, is

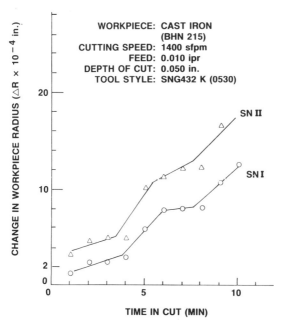

FIG. 11. The wear of silicon nitride ceramic composite metal cutting tools is presented where tool wear, as measured by the increase in workpiece radius, is shown to be lowered by the higher reliability of the SN-I composition. This composition's reliability is greater because the iron, seeded in SN-II, is absent in SN-I.

expected to decrease with increased impurity concentration. Consequently, the strength of the composite seeded with iron (SN-II) shows a reduction in strength at 1000°C and a considerable decrease at 1200°C.

The silicon nitride–titanium carbide composite is manufactured for several commercial applications. To illustrate the role microstructure plays in reliability, the machining performance of metal cutting tools from the high purity SN-I composite is compared with that of tools from the composite with iron particles (SN-II), in Fig. 11. The tool wear, as measured by the change in bar radius, is greater for the SN-II tools versus the higher purity or more optimum microstructure SN-I tools. These data are typical of GTE's experience with ceramic composites; that is, the greater component reliability is associated with the improved microstructure which results from proper process control and high purity powder handling practices.

4. COMPONENT FABRICATION

The introduction stated the need for improved reliability in ceramic components with properties approaching those achieved in research programs. The earlier sections stressed the causes for and elimination of processing defects. This section will briefly illustrate the realization of high strength, high Weibull modulus properties in injection-molded components. The injection-molding process has been reported in detail[14] and will not be presented here.

4.1. Small Cross-section Parts

The CATE (Ceramic Applications in Turbine Engines) axial turbine blades are typical examples of small cross-section parts, shown in Fig. 12. Properties of injection-molded and sintered CATE axial turbine blades have been discussed in great detail in earlier publications.[17] A brief summary is presented here. The CATE blades, typical of small intricate shapes, were sintered to high density (>98% TD) with a high degree of dimensional control. For example, the geometric tolerances of the air-foil contour and the setting angle in the CATE blades were maintained within ±0·002 in and ±1·0°, respectively. Eighteen of these CATE blades were cold spin tested to failure at Allison Gas Turbine Division of General Motors. The results summarized are: (1) The average failure speed was 50 310 rev per min, equivalent to 136%

FIG. 12. Injection-molded turbine components as molded (left) and after sintering (right). The CATE axial turbine blades are in the foreground while the AGT gas turbine rotor is in the rear.

of design speed and (2) the lowest speed failure was 45 200 rev per min or 122% of design speed.

The Weibull modulus of this test distribution is 18, essentially the same as reported for the strength data in Fig. 5. The similarity of the Weibull moduli is interpreted as indicating the achievement of high reliability in components, similar to that measured in the simple strength bars.

4.2. Large Cross-section Parts

Silicon nitride gas turbine rotors have been injection-molded and sintered with good maintenance of geometry from the AY6 composition. Other injection-molded rotor components were consolidated by hot isostatic pressing. The cold spin test data, as obtained by the G.M. Division, from five injection-molded and HIP'd rotors indicated that four of the five rotors were spun to or exceeded the design speed (86 240 rev per min). To ascertain the failure cause for the rotor which burst below the design specification, a fractographic analysis was initiated.

Indications were that small alumina milling media chips were present in that rotor. The corrective action was to utilize silicon nitride media (AY6) whose inherent toughness had been demonstrated to prevent such chip formation. The strength and strength distribution of MOR bars cut from two HIP'd rotors are shown in Fig. 13. The low average strength and Weibull modulus noted in rotor 549 was a result of the alumina particles, detected at the fracture origins of individual low strength bars. These inclusions are not fundamental to Si_3N_4 material and were eliminated by improved processing with Si_3N_4 media.

The rotor 615 was processed from powder milled with the silicon nitride media with all other processing schedules identical to that of rotor 549. The strength data indicate a modest gain to 122 000 psi with the elimination of the alumina inclusions, but the reliability as measured by the Weibull modulus is trebled to 21·5. This significantly improved reliability has been maintained with the gains realized by using the high toughness, silicon nitride milling media. Excellent high temperature strength and oxidation resistance were determined for rotor materials processed by the schedule utilized for the 615 rotor. This again demonstrated that the reliability improvement inherent in the injection-molded green microstructure (Fig. 5), can be realized in

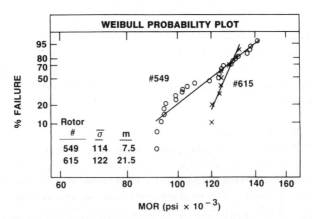

FIG. 13. A Weibull probability plot showing the strength (MOR) of rotor 549 samples, lowered by Al_2O_3 media chips, contrasted with the higher reliability processing schedule for rotor 615 which shows a much improved modulus.

full-scale components. It also identifies the difficulty associated with the translation from research specimen to component fabrication. The alumina milling media limited reliability while the corrective action with silicon nitride media regained the property level associated with the research results.

5. CONCLUSIONS

Silicon nitride ceramic and ceramic composite compositions were fabricated by several shapemaking and consolidation techniques. Their ambient temperature strength and reliability behavior can be summarized: strength is controlled by microstructural defects; these flaws originate in the green-body microstructure and evolve during sintering; and the flaws are density inhomogeneities or metallic inclusions. Fabrication by dispersed particle processes, slipcasting and injection-molding was identified as the means to control green microstructure while metallic inclusions must be excluded with appropriate processing equipment safeguards. It was shown that injection-molded turbine components and metal cutting tools can be produced with the same level of reliability and properties as found in the laboratory research program results.

ACKNOWLEDGEMENTS

The authors extend grateful appreciation to their many colleagues whose experimental expertise was key to the research achievements discussed in this paper. The turbine component fabrication development was conducted with Department of Energy support under sub-contract from the Allison Gas Turbine Division, G.M.

REFERENCES

1. SMITH, J. T. and QUACKENBUSH, C. L. In *Factors in Densification and Sintering of Oxide and Non-oxide Ceramics,* Somiya, S. and Saito, S. (Eds), KTK Scientific Publishers, Tokyo, p. 426, 1979.
2. BALDONI, J. G., BULJAN, S. T. and SARIN, B. K. In *Proceedings of the Second International Conference on the Science of Hard Materials,* Rhodes, 1984, Plenum Press, New York, 1985, in press.

3. SARIN, V. K. and BULJAN, S. T. SME Technical Paper No. MR83-189, 1983.
4. RICE, R. W. and McDONOUGH, W. J. *Journal of the American Ceramic Society*, **58** (1975) 264.
5. CLARKE, D. R. and THOMAS, G. *Journal of the American Ceramic Society*, **61** (1978) 114.
6. JACK, K. H. In *Ceramics for High Performance Applications*, Burke, J. J., Gorum, A. E. and Katz, R. N. (Eds), Brook Hill Publishing Co., Chestnut Hill, Mass., p. 265, 1974.
7. WILD, S., GRIEVESON, P., JACK, K. H. and LATIMER, M. J. In *Special Ceramics*, Vol. 5, Popper, P. (Ed.), British Ceramic Research Association, Stoke-on-Trent, p. 377, 1972.
8. SMITH, J. T. and QUACKENBUSH, C. L. *American Ceramic Society Bulletin*, **59** (1980) 529.
9. BALDONI, J. G., BULJAN, S. T. and SARIN, V. K. presented at *Basic Science and Electronics Joint Fall Meeting, American Ceramic Society*, Cambridge, MA, September, 1982, to be published.
10. QUACKENBUSH, C. L., NEIL, J. T. and SMITH, J. T. ASMP Paper 81-GT-220, March, 1981.
11. QUACKENBUSH, C. L. and SMITH, J. T. *American Ceramic Society*, **59** (1980) 533.
12. BULJAN, S. T. and SARIN, V. K. In *Proceedings of the Second International Conference on Science of Hard Materials*, Rhodes, 1984, Plenum Press, New York, 1985, in press.
13. PASTO, A. E., NEIL, J. T. and QUACKENBUSH, C. L. In *Proceedings of the International Conference on Ultrastructure Processing of Ceramics, Glasses and Composites*, Hench, L. L. and Ulrich, D. R. (Eds), John Wiley, New York, p. 476, 1984.
14. QUACKENBUSH, C. L., FRENCH, K. and NEIL, J. T. *Ceramic Engineering Science and Proceedings*, **3** (1982) 2034.
15. PASTO, A. E. *Journal of the American Ceramic Society*, **67** (1984) C-178.
16. EVANS, A. G. and CHARLES, E. A. *Journal of the American Ceramic Society*, **59** (1976) 179.
17. NEIL, J. T., FRENCH, K. W., QUACKENBUSH, C. L. and SMITH, J. T. ASME Paper No. 82-GT-252, presented at the *International Gas Turbine Conference*, London, April, 1982.

COMMENTS AND DISCUSSION

Chairman: R. W. DAVIDGE

R. T. Cundhill: When you look at hot isostatically pressed materials do you find that you do not have the white spot defects and are left with only the metallic inclusions.

J. T. Smith: Initially, in processing, we eliminate the metallic inclusions which anyone can do. There is an open debate within our own group as to the benefit of hot isostatic pressing. Our present feeling is that hot isostatic pressure will eliminate the sintering defects, places where the liquid, for instance, does not penetrate properly, but if you have agglomerates in your green microstructure, it appears that hot isostatic pressing does not force the liquid between the particles of those agglomerates to improve the reliability, so correct processing through a green microstructure is essential; HIP can not eliminate the errors of the green microstructure.

P. Popper: Can the metallic inclusions be removed by some conventional technique such as sieving?

Smith: They are so fine and so small, they all come from processing, so there is no practical way to take out 10–20 μm particles. The practical way is not to let them get in, in the first place.

M. J. Pomeroy: You mentioned that your silicon nitride/titanium carbide composites show increases in wear resistance. Is that due to fracture toughness or hardness and have you a microstructural or compositional explanation for this?

Smith: I think we have got two factors confused here. What I was trying to show was that in some modelling work, which I did not report, the abrasion wear resistance is a function of both the fracture toughness and the hardness. Now in the case of the particular silicon nitride composite I reported, we find the fracture toughness to be the same as the monolithic body and therefore we attribute the abrasion wear resistance to the addition of titanium carbide which raises the hardness.

29

Panel Discussion: Aspects of Reliability

Chairman: W. LONG

R. W. Davidge: Perhaps I can define the problems as I see them and then we can open up the discussion. Reliability as a subject has not really been discussed at this conference other than two papers. Nevertheless it is an important problem and it does need solutions.

If you are manufacturing components, some of them will have defects, others will not. The basic problem is this—you require some technique or methodology, which I will refer to as quality control which will tell you whether the components that you are making are good ones or bad ones. Now if you have developed a technique which works, then you can fit a good component into a car and drive off! If the component is a bad one you need a dustbin to throw it in! So the question is 'what is quality control?' What is the magic wand you can wave over a component to say good or bad for whatever purpose you need it.

There are two main methods. The first one is NDT. We have had one paper on this which discussed the limits of flaws you can see. If you are concerned with green compacts before firing you might be looking for defects which are perhaps 100 μm or a few millimeters in size and as we have heard from several papers the sort of defect in a fired ceramic that controls the strength is in the region of 1000–1200 μm. There are a number of techniques such as X-ray, microfocus, ultrasonic techniques, thermography and generally these can detect defects that are greater than 25–100 μm. The resolution will vary depending on the defect—for example whether it is an inclusion of tungsten carbide or a very thin crack. But these techniques are not sensitive enough to tell you what the strength of the material is without breaking it. So NDT generally will not tell us all we need to know.

Secondly, you can assess the mechanical properties. You can look at statistical and time dependent, size dependent, stress state dependence of strength and plot strength–probability–time diagrams or carry out proof testing and look at proof test diagrams. Using these, then, for low to moderate temperatures where the original flaws in the material remain but no additional flaws are created by testing, the tests are good and reliable and have been demonstrated for quite a number of components, from hip joints to grinding wheels, with good agreement.

On the other hand, at higher temperatures (1000°C) which is the area of interest for gas turbines, then new flaws are created which can grow in size to form a new critical flaw (see Fig. 1).[1] Example 4 shows the sequence of events.[2] Small voids are nucleated at grain intersections and these will link together to form fractures along particular grain boundary facets leading to critical flaws.

Example 5 shows chemically induced high temperature behaviour such as oxidation processes in silicon nitride/carbide materials forming a silica glaze on the surface. However, cracks formed in the glaze can permeate into the underlying ceramic causing a strength reduction.[3]

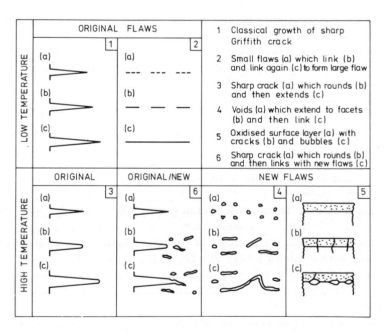

FIG. 1. Generation of flaws in engineering ceramics at low and high temperatures.

CONCLUSIONS

At low to moderate temperatures, reliability prediction is possible using mechanical property assessment but not, generally, NDT techniques.
At high temperatures, prediction techniques need development mainly because new flaws are generated by either creep or chemical effects which confuse the situation.

K. H. Jack: In quality control, you have to have a feed-back to processing. You can go on 'quality controlling' ad infinitum but the problem is you must relay back the information to improve processing.
Davidge: Yes, that is true. Quality control is useless if it rejects everything. The materials people will say they need better design and the engineers will ask for better materials. Both have to be improved.
W. G. Long: But you also need good quality control on the raw materials to start with. You need to know that the materials you are working with today are the same as the ones you had last month and the same as the ones used for the prototypes that the engine manufacturer did the expensive testing on. So it has to be quality assurance all the way from raw materials through processing to finished component. But at what cost?
Jack: I see one problem in increasing sophistication and that is increasing costs and there will be no future for these materials.
W. Grellner: Well I think there is one possibility of bringing down these high costs. You need quality control and the most important time is during the development of the process and if you have developed a process you need to control the process. Of course, you can carry out QC on some of your finished components but it is very expensive to do it on all components, so you need to control the processing. You can use NDT and other techniques. But the earlier you detect problems and faults, the less costly will be the process.
Davidge: Yes, it is no good making a product for $10 000 and after testing, it breaks, if you can spend that money on more research early on in the process which gives much more detail that can then be applied to a large number of pieces.
J. T. Smith: In our own approach we have taken the time to do the process research and you have to have quality control in place at each

step in the process. Take injection-molding—we do QC after molding, after burn-out and after sintering. I disagree with some of your remarks about NDT in that in the cross-section we are using, you can in fact find flaws below 200 μm and metallic inclusions below 100 μm by using micro-focus techniques if you also take advantage of the projection capability of that technique. When you actually look at the fracture mechanics and relate it to idealized flaw sizes and then look at real materials, you will find that if you can find flaws below 200 μm for the current design you will have the reliability you need.

Davidge: I was only making broad generalisations.

Smith: Well as a matter of fact, Rolls has some of the best projection micro-focus, X-ray capabilities in the world. You should take advantage of it!

Long: Do you think the engine manufacturers are going to pay for this extra reliability and increased cost?

R. T. Cundhill: Well no-one here has mentioned specifications. In normal industrial practice you have quality of performance and quality of conformance, and the quality of conformance means that the article conforms to specifications and standards. We have not got these yet.

Grellner: It is usual to have specifications for ceramic articles too but it depends on what the customer wants to pay. For the nuclear ceramics programme there are very severe specifications.

Jack: But there must be applications where specifications are not very high. I was impressed by Dr Kennedy[4] who got a higher reliability by degrading the material.

Smith: The quality control that we carry out with cutting tools includes density, fracture toughness, X-ray diffraction, microstructure and hardness. The point is that a cutting tool, when you put it against an expensive workpiece such as the nickel-based alloys, has to have every bit as much reliability as a more sophisticated component. It is an easy shape to make so we have to start getting involved in these easy shapes and learn how to treat them. It is easy to make one or two cutting tools but, when you start loading up furnaces with them, all the thermodynamics that you think you understand change; the inter-reactions change, the powder beds that you use may or may not be applicable. So I have to make the point again that the research basis you use has to be applied in production.

Jack: Yes, but cutting tools have the advantage that they are very small components.

Smith: OK, but you make them in bulk. That is the point that is different.

Long: I think that the success that has been achieved with cutting tools should give encouragement to the further development of these high performance ceramics in more severe environments.

REFERENCES

1. DAVIDGE, R. W. In *Mechanically, Chemically and Thermally Induced Failure in Engineering Materials,* Bolton, J. D. and Hampshire, S. (Eds), Materials Research Publications, NIHE, Limerick, p. 291, 1984.
2. EVANS, A. G. and BLUMENTHAL, W. In *Fracture Mechanics of Ceramics,* Vol. 6, Bradt, R. C., Hasselman, D. P. H. and Lange, F. F. (Eds), Plenum Press, New York, p. 423, 1983.
3. EVANS, A. G. and DAVIDGE, R. W. *Journal of Materials Science,* 5 (1970) 314.
4. KENNEDY, P. Effect of Microstructural Features on the Mechanical Properties of REFEL Self-bonded Silicon Carbide, this volume, pp. 301–317.

List of Participants

M. Ahmed	NIHE, Limerick, Ireland.
K. Allen	Dundalk Regional Technical College, Ireland.
M. Austin	Trinity College, Dublin, Ireland.
E. A. Belfield	Advanced Materials Engineering Ltd, Stourport on Severn, UK.
M. Billy	UER des Sciences, Limoges, France.
P. A. Blackham	Dynamit Nobel (UK) Ltd, Slough, UK.
S. Bosković	'Boris Kidric' Institute of Nuclear Sciences, Belgrade, Yugoslavia.
R. Brage	KemaNord Industrikemi, Ljungaverk, Sweden.
A. Briggs	AERE, Harwell, UK.
L. Brosnan	Radiac Abrasives Ltd, Castleisland, Ireland.
R. J. Brook	University of Leeds, UK.
D. Broussaud	Ecole des Mines de Paris, Evry, France.
F. Cambier	Centre Recherches de l'Industrie Belge de la Céramiques, Mons, Belgium.
D. Campos-Loriz	Sohio Engineering Materials, Niagara Falls, USA.
E. Cashell	Cork Regional Technical College, Ireland.
P. Chagnon	Rhone-Poulenc Recherches, D'Aubervilliers, France.
N. Cotterill	TRW Ceramics Ltd., Derby, UK.
R. T. Cundhill	SKF Engineering & Research Centre, Nieuwegein, The Netherlands.
R. W. Davidge	AERE, Harwell, UK.
D. Devenny	Consarc Engineering Ltd, Bellshill, UK.
T. Ekström	AB Sandvik Hard Materials, Stockholm, Sweden.
Q. Fan	University of Leeds, UK.
J. Fardoe	Morganite Special Carbons Ltd, London, UK.
M. Farmer	University College, Dublin, Ireland.
R. Forham	JRC, Petten, The Netherlands.
D. Fricker	Howmedica International Inc., Limerick, Ireland.

C. Galassi	Research Institute for Ceramics Technology, Faenza, Italy.
R. C. Gambie	Norton Industrial Ceramics, Weybridge, UK.
G. Gee	Boart Hardmetals (Europe) Ltd, Shannon, Ireland.
R. Gilissen	SCK, Mol, Belgium.
W. Grellner	ESK-GmbH, Kempten, West Germany.
E. Gugel	Forschungsinstitut der Cremer-Gruppe, Roedental, West Germany.
D. A. Gunn	GR-Stein Refractories Ltd, Worksop, UK.
S. Hampshire	NIHE, Limerick, Ireland.
A. Hendry	University of Newcastle upon Tyne, UK.
S. Heyez	S. A. Solvay Laboratoire de Recherche, Brussels, Belgium.
G. Higgins	BHP Laboratories, Limerick, Ireland.
I. Higgins	University of Newcastle upon Tyne, UK.
B. C. Hobson	TAC Engineering Materials (T & N), Manchester, UK.
D. Holmes	British Ceramic Research Association, Stoke on Trent, UK.
S. A. Horton	University of Leeds, UK.
K. H. Jack	University of Newcastle upon Tyne, UK.
A. Johnsson	AB Volvo, Goteborg, Sweden.
H. O. Juchem	De Beers Industrial Diamond Division, Ascot, UK.
A. Kennedy	Institute for Industrial Research & Standards, Dublin, Ireland.
P. Kennedy	Springfields Nuclear Power Development Laboratories, Preston, UK.
E. Kensella	National Board for Science & Technology, Dublin, Ireland.
M. S. Lacey	University College, Dublin, Ireland.
W. Lang	National Engineering Laboratory, East Kilbride, UK.
C. Leblud	JRC, Petten, The Netherlands.
S. J. Lee	THOR Ceramics Ltd, Dunbartonshire, UK.
A. Leriche	Centre Recherches de l'Industrie Belge de la Céramiques, Mons, Belgium.
W. G. Long	Babcock & Wilcox, Lynchberg, USA.
H. Luthe	Hermann C. Starck Berlin, West Germany.
L. McDonnell	Cork Regional Technical College, Ireland.
S. Mason	University of Warwick, UK.
H. Moeller	Babcock & Wilcox, Lynchberg, USA.
H. Mostaghaci	University of Leeds, UK.
J. N. Mulcahy	University College Dublin, Ireland.
A. Muller-Zell	Hutschenreuther AG, Selb, West Germany.
T. Murakami	Komatsu Ltd, Oyama-shi Tochigi-ken, Japan.
D. Murphy	Industrial Development Authority of Ireland, Dublin, Ireland.
W. Mustel	Ceraver Ltd, Tarbes, France.
B. C. Mutsuddy	Battelle Columbus Laboratories, USA.
R. Nelissen	Royal Sphinx, Maastrisht, The Netherlands.

List of Participants

M. O. Nicolls	De Beers Industrial Diamond Division, Ascot, UK.
H. Nosbusch	CEC, Brussels, Belgium.
K. Notter	De Beers Industrial Diamond Division, Ascot, UK.
K. O'Reilly	NIHE, Limerick, Ireland.
K. Parkinson	BNFL, Warrington, UK.
J. M. Perrin	University of Leeds, UK.
E. Petty	NIHE, Limerick, Ireland.
M. Peuckert	Hoechst AG, Frankfurt, West Germany.
H. Pickup	University of Leeds, UK.
R. C. Piller	AERE, Harwell, UK.
M. J. Pomeroy	NIHE, Limerick, Ireland.
P. Popper	Newcastle, UK.
M. Redington	Thomond College of Education, Limerick, Ireland.
D. Reid	De Beers Industrial Diamond Division, Shannon, Ireland.
P. Robyn	Glaverbel, Jumet, Belgium.
A. Roos	Royal Sphinx, Maastrisht, The Netherlands.
G. Roult	Centre d'Etudes Nucléaires, Grenoble, France.
K. Salmon	Advanced Materials Engineering Ltd, Stourport on Severn, UK.
B. Saruhan	NIHE, Limerick, Ireland.
G. Schwier	Hermann C. Starck Berlin, Goslar, West Germany.
P. Selgert	Hoechst, AG, Frankfurt, West Germany.
B. Shaw	Dyson Refractories Research & Development Laboratories, Sheffield, UK.
S. Slasor	University of Newcastle upon Tyne, UK.
N. Smith	De Beers Industrial Diamond Division, Shannon, Ireland.
J. T. Smith	GTE Labs. Inc., Waltham, USA.
T. Soma	NGK Insulators Ltd, Nagoya, Japan.
C. J. Spacie	Morgan Materials Technology Ltd, Stourport on Severn, UK.
A. Spooner	De Beers Industrial Diamond Division, Shannon, Ireland.
A. Szweda	Lucas Cookson Syalon Ltd, Solihull, UK.
B. A. Thiele	Kernforschungsanlage Juelich (KFA), Juelich, West Germany.
D. P. Thompson	University of Newcastle upon Tyne, UK.
J. Tooher	Trinity College, Dublin, Ireland.
W. Tredway	University of Illinois at Urbana-Champaign, USA.
H. M. Verhoog	Hoogovens Group, Igmuiden, The Netherlands.
R. Wallace	Crane Packing, Slough, UK.
C. Weber	NIHE, Limerick, Ireland.
A. J. Wickens	Morgan Materials Technology Ltd, Stourport on Severn, UK.
I. Wilson	TRW Ceramics Ltd, Derby, UK.
H. Wopler	De Beers Industrial Diamond Division, Ascot, UK.
G. Wötting	Institut für Werkstoff-Forschung, Köln, West Germany.

Index of Contributors

Austin, M., 255
Balkwill, K. P., 341
Bigay, Y., 149
Bosković, S., 165
Briggs, A., 341
Brook, R. J., 41, 53
Brossard, M., 191
Broussaud, D., 281
Buljan, S. T., 409

Cambier, F., 53
Cashell, E. M., 213

Davidge, R. W., 341
Denape, J., 281
Dowson, D., 281

Eisele, U., 41
Ekström, T., 231
El-Baradie, M., 255

Fan, Q., 149
Farmer, M. H., 375
Ferguson, P., 97
Fessler, H., 319
Fricker, D. C., 319

Gilbart, E., 41
Goursat, P., 191

Hampshire, S., 69
Hendry, A., 119
Higgins, I., 119
Horton, S. A., 281
Hoven, H., 299

Ingelström, N., 231

Jack, K. H., 1

Kanka, B., 83
Kennedy, P., 301
Koizlik, K., 299
Kostić, E., 165

Labbe, J. C., 191
Lacey, M. S., 375
Leriche, A., 53
Lewis, M. H., 175
Linke, J., 299

Mason, S., 175
Matsui, M., 361
McDonnell, L. D., 213
Monaghan, J., 255
Mostaghaci, M., 149
Mulcahy, J. N., 375
Mutsuddy, B. C., 397

Neil, J. T., 409

443

Oda, I., 361

Pasto, A. E., 409
Pickup, H., 41, 53
Piller, R. C., 341
Pomeroy, M. J., 69

Rae, A. W. J. M., 97
Riley, F. L., 149, 281
Risbud, S. H., 203
Roult, G., 191

Saruhan, B., 69
Siddiqi, S. A., 119
Slasor, S., 223
Smith, J. T., 409
Soma, T., 361
Spacie, C. J., 133

Sun, W. Y., 105
Szweda, A., 175

Thiele, B. A., 299
Thompson, D. P., 105, 223
Tooher, J., 255
Torre, J. P., 149
Tredway, W. K., 203

Vandeneede, V., 53

Wallbridge, N., 281
Walls, P. A., 105
Wallure, E., 299
Wötting, G., 83

Ziegler, G., 83
Zilberstein, G., 409

Subject Index

Activation energies, densification processes kinetics, 51
Åkermanite, 4
Aircraft engines
  problems of ceramics in, 32
  relative cost factors in, 38
Alcoa A17 alumina, 109, 227
Alumina
  cutting tool use, 256
  performance data, 14, 249
  hot hardness data, 15, 250
  reaction with α'-sialons, 112–16
  reaction with silicon nitride, 108
  replacement by β'-sialon, 19
  silicon nitride affected by, 88–9
Aluminate oxynitride glasses, 207–8
Aluminium nitride, 299
  reaction with silicon oxynitride, 109
  see also Syalon, 404
Aluminosilicates, sialons produced from, 121–5, 134–41, 149–53
Amborite, 257, 258
  cost of, 277
  cutting performance of, 262–3, 268, 269, 272, 273, 275, 276, 277
Apatite, 3, 8, 74, 79, 80
Aspect ratios
  sialons, 238
  silicon nitride, 67, 78, 85, 87, 94

Auger electron spectroscopy (AES) PEAT microscope, use of, 214, 216, 218
Automotive components, 17, 18, 32–3, 433
  cost factors for, 37

Ball bearings, sialons used in, 16, 17
Barium sialon glasses, 205, 206–7
  effect of nitrogen, 206
Basic Research for Industrial Technology for Europe (BRITE) programme, 379
Behaviour diagrams
  magnesia–neodymia–silica, 77
  silicon–aluminium–oxygen–nitrogen, 6
  yttrium–silicon–aluminium–oxygen–nitrogen, 8, 9, 10, 23
  yttrium–silicon–oxygen–nitrogen, 3
  see also Phase diagrams
Bend tests, 388–91
  silicon nitride, 347, 349, 363, 367, 371–2
  thickness effects on, 389, 390
  validity to fracture (design) strength, 392

445

Binders
  properties of, 399
  removal profile, 407–8
  removal study, 397–408
    experimental procedures, 398–9
    thermogravimetric analysis results, 399–404
    volatile-products analysis, 404–7
    time for removal, 398
Black ash, 18, 120
  see also Rice husk ash
Boron nitride
  nuclear fusion reactor use, 300
  see also Cubic boron nitride (CBN)
Borzan (BZN), 257, 258
Brinell hardness. See Hardness data
Butyl oleate binder
  properties of, 399
  thermal degradation of, 400–2

Calcium α'-sialons, 113, 114, 115, 116
Carbon black, 120
Carbo-oxynitride glasses, 210, 212
Carborundum Resitant Materials Company, 378
Carbothermal reduction processes, 18, 120, 121, 133, 134, 146
Cast iron
  machining of, 268–77
  sialons used in machining, 14, 248, 249
Castor oil, properties of, 399
CATE (Ceramic Applications in Turbine Engines) axial turbine blades, 426, 427
Ceramic alloying, 4
Chebyshev quadrature formulae, 338
China, research funding, 38
Clay, 121–2, 134, 136, 156
  pot-clay experimental specimens, 387, 389
Clay–coal mixtures, thermal reaction of, 18, 122–5, 135–41
Coal
  analytical composition of, 122, 134
  carbon from, 18, 120, 122, 134

Commercialisation aspects, 31–9
Component fabrication, 2, 426–9
Composite components, 395
Composites
  silicon nitride–titanium carbide, 421–6
  $\alpha'-\beta'$, 20, 27, 110, 112
  $O'-\beta'$, 22, 27, 109
Cooperation projects, 36, 39
Cost considerations
  automotive components, 37
  ceramic cutting tool inserts, 277–8
Creep data
  bend tests, 175, 185, 186, 187, 188
  commercial ceramics, 12, 29, 185
  Compressive tests, 175, 185, 186, 187, 188
Cubic boron nitride (CBN), 256–9
  commercially available materials, 257–8
  cost of, 277
  cutting performance of, 262–3, 268, 269, 272, 273, 275, 276, 277
  hardness of, 258
Cutting tools
  bulk manufacture of, 436
  ceramics as, 13, 14, 256–7
  edge wear mechanisms, 242, 247
  flank wear curves
    alumina on cast iron, 249
    alumina-coated carbide on cast iron, 271
    cubic boron nitride
      cast iron machining, 272
      Vitallium machining, 262, 268
    sialons
      cast iron machining, 249, 272
      Incoloy 901 machining, 248, 251
      Inconel 718 machining, 251
      Vitallium machining, 267
    tungsten carbide
      cast iron machining, 271
      Vitallium machining, 261, 266
  market sector, 32, 36
  performance data, 14, 248–50, 259–79

Cutting tools—*contd.*
  reliability of, 436
  sialons used as, 13, 14, 100, 104,
    232, 247–50, 251, 267, 272,
    275–7
  speeds used, 104, 247
  V–T diagrams
    cast iron turning, 273
    Vitallium machining, 263, 269
Cymrite, 206, 207

Dektak surface profile traces, 343,
  344, 345
Densification
  kinetics
    density effects on, 46, 47, 59–61
    illite-derived sialons, 159–62
    kaolinite-derived sialons, 159–62
    silicon nitride, 41–6, 77
    model, 126–7
    processes, silicon oxynitride, 125–7
Devitrification heat treatments
  sialons, 9, 116, 177, 178, 179
  silicon nitride, 75, 77
  silicon oxynitride, 130
Die inserts, 33
  sialons used in, 16, 17, 100
Diesel engine components, 17, 32–3
  European research, 376–92
  Japanese research, 37
Dog-bone tensile specimen, 388
  disadvantages of, 391
Dovetails, turbine blade root, 323
Dublin, University College (UCD)
  co-ordination role in ceramic
    engine research, 384–5
  cylinder-bursting test programme,
    380–5
  engine research at, 376–9
  flat-specimen test programme,
    385–92
  seminar on ceramic engine com-
    ponents, 379–80
Dysprosium–aluminium garnet
  (DyAG), 169, 174

Economic considerations. *See* Cost
  considerations
Edge-stress integrals
  notched components
    shear loading cases, 331
    tension loading cases, 329
    numerical evaluation of, 337–8
  shouldered components, 324, 326
EEC
  basic technological research
    programme, 379
  cooperation projects, 36, 39
  research/design contracts, 377, 379
Electrical properties, sialons, 12,
  22–3
Electron beam tests, 299–300
Electron-beam microanalysis, silicon
  oxynitride, 129
Electronics components, market,
  31–2, 36
Elongated grain structure, silicon
  oxynitride, 129, 130
Europe. *See* EEC

Failure probability
  addition of characteristic volumes,
    333
  plots
    silicon nitride ceramics, 345, 346,
      417
    silicon nitride components, 428
  rigorous evaluation of, 321–2
Faith, manufacturers, 35, 39
Fibre composites, 38–9, 395
Fibrous microstructure, sintered
  silicon nitride, 78, 81
Flaws
  generation of, 434
  silicon nitride, 373, 412
  size limitation for detection of, 436
Fluoronitride glasses, 209
Fluxing agents, 2
  see also Sintering aids
Fosterite, 107, 212
Fracture toughness. *See* Toughness
Friction coefficient, sialons, 12

Gamma function
  equation for, 43, 62
  silicon nitride, 46, 62, 63, 370
Garnet, formation of, 8, 169, 171
Gas turbines
  advantages of ceramics in, 2, 32
  requirements for ceramic materials in, 176
  rotors, 426–7
Gehlenite, 111
Glass phase grain boundary, 3, 23, 70, 75, 79, 106, 107, 171, 177
Glass transition temperatures
  effects of carbon
    Mg–Si–Al–O–C glasses, 209
  effects of nitrogen
    aluminium oxynitride glasses, 208
    Ba–Si–Al–O–N glasses, 206
    silicon nitride containing magnesia, 349
Globularization processes, silicon nitride sintering, 91–3
Grain boundary phase, 3, 23, 70, 75, 79, 106, 107, 171, 177, 411
Grain growth rate, ratio to densification rate, 42, 46, 49
Graphite, thermal shock resistance of, 299
Grey cast iron. *See* Cast iron
Griffith flaw criterion/equation, 303, 373

Hardness data
  cubic boron nitride, 258
  grey cast iron, 270
  loading rate used, 174
  Mg–Si–Al–O–C glasses, 209
  plotted against toughness, 242, 246
  sialons, 12, 13, 15, 20, 21, 171, 172, 229, 239, 244, 283
  silicon carbide, 283
  silicon nitride–titanium carbide composites, 424
  Y–Si–Al–O–N glasses, 25

Heat treatment effects
  glasses, 204
  sialons, 177, 184–5
  silicon nitride, 75, 77
  silicon oxynitride, 130
Heat-resistant alloys
  sialons used in machining, 14, 247, 248, 250, 251
  wear patterns on machining, 242, 247
Hexacelsian 206, 207
High-speed machining, sialons used in, 104, 247, 248
Hitachi aluminium nitride/silicon carbide material, 300
Homogeneity, research into, 34
Hot hardness data
  alumina, 250
  sialons, 13, 15, 23, 250
  indentor type used, 28
Hot-isostatically pressed (HIP) silicon nitride, 4, 414, 417, 430–1
Hot-pressed silicon carbide
  creep data for, 12, 29
Hot-pressed silicon nitride (HPSN), 2, 53–4, 341, 342, 344, 346, 349, 350, 351, 353, 355
  bend test load/displacement curves for, 351
  chemical composition of, 342
  compared with pressureless-sintered material, 58–66
  fracture surfaces of, 349, 350, 355
  moduli of rupture listed for, 346
  molten steel immersion results, 145, 146
  reaction sequences with magnesia, 106–7
  scanning electron micrographs of, 344, 355
  sodium hydroxide effects on, 146
  stepped temperature stress rupture results, 353
  surface finish of, 344, 359
  surface profile traces for, 344
  temperature dependence of MOR strength, 352
  Weibull moduli listed, 346

Subject Index 449

Idiomorphic crystallisation, 85
Illite
 analytical composition of, 152
 sialons produced from, 149–53
Incoloy, 901
 sialons used in machining of, 14, 248, 250, 251
Inconel, 718
 cubic boron nitride used in machining of, 258
 sialons used in machining of, 250, 251
Injection-molded components
 binders used, 398, 399
 quality control of, 435
 rotors, 427
 silicon nitride ceramics, 416, 417, 418
Integrals, numerical evaluation procedures
 double integrals, 338
 single integrals, 337–8
Internal combustion (IC) engines
 advantages of ceramics for, 376
 ceramic components, 17, 32–3
 testing of, 376–92
 debris after failure, 377–9
Investment return, Japanese attitude to, 35, 36
Iscanite, 257
 cost of, 277
 cutting performance, of, 267, 269, 272, 273, 276, 277
Iscar IC20 carbide
 cost of, 277
 cutting performance of, 261, 263, 266, 269
Isopressed silicon nitride, 414–16

Jänecke prism behaviour diagrams
 Mg–Nd–Si–O–N system, 78
 Mg–Si–Al–O–N system, 24
 Y–Si–Al–O–N system, 8, 9, 178
Japan
 diesel engine components, 37
 return on investment, 35, 36
 steel companies diversification, 34–5

Kaolinite
 analytical composition of, 122, 134, 152
 β'-sialon produced from, 121–5, 134–41
 sialons produced from, 151–3
 thermal decomposition of, 122
Kingery densification model, 126–7
Knoop microhardness. See Hardness data

Laser illumination, lateral cracks detected by, 219–20
Lignite clay
 analytical composition, of, 140
 effect on sialon production, 141
Liquid-phase sintering, 3, 49, 77, 78, 79, 84, 92, 125, 167, 176
Lithium–silicon nitride, 22–3
Low-atomic-number ceramics, 299–300
Low-cost ceramics, 19, 37, 121–30, 133–41

Magnesia, 2, 43, 53
 effects on silicon nitride densification, 3, 70, 71–7, 106–7
Magnesia–neodymia–silica system, phase relationships, 77
Magnesium sialon glasses, 24, 205, 206, 211
Market breakdown, engineering ceramics, 31–2, 36
Mathematical modelling, UCD co-ordination of, 385
Melilite, 3, 8, 74
 effect on densification of α'-sialons, 228–9, 230
 oxidation of, 4
Micro-focus X-ray techniques, 436
Microstructural characterization
 aluminosilicate produced sialons, 153, 157–8
 sialons, 235–9
 intergranular phase, 235, 237, 238

Microstructural characterization—contd.
  silicon nitride powders, 44, 58
  sintered silicon nitride, 65–6, 76, 78, 84–93
  Syalon ceramics, 177–83
Modulus of rupture (MOR)
  experimental method for determination, of, 305
  sialons, 12, 229, 283
  silicon carbide, 283, 312, 314, 315
  silicon nitride, 345–7
  titanium carbide composites, 424
Molten metal resistance, sialon ceramics, 17–18, 100, 145–6
Mullite, 122, 139, 177
Multianion glasses, 203–12
  aluminate oxynitride glasses, 207–8
  fluoronitride glasses, 209
  heat treatments applied, 204–5
  Nd–Si–Al–O–N glasses, 25
  oxycarbide glasses, 208–9
  oxyfluoronitride glasses, 209–10
  oxynitride glasses, 23–6, 205–7
  Y–Si–Al–O–N glasses, 24, 25
Murakami's reagent/etchant, 282

National Coal Board (NCB) anthracite, 122, 134
Near-net-shaping, 176
Neodymia (sintering aid)
  effects on silicon nitride densification, 71–7
  preparation of α'-sialon, 228
Neodymium–silicon–aluminium–oxygen–nitrogen system
  densities of glasses, 25
Neuber notch theory, 323–4
  bending theory, 328
  notation for, 322
  shear theory, 330
  stress distributions, 327
  tension theory, 327–8
Neutron diffraction spectrometry, 192–201
  see also Time-of-flight neutron diffraction spectrometry

Nickel-based alloys
  sialons used in machining, 247, 248, 250, 251
Nitrogen glasses, 23–6
NMR spectroscopy
  structural determination of oxycarbide glasses, 211, 212
Nomarski contrast micrograph, sialon wear tests, 293
Non-destructive evaluation (NDE) techniques
  PEAT microscope, 214–20, 221
Non-destructive testing (NDT)
  silicon nitride, 374
  size limitations of, 433
Norton Lucas HS130 silicon nitride, 11, 12, 29, 30
Norton NC132, creep data, 29, 185
Norton NC350 reaction-bonded silicon nitride, 356
Norton NC430, creep data, 29
Notched components, failure probability of, 327–31
Nuclear fusion reactors
  components, 299
  in-pile testing of materials for, 300

Organic binders, oxidative removal of, 399–407
Orthopaedic implants, 281
Oswald ripening, 47, 49
Owen Illinois, 377
Oxycarbide glasses, 208–9
Oxycarbonitride glasses, 210, 212
Oxyfluoronitride glasses, 209
Oxynitride ceramics, disadvantages of, 176
Oxynitride glasses, 23–6, 205–8, 212
Oyez First European Symposium on Engineering Ceramics, 31
  Japanese comment at, 35

Particle size
  distribution
    effect on densification kinetics, 47, 51

Particle size—*contd.*
distribution—*contd.*
 sialons, 159
 effect on sialon densification, 154, 155, 163–4
 values quoted for sialons, 152, 153
Phase diagrams
 magnesia–neodymia–silica, 77
 silicon–carbon–oxygen–nitrogen system, 121
 silicon nitride–silica–alumina–aluminium nitride, 108
 yttria–alumina–silica system, 90
Photo Electro Acoustic Thermal (PEAT) microscope, 214–18
 analytical techniques incorporated, into, 214
 applications for, 218–20
 flowing gas treatment in, 221
 specimen size in, 221
Photo-acoustic spectroscopy (PAS)
 PEAT microscope use of, 214, 215
Poisson's ratio, sialons, 12
Polyethylene binder
 properties of, 399
 thermal degradation of, 400–2
Polytypoid phases, 9, 108
 2H, 300
 12H, 6, 108, 111, 227
 15R, 6, 108, 113, 115, 227
 21R, 6, 98, 103, 108
Post-sintered reaction-bonded silicon nitride (PSRNSN), 4
Powder bed, 4, 70, 81, 166
Powders, research into, 34, 53, 97
Pressureless-sintered ceramics, advantages of, 176–7
Pressureless-sintered sialons, 7, 69–70
Pressureless-sintered silicon nitride
 compared with hot-pressed material, 58–66
 sintering study, 69–81
  effect of additive on densification, 71–4
  experimental procedures, 70–1
  heat treatment effects, 75, 77
  microstructural features observed 75, 76

Pressureless-sintered silicon nitride—*contd.*
 sintering study—*contd.*
  secondary phases observed, 74
Processing, ceramics, 18, 33, 410, 435

Quality control, 433, 435–6
 cost aspects of, 435

RB211 aeroturbine turbine discs, machining of, 13
Reaction sequences
 effects of sintering aids on, 168
 hot-pressed silicon nitride, 106–7
 neodymium α'-sialon, 228–9
 α'–β' sialons, 110–12
 β'-sialons, 107–9
 O'–β' sialons, 109–10
 silicon carbide, 303–4
 yttrium α'-sialon, 227–8
Reaction-bonded silicon carbide, 303
 phase diagram for, 304
 production process for, 303–4
Reaction-bonded silicon nitride (RBSN), 3–4, 53, 341, 342, 343, 346, 347, 348, 350, 351, 353, 354, 356, 358
 bend test load/displacement curves, 351
 chemical composition of, 342
 creep data for, 12, 29
 densification behaviour, 61
 failure probability for shouldered components, 333, 338–9
 fracture surfaces, 347, 348
 moduli of rupture listed, 346
 scanning electron micrographs of, 343, 347, 354
 stepped temperature stress rupture SEM micrographs, 352, 354
 stepped temperature stress rupture test, 350, 353
 surface profiles of, 343
 temperature dependence of MOR strength, 352
 Weibull moduli listed, 346

REFEL silicon carbide, 302
  composition of, 305
  creep data for, 12, 29
  cylinder-bursting tests, 382, 384
  hardness affected by free silicon
    content, 309
  strength affected by
    free silicon content, 309, 311
    grain size, 306, 307, 313
    lapping operations, 316
    surface oxidation, 317
    temperature, 305–6
  study of properties of, 305–317
    experimental procedure, 305
    results for, 305–13
  surface energy affected by free
    silicon content, 311
  surface finish effects on strength,
    312, 313, 314, 315
  toughness affected by free silicon
    content, 306–9, 310, 313
  Young's modulus affected by free
    silicon content, 308, 311
Refractories, sialons used with silicon
  carbide, 100, 103–4
Reliability, 37, 409, 433–7
  improvement in
    silicon nitride ceramics, 417
    silicon nitride composites, 428–9
Research effort, advanced ceramics,
  33, 34
  funding for, 33, 38
Rice husk ash
  composition of, 18, 120, 126
  impurities in, 126
  silicon oxynitride produced from,
    126, 127–30
Rockwell hardness. See Hardness
  data
Roller bearings, sialons used in, 16,
  17

Saggars, β'-sialons produced in,
  135–6, 137
Scanning Auger microscopy (SAM)
  PEAT microscope, use of, 214

Scanning electron micrographs
  barium–silicon–aluminium–
    oxygen–nitrogen system,
    207
  illite-derived sialon ceramics, 157
  kaolinite-derived sialon ceramics,
    158
  sialon wear test piece, 292
  α'-sialon–alumina reaction
    couples, 114
  α'–β' sialon composites, 236–7,
    243
  β'-sialons, 240–1
  silicon carbide wear test piece, 294
  silicon nitride ceramics, 65–6, 76,
    86, 91, 343–4, 347–9, 354–5
    flaws, 412, 413, 415, 416
    green body, 418–19
  silicon nitride powders, 44, 58
  silicon oxynitride, 129, 130
Scanning electron microscopy (SEM)
  experimental procedure, for, 233
  PEAT microscope, use of, 214,
    216, 218
Scissors, ceramics used for, 33, 35
Secrecy aspects, 35
Shear modulus
  sialons, 283
  silicon carbide, 283
Shouldered components
  edge stress distribution in, 325
  failure probability of, 324–7
Sialons
  acid leaching prior to sintering,
    141, 142
  aluminosilicate produced, 151
  hot-pressing of, 153
  microstructural studies for, 153,
    157–8
  powder milling of, 151–3
  applications of, 232
  aspect ratios of, 238
  characterisation studies
    experimental procedures, 232–4
    microstructural studies, 235–9,
      240–1, 243
    phase compositions listed, 234–5

Sialons—contd.
  clay–coal produced, 121–5,
    134–41, 150–3
  properties of, 143–6
  commercial ceramics, 11–18
  compositions as equivalents O and
    Al, 232–3
  creep data, 12, 185
  cutting performance predicted for,
    248–50
  devitrification heat treatments, 9,
    116, 177
  effect of raw materials on, 239, 243
  effect of sintering temperature on,
    238–41
  grain growth affected by temperature and time, 238–9
  hardness of, 15, 171, 172, 239,
    242–5, 252
  high temperature deformation and
    fracture, 184
  hot hardness of, 250
  intergranular phase, 235, 237, 238,
    284
  lattice parameters of, 112, 113, 169
  mechanical properties of, 7, 229,
    283
  metal-cutting performance of, 242,
    247–50
  microstructural studies of, 235–9,
    240–1, 243
  modulus of rupture for, 229, 283
  molten metal resistance of, 17–18,
    145–6
  oxidation of, 145, 182
  α'/β' phase ratio, 238, 244
  physical properties of, 239–42
  production methods listed, 99, 150
  sintering aid effects on, 168–72
  sintering of, 141–6
  slag attack on, 17, 146
  sodium hydroxide effects on, 146
  thermal diffusivity of, 242, 246,
    252, 252–3
  toughness of, 171, 172, 242, 244,
    245
  wear debris from, 293, 295

Sialons—contd.
  wear factor results, 290, 291, 296–7
  X-phase, 107–8, 123, 135
  X-ray diffraction studies of, 137,
    139, 140
  see also main entries α'-Sialons;
    $\alpha'-\beta'$ Sialons; β'-Sialons;
    O'-Sialons; O'$-\beta'$ Sialons;
    X-Sialons; Syalon
α'-Sialons, 19–21
  crystal structure of, 19
  general formulae for, 224
  hot hardness of, 21, 30
  mechanical properties of, 229
  modulus of rupture for, 229
  phase relationships for, 20, 225
  production from nitrides, 19
  production from oxides, 20
  reaction sequences for, 111, 112
  unit cell dimensions for, 112, 113
  see also Neodymium α'-sialon;
    Syalon 101; Yttrium
    α'-sialon
$\alpha'-\beta'$ Sialons, 21, 30
  cutting performance of, 247, 248,
    249, 251
  effect of raw materials on, 239, 243
  hardness of, effect of alpha-phase
    on, 244, 252
  reaction sequences for, 111, 112
  thermal diffusivity of, 242, 246
  toughness of
    effect of alpha-phase on, 244,
      245
    hardness plots, 246
β'-Sialons, 5–7
  applications for, 11, 223
  commercial powders available,
    98–101
  comparison with beta silicon
    nitride, 198
  composition range of, 6, 97, 150,
    176, 191, 234
  creep of, 12, 29
  crystal structure of, 5
  cutting performance of, 14
  densification of, 69–70, 79

β'-Sialons—contd.
  electrical conductivity of, 12
  engineering applications for, 13–18
  forming of, 13
  hardness of, 12
  high-pressure behaviour of, 196–8, 200
  high-temperature behaviour of, 184–7, 196–9, 201
  hot hardness of, 13, 15, 21, 30
  kaolin used for production of, 121–5
  machining of, 13
  mechanical properties of, 7, 283
  microstructural development, 177–9
  modulus of rupture for, 12, 283
  oxidation of, 11, 182, 190
  phase relationships with α'-sialons, 20, 111
  physical properties of, 7, 12, 143, 239, 242, 244–5
  Poisson's ratio of, 12
  preparation from silicon nitride and alumina, 195–6
  pressureless-sintering of, 7, 69–70, 175
  processing routes for, 18–19
  production methods, 18, 223–4
    clay–coal mixtures, 121–30, 135–41
    saggars, 135–6, 137
    vertical tube furnaces, 136, 138, 139
  raw materials for, 18–19, 122, 134
  reaction sequences for, 107–9, 110, 111, 112
  relationship with silicon nitride, 7
  shear modulus for, 283
  sintering of, 13
  structural determination of, 193–5
  tetrahedral angles in, 196, 198, 199
  thermal properties of, 7, 12, 242, 246
  toughness of, 12, 184
  transmission electron micrograph of, 284
  unit cell dimensions for, 112, 113

β'-Sialons—contd.
  Weibull modulus of, 12
  YAG bi-phase composites, 9, 10
    creep data for, 12, 185–7
    heat treatment to form, 7, 8, 177
    microstructure of, 179, 181, 187
    oxidation of, 11, 182, 190
    toughness of, 184
  see also Syalon, 201
  Young's modulus for, 12, 283
O'-Sialons, 22–3
  electrical properties of, 22–3
  production of, 22
  reaction sequence for, 110
O'–β' Sialons, 22, 183, 187, 188, 189
  preparation of, 109–10
Silicon–aluminium–oxygen–nitrogen system
  behaviour diagram for, 6
  phase relationships, 6, 108
Silicon carbide
  aluminium nitride modified, 300
  densification of, 303–5
  mechanical properties of, 283
  modulus of rupture for, 283
  shear modulus for, 283
  sialons used with, 100
  wear debris from, 295
  wear factor results, 290, 291, 296, 297
  Young's modulus for, 283
Silicon–carbon–oxygen–nitrogen–system, 120–1
  phase stability diagram for, 121
Silicon nitride
  alpha–beta phase transformation, 71, 72, 73, 74, 75, 76, 77, 78, 79, 95, 106, 167, 411
  alpha-phase, effect of, 84–5
  beta-phase
    content of, 57, 64
    importance of, 70
    stability of, 121
  binders used, 399
    thermal degradation of, 400–4
    volatile-products analysis of, 404–7
  chemical characteristics listed, 56

Silicon nitride—*contd.*
  chemistry of, 417–21
  crystal structure of, 5
  cutting tool use, 257, 264
  densification, 2, 8–11, 53–67
    kinetics, 41–8, 77–9
      experimental method, 42–3
      results, 43–6
    density values quoted, 59, 71, 73, 74, 88, 89, 167, 168
    elongated particles, 57, 58
    grain growth mechanisms, 42, 47–8, 49–50
    grain refinement of, 106
    liquid-phase sintering of, 3, 49, 77, 78, 79, 84, 92, 167
    mechanical properties
      microstructural effects on, 92, 93, 96
      processing effects on, 64
  mechanical properties, of, 165
  metallic inclusions in, 412, 417–21
  microstructural characterisation of, 86–93
  microstructural development during sintering of, 84–5
  microstructure of, 410–17
  physical characteristics listed, 2, 56, 57
  pit/white spot flaws in, 411–14
  pressureless sintering of, 58, 69–79
  production methods for, 54, 120
  relationship with β'-sialon, 7
  scanning electron micrographs of, 44, 58, 65–6
  secondary phases, 3, 74
  shaping techniques used, 414
  sinterability study, 53–67
    experimental procedures, 54–5
    results, 55–67
  sintering aids used, 4, 43, 45, 51, 53, 69–82, 83, 88, 106, 165
  solution–reprecipitation processes, 3, 49, 81, 84, 96, 107, 167, 411
  tensile strength of, 364–7
    experimental procedure, 362–3
  titanium carbide composites, 421–6

Silicon nitride—*contd.*
  titanium carbide composites—*contd.*
    effect of impurities on, 423–4
    mechanical properties of, 424
    microstructure of, 421–3
    wear of cutting tools, 425
  YAG composite, 172
  see also Hot-pressed, Pressureless-sintered, and Reaction-bonded silicon nitride
Silicon oxynitride
  crystal structure of, 23
  densification of, 125–7
  hot pressing of, 127–30
  solubility of alumina in, 6, 22
Single-crystal pulling die, 16, 18
Sinterability study, silicon nitride powders, 53–67
Sintering aids, 4, 43, 45, 51, 53, 69–82, 83, 88, 106, 165–73
  see also Alumina; Magnesia; Neodymia; Yttria
Size effects
  strength of ceramics, 361, 367, 370
    experimental studies, 362–73
    silicon nitride, 371
  testing of ceramics, 389–91, 392–3
Slag, sialons attacked by, 17, 146
Sliding contact wear tests, 286–97
Sliding wear mechanisms, 296
Slip-cast silicon nitride ceramics, 416, 417, 418
Solution–reprecipitation processes, silicon nitride, 3, 49, 81, 84, 96, 107, 167, 411
Space groups, sialons, 193–4
Specifications, 436
Steel companies, diversification into ceramics, 34–5
Stepped temperature stress rupture (STSR) tests, 349–50
  silicon nitride, 349–50, 353
    SEM micrographs, 352, 354–5, 356
Strength (of ceramics), factors affecting, 302–3, 361, 367, 370
Stress concentration factors, 322

Stress-volume integrals, 324
  notched components
    shear loading case, 332
    tension loading case, 330, 336–7
    numerical evaluation of, 337–8
Structural engineering components market, 32
Sub-critical crack growth (SCG), β'-sialons, 184
Sub-surface defects, location by PEAT microscope, 216, 218–20
Supersaturation, silicon nitride affected by, 85, 94, 95
Surface finish, 274–7
  milling of Vitallium, 275–6
  silicon carbide, 312, 313, 314, 315
  silicon nitride, 343, 344, 359
  turning of cast iron, 276–7
  turning of Vitallium, 274, 275
Surface microstructure, 179, 214–20
Surface profiles
  sialon wear test specimens, 285
  silicon nitride, 343–4
Surgical implants, machining of, 259–68
Syalon (trade name) ceramics, 97
  component fabricating techniques, 100
  creep data for, 12, 185
  engineering grades, 98, 99–100
  high-nitrogen, 178–9, 180
  high-temperature deformation, 184–8
  mechanical properties of, 229
  microstructural development, 177–83
  powder manufacturing processes, 98, 99
    addition to zirconia, 101
  refractory grades, 98, 100–1
  surface microstructure of, 179, 182–3
  see also Sialons

Talysurf traces, sialon wear test specimens, 285

Taylor tool-life curves, 263, 264, 268, 269, 273–4
  see also Cutting tools, V–T diagrams
Tensile tests
  bend tests compared with, 373
  bending stress component in, 364
  disadvantages of, 392
  grip configuration for, 363
  silicon nitride, 364–6
  specimen configurations/geometry/dimensions, 362, 388
Thermal conductivity, β'-sialons, 12
Thermal diffusivity, sialons, 242, 246, 252
Thermal expansion coefficient, β'-sialons, 7, 12
Thermal shock resistance
  importance in nuclear fusion reactors, 299
  sialons, 12
Thermal wave microscopy (TWM)
  advantages of, 216
  detection systems for, 217
  PEAT microscope, use of, 214, 216–17, 218, 219–20
  resolution limit of, 216–17
Thermogravimetric analysis, organic binders, 399–404
Time-of-flight neutron diffraction spectrometry
  advantages of, 192
  data analysis, 192–3
  experimental method, 192–3
  β'-sialon results, 193–201
Timoney variable compression ratio mechanism, 377
Titanium carbide, composites with silicon nitride, 421–6
Tool wear curves. See Cutting tools, flank wear curves
Toughening, research into, 34, 38–9
Toughness data
  plotted against hardness, 242, 246
  sialons, 12, 184, 229, 242, 245, 253, 283
  silicon carbide, 283

Toughness data—*contd.*
 silicon nitride–titanium carbide composites, 424
Tribology, 281
Tungsten carbide
 cutting tool use, 257
 cost considerations, 277
 performance data, 14, 260–2, 266, 269, 271, 273, 275–7
 hot hardness data, 15
Turbine components, sialons used in, 26, 27

Vickers hardness data
 indentor used, 28–9
 loading rate used, 174
 see also Hardness data
Vitallium
 chemical composition of, 259
 milling of, 264–8
 surface finish for machining of, 274–6
 turning of, 260–4
Volcanic ash (sialon raw material), 18

Waspalloy, machining of, 258
Wear tests, 282–97
 cutting tools. *See* Cutting tools, flank wear curves debris examination in, 290, 293, 295
 effects of loading, 291–2, 296
 effects of sliding speed, 291–2, 297
 experimental procedures, 282–90
 measurement of wear rates, 288
 plate-on-rotating-cylinder test, 286, 288
 results for, 291–2
 reciprocatory pin-on-plate test, 286, 287
 results for, 289, 290
 results, sialons and silicon carbide, 290–5
 surface preparation for, 284–6
 temperature effects on, 298

Wear tests—*contd.*
 tri-pin-on-disc test, 286, 287
 results for, 290–1
 worn-surface characterisation for, 288, 290, 292–5
Weibull equation, 321
Weibull failure probability plots
 silicon nitride ceramics, 345, 346, 417
 silicon nitride components, 428
Weibull modulus
 calculation of, 394–5, 414
 effect on Neuber notched bar behaviour, 329–32
 sialons, 12
 silicon nitride, 346, 370, 371, 373
Weibull statistics, 320, 371
Weibull weakest link theory, 367, 370
Welding operations, sialons used in, 13, 15, 17
Wollastonite, 8, 9, 169
World market, advanced ceramics, 33, 34, 36

X-phase sialons, 107–8, 123, 135, 141
X-ray diffraction studies
 clay–coal reaction products, 125
 sialon, 115, 137, 139, 140
  alumina reaction couples, 113, 115
  production, 169, 170, 171
  wear tests, 293
 silicon nitride, 71, 420

Young's modulus
 experimental determination of, 305
 β'-sialon, 12, 283
 silicon carbide, 283, 311
Yttria, 2, 43, 53, 153
 alumina–silica phase diagram, 90
 sialon densification, 7, 8–11, 155, 161–2
 silicon nitride densification, 3, 70, 88–9

Yttrium α'-sialon
  densification of, 227–8
  mechanical properties of, 229
  phase relationships for, 225–7
  reaction sequences in formation of, 227–8
Yttrium–aluminium garnet (YAG), 8, 22, 70, 168, 171, 173
  composite with silicon nitride, 172
  crystallisation of, 8, 70, 171, 177
Yttrium–silicon–aluminium–oxygen–nitrogen system, 9, 10, 23
  behaviour diagrams for, 8, 9, 10, 23, 178

Yttrium–silicon–aluminium–oxygen–nitrogen system—*contd.*
  glass-forming regions in, 9, 10
  hardness of glasses, 25
  viscosities of glasses, 24
Yttrium–silicon–oxygen–nitrogen system
  behaviour diagram for, 3
Yttro-garnet. *See* Yttrium–aluminium garnet

Zirconia
  effects of sialon addition, 101
  stabilised, 43, 101